AF788529

Light Scattering Reviews 7

Radiative Transfer and Optical Properties of Atmosphere and Underlying Surface

Alexander A. Kokhanovsky (Editor)

Light Scattering Reviews 7

Radiative Transfer and Optical Properties of Atmosphere and Underlying Surface

Springer

Published in association with
Praxis Publishing
Chichester, UK

PRAXIS

Dr Alexander A. Kokhanovsky
Institute of Environmental Physics
University of Bremen
Bremen
Germany

SPRINGER–PRAXIS BOOKS IN ENVIRONMENTAL SCIENCES (*LIGHT SCATTERING SUB-SERIES*)
EDITORIAL *ADVISORY BOARD MEMBER*: Dr Alexander A. Kokhanovsky, Ph.D., Institute of Environmental Physics, University of Bremen, Bremen, Germany

ISBN 978-3-642-21906-1 ISBN 978-3-642-21907-8 (eBook)
DOI 10.1007/978-3-642-21907-8
Springer Heidelberg New York Dordrecht London

Library of Congress Control Number: 2012945183

© Springer-Verlag Berlin Heidelberg 2013
This work is subject to copyright. All rights are reserved by the Publisher, whether the whole or part of the material is concerned, specifically the rights of translation, reprinting, reuse of illustrations, recitation, broadcasting, reproduction on microfilms or in any other physical way, and transmission or information storage and retrieval, electronic adaptation, computer software, or by similar or dissimilar methodology now known or hereafter developed. Exempted from this legal reservation are brief excerpts in connection with reviews or scholarly analysis or material supplied specifically for the purpose of being entered and executed on a computer system, for exclusive use by the purchaser of the work. Duplication of this publication or parts thereof is permitted only under the provisions of the Copyright Law of the Publisher's location, in its current version, and permission for use must always be obtained from Springer. Permissions for use may be obtained through RightsLink at the Copyright Clearance Center. Violations are liable to prosecution under the respective Copyright Law.
The use of general descriptive names, registered names, trademarks, service marks, etc. in this publication does not imply, even in the absence of a specific statement, that such names are exempt from the relevant protective laws and regulations and therefore free for general use.
While the advice and information in this book are believed to be true and accurate at the date of publication, neither the authors nor the editors nor the publisher can accept any legal responsibility for any errors or omissions that may be made. The publisher makes no warranty, express or implied, with respect to the material contained herein.

Cover design: Jim Wilkie
Project copy editor: Mike Shardlow
Author-generated LaTex, processed by EDV-Beratung, Germany

Printed on acid-free paper

Springer is part of Springer Science+Business Media (www.springer.com)

Contents

List of Contributors

Teruo Aoki
Meteorological Research Institute
1-1 Nagamine
Tsukuba
Ibaraki 305-0052
Japan
teaoki@mri-jma.go.jp

Larry K. Berg
Atmospheric Science & Global Change Division
Pacific Northwest National Laboratory
PO Box 999, MSIN K9-24
Richland, WA 99352
USA
Larry.Berg@pnnl.gov

Evgueni Kassianov
Atmospheric Science & Global Change Division
Pacific Northwest National Laboratory
PO Box 999, MSIN K9-24
Richland, WA 99352
USA
Evgueni.Kassianov@pnnl.gov

Alexander D. Klose
Department of Radiology
Harkness Pavilion, 3rd Floor
180 Fort Washington Avenue
New York, NY 10032
USA
ak2083@columbia.edu

Pavlos Kollias
McGill University
Department of Atmospheric and Oceanic Sciences
805 Sherbrooke Street West
Montreal, Quebec H3A 0B9
Canada
pavlos.kollias@mcgill.ca

Andres Kuusk
Tartu Observatory
61602 Tõravere
Estonia
andres@aai.ee

Joel Kuusk
Tartu Observatory
61602 Tõravere
Estonia
joel@aai.ee

Dana E. Lane-Veron
University of Delaware
College of Earth, Ocean, and Environment
114 B Robinson Hall
Newark, DE 19716
USA
dveron@udel.edu

Mait Lang
Tartu Observatory
61602 Tõravere
Estonia
lang@aai.ee

Casey C. Moore
WET Labs, Inc.
620 Applegate Street
Philomath, OR, 97370
USA
casey@wetlabs.com

Mikhail Ovchinnikov
Atmospheric Science & Global Change Division
Pacific Northwest National Laboratory
PO Box 999, MSIN K9-24
Richland, WA 99352
USA
Mikhail.Ovchinnikov@pnnl.gov

Elena V. Petrova
Space Research Institute of RAS
Profsoyuznaya 84/32
117997, Moscow
Russia
epetrova@iki.rssi.ru

James M. Sullivan
WET Labs, Inc.
70 Dean Knauss Road,
Narragansett, RI 02882
USA
jims@wetlabs.com

Victor P. Tishkovets
Institute of Radio Astronomy of NASU
4 Chervonopraporna Str.
Kharkov, 61002
Ukraine
tishkovets@ri.kharkov.ua

Michael S. Twardowski
WET Labs, Inc., Department of Research
70 Dean Knauss Drive
Narragansett, RI 02882
USA
mtwardo@wetlabs.com

J. Ronald V. Zaneveld
WET Labs, Inc.
620 Applegate Street
Philomath, OR 97370
USA
ron@wetlabs.com

Xiaodong Zhang
Earth System Science and Policy
University of North Dakota
Grand Forks, ND 58202
USA
zhang@aero.und.edu

Notes on the contributors

Teruo Aoki graduated from Meteorological College, Kashiwa, Japan, in 1981. He received his PhD in Polar Science from the Graduate University of Advanced Studies (National Institute of Polar Research), Japan, in 2000. He is currently a head of the 3rd Research Laboratory at Physical Meteorology Research Department in Meteorological Research Institute (MRI) of Japan Meteorological Agency (JMA). His research interests are optical properties of snow, satellite remote sensing of cryosphere, and aerosol-snow/ice interaction and the relevant climate impact. His experiment areas are Antarctica, Alaska, the Himalayas, and Hokkaido in Japan. He received the Science Award of the Japanese Society of Snow and Ice in 2005 and the Award of the Meteorological Society of Japan in 2008.

Larry K. Berg received his MSc and PhD degrees in Atmospheric Science from the University of British Columbia. He joined the staff at Pacific Northwest National Laboratory in June 2002. His research interests include cloud parameterizations, boundary-layer meteorology, turbulence, cloud–aerosol interactions, mesoscale modeling and atmospheric dispersion. Dr Berg has helped lead a number of different field studies supported by the Departments of Energy, Homeland Security, and Defense. To date, he has published more than 15 peer-reviewed journal articles and has contributed to several book chapters.

Evgueni Kassianov received his PhD in 1995 from the Institute of Atmospheric Optics (Tomsk, Russia). This work was focused on the visible and infrared radiative transfer in inhomogeneous clouds. In 1997, he joined the University of Oklahoma (Norman, USA) and worked there as a visiting scientist for 2 years. Currently, he is a senior research scientist at the Pacific Northwest National Laboratory in Richland (Washington, USA). He published over 40 peer-reviewed papers, dealing mostly with radiative transfer in cloudy atmosphere. His research activities include the remote sensing of clouds and aerosols and improved representation of their small-scale variations in large-scale models.

Alexander D. Klose graduated from the Technical University of Berlin, Germany, and received a Diploma in Physics in 1997, while working on optical tomography of rheumatoid arthritis in human finger joints. He earned a PhD in Physics from the Free University of Berlin, Germany, in 2001, and devoted his time to the development of the first transport-theory-based optical image reconstruction method used in clinical research. During his graduate studies, he was a visiting scientist at Los Alamos National Laboratory and at the State University of New York. In 2002–2004, he was an Ernst-Schering Postdoctoral Fellow at the Department of Biomedical Engineering of Columbia University in New York and started working on fluorescence tomography using the radiative transfer equation. In 2006, he developed in collaboration with Edward Larsen (University of Michigan, Ann Arbor) a high-order radiative transfer model based on the simplified spherical harmonics equations for visible light in strongly scattering tissue. He is currently an Assistant Professor at the Department of Radiology of Columbia University Medical Center. His research focuses on numerical solutions of light propagation models in tissue and on image reconstruction of fluorescent and bioluminescent sources for preclinical imaging of small animals. He has developed various image reconstruction algorithms for multispectral fluorescence and bioluminescence tomography.

Pavlos Kollias received his PhD in Meteorology from the Rosenstiel School of Marine and Atmospheric Sciences, University of Miami, in 2000. He is currently an Associate Professor at the Department of Atmospheric and Oceanic Sciences at McGill University. He is also a Tier-II Canada Research Chair in Radar Applications in Weather and Climate. He is an international leader in the application of short-wavelength radars for cloud and precipitation research from ground-based and space-based platforms. He has published over 50 scientific articles in peer-reviewed literature in the areas of millimeter-wavelength radar research, cloud and precipitation physics and dynamics.

Andres Kuusk graduated from Tartu University, Estonia, as a physicist in 1970. He received his PhD in geophysics from the Main Geophysical Observatory, St Petersburg, in 1979 and DSc in geophysics from the University of Tartu in 1991. He is currently working as a senior researcher and head of the Department of Atmospheric Physics in Tartu Observatory, Estonia. His research interests are radiative transfer in vegetation canopies, canopy reflectance models, and remote sensing of vegetation canopies in optical spectral domain.

Joel Kuusk received his BSc in Physics in 2003, MSc in 2005, and PhD in 2011 from the University of Tartu, Estonia. In 2006–2011 he worked as an engineer in the Tartu Observatory and since 2011 has worked as a research associate. His research interests are optical radiometry and vegetation remote sensing, particularly top-of-canopy spectral reflectance measurements above forests from manned and unmanned helicopters.

Dana E. Lane-Veron received her PhD in Oceanography from the Scripps Institution of Oceanography at the University of California, San Diego. She is currently part of the research faculty at the University of Delaware, in the College of Earth, Ocean, and Environment. Dana's research interests include climate change, cloud physics, cloud–aerosol–radiation interactions, stochastic radiative transfer, global energy balance, land surface–atmospheric interactions, sea breeze circulation, offshore wind resource assessment, and numerical modeling and observation of the atmospheric boundary layer. Her research has been supported by grants from the US Department of Energy (DOE), the National Science Foundation (NSF) and the National Oceanic and Atmospheric Administration (NOAA).

Mait Lang graduated from Kaarepere technical school, Luua, Estonia, as a forester in 1992. He received his PhD in forestry from the Estonian University of Life Sciences in 2006. He is currently working as a senior researcher at the Institute of Forestry and Rural Engineering of Estonian University of Life Sciences and as a senior researcher at the Department of Atmospheric Physics of Tartu Observatory, Estonia. His research interests are the application of remote sensing to forest inventories and the modeling of forest canopy structure.

Casey Moore received his BS from Lewis and Clark College (Physics) in 1980, and MS from the University of California at Santa Barbara (Scientific Instrumentation) in 1983. He is President and founder of WET Labs, Inc. His research interests are in inherent optical properties, relationships between particles and optical characteristics, and the design of optical instrumentation.

Mikhail Ovchinnikov graduated from the Moscow Institute of Physics and Technology, Moscow, Russia, in 1989 with an MS degree in Applied Mathematics. He received his PhD in Meteorology from the University of Oklahoma (OU), USA, in 1997 earning OU's Outstanding Ph.D. Dissertation Award. He is currently a senior research scientist at the Pacific Northwest National Laboratory in Richland, Washington, USA. His research interests include cloud physics, numerical modeling and remote sensing of atmosphere, and aerosol–cloud–radiation interactions and their role in climate. His research has been supported by grants from the US Department of Energy (DOE) and National Aeronautic and Space Agency (NASA).

Elena V. Petrova graduated from the State Pedagogical University (Department of Physics), Moscow, Russia, in 1972. Since 1974, she has been with the Space Research Institute of the Russian Academy of Sciences, and she is currently a senior scientist at the Planetary Physics Department of this institute. She took part in the preparation and realization of spectrophotometric and polarimetric experiments for the space missions to Venus and Mars and gained her PhD degree in 1993 for a thesis on the properties of the atmosphere and surface of Mars from the Phobos-2 spectrophotometric data. She is the author/co-author of more than 100 scientific publications including several peer-reviewed book chapters. Her research interests include light scattering by irregular particles and aggregates and remote sensing of atmospheres and surfaces of the solar system bodies. Dr Petrova is the recipient of two Bronze Medals of the Exhibition of Economic Achievements of the USSR (Moscow, 1976–1979).

Jim Sullivan received his PhD from the University of Rhode Island, Graduate School of Oceanography (URI-GSO) in 2000. He accepted a research faculty position at URI-GSO and is currently a senior research scientist for WET Labs, Inc. His current research focus is on ocean optics and sources of optical backscattering, understanding the biological and physical mechanisms that control the spatial-temporal distribution of optical fine-structure and phytoplankton populations in coastal oceans, and oceanographic instrument development (including in situ holographic devices and autonomous sampling platforms).

Victor P. Tishkovets graduated from the Physics Department of Kharkov State University in 1973. He gained his PhD degree in 1982 from the Main Astronomical Observatory in Kiev and a Habilitation Doctoral degree in 2009 from the Kharkov V. N. Karasin National University. He worked at the Institute of Astronomy of the Kharkov University from 1975 to 2003 and has been at the Institute of Radio Astronomy of the National Academy of Sciences of Ukraine in Kharkov thereafter. Dr Tishkovets is the author/co-author of more than 80 scientific publications including several peer-reviewed book chapters and a monograph. His research interests have included electromagnetic scattering by aggregates of particles and particulate media and remote sensing.

Michael S. Twardowski received his BS from Trinity University (Biology) in 1992, and PhD from the University of Rhode Island, Graduate School of Oceanography in 1998. He is Vice-President, Director of Research at WET Labs, Inc. His research interests are in inherent and apparent optical properties, relationships between particles and optical properties, the design of optical instrumentation, and using optical sensing techniques such as backscattering and remote sensing as proxies to investigate the biogeochemistry of natural waters.

J. Ronald V. Zaneveld received his MS from the Massachusetts Institute of Technology (Physics) in 1966, and PhD from Oregon State University (Oceanography) in 1971. He is Professor Emeritus at Oregon State University and Senior Oceanographer for WET Labs, Inc. His research interests are in radiative transfer theory, inherent and apparent optical properties, relationships between particles and optical characteristics, and the design of optical instrumentation. He received the prestigious Jerlov Award from The Oceanography Society (TOS) in 2006 for his achievements in ocean optics research.

Xiaodong Zhang graduated from the Department of Computer Science of Nanjing University, China in 1989. He received MS and PhD degrees, both in Oceanography, from Dalhousie University, Canada, in 1998 and 2001, respectively. His graduate research was focused on evaluating the optical properties of oceanic bubbles and their contribution to light scattering in the ocean through theoretical modeling, laboratory experiments and field observation. He is the founding and faculty member of the Department of Earth System Science and Policy at the University of North Dakota, where he is actively pursuing research in aquatic optics and other related fields. His current research interests include molecular scattering, inversion of volume scattering functions, and retrieval of vapor flux through scintillations.

Preface

This volume of *Light Scattering Reviews* is aimed at the presentation of recent advances in theoretical and experimental research in the general field of light scattering and radiative transfer. The first chapter of the volume prepared by V. Tishkovets and E. Petrova is devoted to studies of the optical properties of densely packed particles for cases where the near-field effects cannot be ignored. In the existing theories of multiple scattering of radiation, it is assumed that the secondary waves propagating from one particle of the medium to another are spherical, which makes these theories inapplicable to closely packed media, where the effects caused by the near field should be important. The mutual shielding of particles in the near field and the field inhomogeneity manifest themselves in some damping of the opposition enhancement in brightness and the strengthening of the negative branch of polarization observed in many atmosphereless celestial bodies. Accounting for the near-field effects allows the absence of the opposition spike in brightness accompanied by the well-pronounced negative branch of polarization observed in low-albedo asteroids to be explained. To interpret correctly the remote sensing data of different objects, methods describing the multiple scattering by discrete media with accounting for the near field are urgently required. The description of methods to deal with this fundamental problem is the main idea of the chapter. The chapter by A. Klose describes multispectral methods in luminescence imaging for the purpose of reconstructing the three-dimensional spatial distribution of light-emitting sources in biological tissue. These source reconstruction methods are covered by the term *multispectral luminescence tomography.* Because multispectral luminescence tomography is performed beyond the diffusion limit of light propagation in scattering tissue at wavelengths between 550 nm and 700 nm, a high-order radiative transfer model based on the simplified spherical harmonics equations (SP_N) needs to be employed. These equations constitute a system of coupled diffusion equations and, therefore, analytical and numerical solutions can easily be found. This chapter gives a detailed overview of the application of the SP_N equations in bioluminescence tomography of luciferase expression in cancer cells, Cerenkov light tomography of radionuclides, and excitation-resolved fluorescence tomography of quantum dots. Furthermore, solving the inverse source problem with an algebraic reconstruction technique, an expectation-maximization method, and gradient-based optimization methods, and with stochastic methods such as an evolution strategy is discussed. The chapter prepared by E. Kassianov et al. provides an up-to-date review of Markovian models introduced independently by the particle transport and atmospheric science communities and applications of these models to the stochastic radiative transfer in a cloudy atmosphere. After an introduction with historical background, the chapter describes different approaches for solving the stochastic radiative transfer equation, including both numerical and analytical averaging. The description is followed by a discussion of advanced passive and active remote sensing techniques that can be used to acquire input parameters of these models. In particular, data provided by next-generation scanning radar with Doppler and polarization capabilities are used for illustration. The relevance of these models to important atmospheric and climate-related issues is discussed throughout. A Kuusk et al. describe a comprehensive database of optical and structural data for the validation of forest radiative transfer models. Three mature stands at the forest test site Järvselja, Estonia, are extensively measured for use as a validation dataset for heterogeneous canopy reflectance models. Individual tree positions and crown dimensions were inventoried. In addition, leaf, needle, stem bark and branch bark visible and near-infrared (VNIR) reflectance spectra, and VNIR reflectance spectra of ground vegetation were measured. This in situ dataset is supported by atmospherically and radiometrically corrected Mode 3 CHRIS reflectance spectra for three view directions, and top-of-canopy VNIR nadir spectra from helicopter measurements. Due to its exhaustive nature the dataset should allow for a faithful reconstruction of the 3D canopy architecture suitable for inclusion in the latest generation of canopy reflectance models. Any radiative transfer simulations thus resulting may furthermore be compared against the remotely sensed observations of the test stands included in the dataset. Instrumentation for estimating the backscattering coefficient by measurement of the volume scattering function over various angular weightings in the backward direction has been in common use for over a decade. The chapter by T. Aoki presents current understanding of optical properties of snow and in particular its reflective characteristics such as snow BRDF and albedo. The relationship between the size of grains and levels of pollution with the snow reflective characteristics is studied. J. Sullivan et al. review different methods used for oceanic backscattering measurements and assess the robustness of theoretical relationships between the backscattering coefficient (spherically integrated volume scattering from 90 to 180 degrees) and the properties that are directly measured. The generic principles of operation, calibration and measurement protocols, and measurement uncertainties for backscattering sensors are reviewed, with an in-depth focus on the commercial WET Labs ECO sensor. The general theory of molecular scattering is reviewed by X. Zhang with a focus on the scattering by water and seawater resulting from fluctuations of density and concentration of the solution and of orientations of the water molecules. The theoretical development was introduced of a refined model estimating scattering by seawater as a function of salinity and temperature. The model results agree with laboratory observations within experimental error.

This volume is dedicated to the memory of a prominent scientist, Kusiel Shifrin (26.07.1918 to 02.06.2011), who made a number of important contributions to various scientific fields including light scattering by small particles, ocean optics, and remote sensing.

Bremen, Germany
October, 2011

Alexander A. Kokhanovsky

Part I

Light Scattering and Radiative Transfer

1 Light scattering by densely packed systems of particles: near-field effects

Victor P. Tishkovets and Elena V. Petrova

1.1 Introduction

The phenomenon of scattering and absorption of electromagnetic waves is actively used in many fields of science and engineering. Among them, there are the remote-sensing techniques for studying different objects and testing the quality of materials. They are extensively used in such areas as radio physics and radiolocation (to study radio-wave propagation and the properties of different objects (Ishimaru, 1978; Bass and Fuks, 1979; Tsang and Kong, 2001; Tsang et al., 2000, 2001)), optics of the atmosphere and ocean, in climatology and ecology (McCartney, 1976; Quinby-Hunt et al., 2000; Mishchenko et al., 2006), and biophysics and optics of solutions and colloids (for sorting cells and suspensions and their non-contact investigation (Horan and Wheeles, 1977; Hoekstra and Sloot, 2000)). In astronomy, up to now, remote-sensing methods have been of importance in the research of the Earth, other planets, their satellites, their atmospheres and regolith, cometary dust, etc. (see, e.g., Mishchenko et al., 2010 and references therein).

The remote-sensing methods are successfully used if the so-called direct and inverse problems can be solved. To solve the direct problem means to determine the characteristics of the scattered radiation from the specified characteristics of the incident radiation and the properties of the object. It is reduced to the solution of the Maxwell equations with specified boundary and initial conditions. To solve the inverse problem means to find the characteristics of the scattering object from the specified characteristics of the incident and scattered radiation. The latter problem is usually ill-posed, and its solution can be obtained only if some portion of a priori information about the object is available. Even in such a case, detailed information about the scattering object can be acquired only if the direct problem can be preliminarily solved. This solution should be sufficiently general, which allows the experimental and theoretical data to be compared in a wide range of the parameters characterizing the object.

The theory of light scattering by individual nonspherical particles of a rather general shape and by confined systems of homogeneous spherical particles has been thoroughly developed. However, in most cases, researchers have to deal with the media containing a very large number of scatterers, while progress in the theory applicable to discrete random media has been substantially slower, especially for particles with sizes comparable to the wavelength of the incident light, λ. In this size range, the wavelength dependence of characteristics of the scattered radiation is particularly strong, which makes this case paramount from the standpoint of interpretation of observational data. The currently available equations describing multiple scattering in media imply that the waves propagating from one particle of the medium to another are spherical (i.e., the scatterers are in the far-field zones of each other) (Mishchenko et al., 2006, 2010). This assumption is valid for sparse media, where the distances between the scatterers are much larger than their sizes and the wavelength. In such media, the wave propagating from particle j to particle s can be considered spherical, and in the vicinity of particle s this wave can be assumed to be a locally homogeneous plane wave. When describing the scattering by particle s, the simplification of a plane wave allows one to invoke such concepts of single-scattering theory as the scattering matrix, the extinction cross-section, etc. Moreover, if the scatterers are randomly positioned throughout the medium, the scattered radiation can be represented as a sum of two parts. One of them corresponds to incoherent (diffuse) scattering, and it is described by the well-known radiative transfer equation (RTE). This part of the scattered radiation depends on the properties of the medium relatively weakly. The second part arises from the interference of pairs of conjugate waves scattered along the same string of particles in the medium but in opposite directions (Watson, 1969; Wolf and Maret, 1985; Akkermans et al., 1988; Barabanenkov et al., 1991; Mishchenko et al., 2006). Constructive interference of the scattered waves manifests itself as a narrow interference peak of intensity centered at exactly the backscattering direction as well as causing the specific behavior of polarization in the backscattering domain. In the literature, this phenomenon is called weak localization (WL) of waves or coherent backscattering (CB) in a particulate medium (see, e.g., Mishchenko et al., 2006). It is currently believed that WL causes the opposition effects in brightness and polarization observed for many atmosphereless bodies of the solar system (Mishchenko et al., 2010). (In this review, we do not consider the so-called strong-localization effect. Remember that the difference between weak and strong localization is determined by the relation λ/l_{mfp}, where l_{mfp} is the mean free path (van Rossum and Nieuwenhuizen, 1999). For WL $\lambda/l_{mfp} \ll 1$, while the strong-localization effects occur if $\lambda/l_{mfp} \sim 1$.)

However, in the densely packed medium composed of scatterers, the sizes of which are comparable to the wavelength, the distances between them can be of the order of λ. In this case, the effects connected with the near field are of particular importance. Even for the homogeneous isotropic medium, the fields at a wavelength scale are strongly inhomogeneous (Tishkovets, 1998, 2008; Tishkovets et al., 1999, 2004a, 2004b; Petrova et. al., 2007; Petrova and Tishkovets, 2011). Because of this, to describe the light scattering by closely packed media is much more difficult than in the approximation of an incident plane homogeneous wave. Since it is a challenge to analyze all these phenomena, they are usually ignored in the current models of multiple scattering by densely packed media. The role of the near field in forming the brightness and polarization characteristics of the scattered radiation, especially in the opposition angular domain, has not been properly studied, which retards the interpretation of the remote sensing data for many objects, particularly, the data of optical observations of atmosphereless celestial bodies.

In the present review, we consider the influence of near-field effects on the characteristics of light scattering by ensembles of spherical particles. An assumption of particle sphericity is not crucial. It only allows us to avoid more complex and cumbersome calculations.

1.2 Scattering of electromagnetic waves by a system of spherical particles. Basic notions and equations

The theory of light scattering by systems (clusters) of spherical particles is considered in detail in many papers (see, for example, Fuller and Mackowski, 2000; Tsang et al., 2000; Tsang and Kong 2001; Mishchenko et al., 2002; Borghese et al., 2002 and references therein). The basic equations of the theory are presented below without derivation or proof.

Consider the scattering of a monochromatic plane wave by a system of homogeneous and isotropic spherical particles with arbitrary sizes and refractive indices. To specify the scattering geometry, we will use a Cartesian coordinate system shown in Fig. 1.1. An incident plane wave propagates in the coordinate system $\hat{\mathbf{k}}_0$ with the z_0-axis directed along the wave vector $\mathbf{k}_0$ ($k_0 = 2\pi/\lambda$, λ is the wavelength). Hereinafter bold letters with carets, $\hat{\mathbf{v}}$, denote the right-handed coordinate systems with the z-axis directed along the vector $\mathbf{v}$. The scattered wave propagates in the coordinate system $\hat{\mathbf{k}}_{sc}$ with the z-axis directed along the wave vector $\mathbf{k}_{sc}$ ($k_{sc} = k_0$). Coordinates of particles are determined by the radius-vectors $\mathbf{R}_j$ ($j = 1 \ldots N$, N is the number of particles in the cluster) in the laboratory coordinate system $\hat{\mathbf{n}}_0$ with axes (x, y, z) and the origin in the center of the cluster. The rotation from the coordinate system $\hat{\mathbf{n}}_0$ to the coordinate system $\hat{\mathbf{k}}_0$ is determined by the Euler angles φ_0, ϑ_0, ψ_0, the rotation from the coordinate system $\hat{\mathbf{n}}_0$ to the system $\hat{\mathbf{k}}_{sc}$ is determined by the Euler angles φ, ϑ, ψ and the rotation from the coordinate system $\hat{\mathbf{k}}_0$ to the system $\hat{\mathbf{k}}_{sc}$ is specified by the Euler angles φ_{sc}, ϑ_{sc}, ψ_{sc}. It is convenient to describe light scattering using the circular polarization (CP) repre-

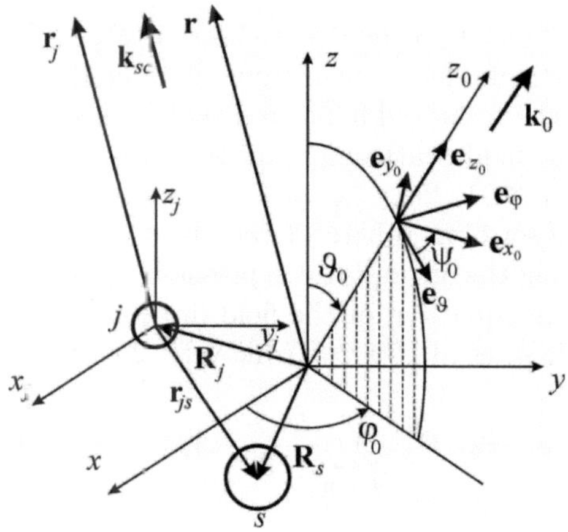

Fig. 1.1. The coordinate systems used for description of light scattering by a cluster.

sentation (Mishchenko et al., 2002). In the CP representation the electric vector of the plane wave can be written as

$$\mathbf{E}^{(0)} = \mathbf{e}_n(\hat{\mathbf{k}}_0) \exp(i\mathbf{k}_0\mathbf{r} - i\omega t) . \tag{1.1}$$

Here $n = \pm 1$, ω is the frequency, $\mathbf{r}$ is the radius vector of the observation point, $\mathbf{e}_n(\hat{\mathbf{k}}_0)$ is a covariant spherical basis vector (Varshalovich et al., 1988) formed by the unit vectors $\mathbf{e}_{x_0}$, $\mathbf{e}_{y_0}$ in the coordinate system $\hat{\mathbf{k}}_0$ (Fig. 1.1). When $n = 1$, the direction of rotation of the vector (1.1) corresponds to the clockwise direction when looking in the direction of the vector $\mathbf{k}_0$. When $n = -1$, the direction of rotation is anticlockwise. In the first case polarization is called right-handed and left-handed otherwise (Mishchenko et al., 2002). The time-dependence of the scattered light is also described by the multiplier $\exp(-i\omega t)$ and hereinafter, time dependencies will be omitted.

In any point outside the particles, the field scattered by the cluster can be represented as superposition of the fields scattered by constituents (Mishchenko et al., 2002, 2006; Tsang et al., 2000)

$$\mathbf{E} = \sum_{j=1}^{N} \mathbf{E}^{(j)} , \tag{1.2}$$

where $\mathbf{E}^{(j)}$ is the field scattered by the jth particle (see, for example, Borghese et al., 2002; Tishkovets, 2008):

$$\mathbf{E}^{(j)} = \sum_{LM} \left[B_{LM}^{(j)} h_L(k_0 r_j) \mathbf{X}_{LM}(\vartheta_j, \varphi_j) + \frac{A_{LM}^{(j)}}{k_0} \nabla \times h_L(k_0 r_j) \mathbf{X}_{LM}(\vartheta_j, \varphi_j) \right] . \tag{1.3}$$

Eq. (1.3) represents the wave scattered by the jth particle in the local coordinate system of the jth particle with the origin in the center of the particle and the axes parallel to those of the laboratory system (Fig. 1.1). Here ϑ_j, φ_j are the angular coordinates of the observation point (of the vector $\mathbf{r}_j$) in this coordinate system, $h_L(x)$ is the Hankel spherical function of the first kind, $\mathbf{X}_{LM}(\vartheta, \varphi)$ are the vector spherical harmonics (Varshalovich et al., 1988; Borghese et al., 2002), and $A_{LM}^{(j)}$, $B_{LM}^{(j)}$ are coefficients determined from the boundary conditions.

The properties of the scattered light are usually considered in the far-field zone. The criteria of the far-field scattering can be written as follows (Mishchenko et al., 2006): $k_0(r - a_m) \gg 1$, $2k_0 r \gg (k_0 a_m)^2$, $r \gg a_m$, where a_m is the radius of the smallest circumscribing sphere of the cluster, and r is the distance to the observation point. Using the asymptotic expression $h_L(x) \approx i^{-L-1} \exp(ix)/x$ ($x \gg L$, $x \gg 1$) the following equation for the field in the far-field zone can be obtained from Eq. (1.3) (Borghese et al., 2002; Tishkovets, 2008):

$$\mathbf{E}^{(j)} = \frac{\exp(ik_0 r)}{-ik_0 r} \exp(-i\mathbf{k}_{sc}\mathbf{R}_j) \sum_{LMp} \frac{2L+1}{2} A_{LM}^{(j)(pn)} D_{Mp}^{*L}(\hat{\mathbf{n}}_0, \hat{\mathbf{k}}_{sc}) \mathbf{e}_p^{(s)}(\hat{\mathbf{k}}_{sc}) . \tag{1.4}$$

Here $\mathbf{e}_p(\hat{\mathbf{k}}_{sc})$ is a covariant spherical basis vector (Varshalovich et al., 1988) in the coordinate system $\hat{\mathbf{k}}_{sc}$, $p = \pm 1$, $D_{Mn}^{L}(\hat{\mathbf{n}}_0, \hat{\mathbf{k}}_{sc}) = D_{Mn}^{L}(\varphi, \vartheta, \psi)$ is the Wigner function (Varshalovich et al., 1988), the asterisk denotes complex conjugation, and

$$A_{LM}^{(j)(pn)} = i^{-L} \sqrt{\frac{1}{2\pi(2L+1)}} \left(A_{LM}^{(j)} + pnB_{LM}^{(j)}\right).$$

Coefficients $A_{LM}^{(j)(pn)}$ are determined from the system of equations (see, for example, Tishkovets, 2008; Tishkovets and Jockers, 2006)

$$\begin{aligned} A_{LM}^{(j)(pn)} &= a_L^{(j)(pn)} \exp(i\mathbf{k}_0\mathbf{R}_j) D_{Mn}^L(\hat{\mathbf{n}}_0, \hat{\mathbf{k}}_0) + \\ &+ \sum_q a_L^{(j)(pq)} \sum_{s\neq j} \sum_{lm} A_{lm}^{(s)(qn)} H_{LMlm}^{(q)}(\hat{\mathbf{n}}_0, \hat{\mathbf{r}}_{js}) \end{aligned} \quad (1.5)$$

Here $\hat{\mathbf{r}}_{js}$ is a coordinate system with the z-axis along the vector $\mathbf{r}_{js}$ (Fig. 1.1), $q = \pm 1$, $a_L^{(j)(pn)} = a_L^{(j)} - pnb_L^{(j)}$, and $a_L^{(j)}$; $b_L^{(j)}$ are the Mie coefficients for the jth particle (Mishchenko et al., 2002).

$$H_{LMlm}^{(q)}(\hat{\mathbf{n}}_0, \hat{\mathbf{r}}_{js}) = \frac{2l-1}{2}(-1)^m \sum_{l_1} i^{-l_1} h_{l_1}(k_0 r_{js}) D_{m_1 0}^{l_1}(\hat{\mathbf{n}}_0, \hat{\mathbf{r}}_{js}) C_{LMl-m}^{l_1 m_1} C_{Lql-q}^{l_1 0} \quad (1.6)$$

are the coefficients of the addition theorem for the vector Helmholtz harmonics (see, for example, Felderhof and Jones, 1987; Tishkovets and Litvinov, 1996; Fuller and Mackowski, 2000). In Eq. (1.6) $m_1 = M - m$ and C are Clebsch–Gordan coefficients (Varshalovich et al., 1988).

Eq. (1.4) represents the transverse spherical wave outgoing from the jth particle. Substitution of Eq. (1.4) into Eq. (1.2) gives the transverse spherical wave propagating from the cluster. (The size of the cluster is assumed to be small as compared to the distance from the observation point. In this case the direction of scattering is the same for all particles (Fig. 1.1).) The wave amplitude is proportional to r^{-1}, where r is the distance from the cluster. In contrast to field (1.4), field (1.3) is not a transverse spherical wave. This field contains the radial component (the component along the vector $\mathbf{r}_j$) and the terms decreasing as $r_j^{-n_0}$ with $n_0 > 1$. The last statement follows from the representation of the Hankel function $h_l(x)$ in the form of a finite power series, containing terms x^{-n_0} with $n_0 = 1, 2, 3, \ldots, l+1$ (Grandshtein and Ryzhik, 1980). In the literature, the terms decreasing as r^{-n_0} with $n_0 > 1$ are associated with the near field (Greffet and Carminati, 1998). Scattering of such a complex field is described by the system of equations (1.5). In this system coefficients (1.6) determine the electromagnetic field between particles, the behavior of which depends on a distance between particles. In the general case coefficients (1.6) contain the terms decreasing faster than r_{js}^{-1}. If one keeps only the terms proportional to r_{js}^{-1} in these coefficients, the coefficients take the form (Tishkovets and Mishchenko, 2004)

$$H_{LMlm}^{(q)}(\hat{\mathbf{n}}_0, \hat{\mathbf{r}}_{js}) = \frac{2l+1}{2} \frac{\exp(ik_0 r_{js})}{-ik_0 r_{js}} D_{Mq}^L(\hat{\mathbf{n}}_0, \hat{\mathbf{r}}_{sj}) D_{mq}^{*l}(\hat{\mathbf{n}}_0, \hat{\mathbf{r}}_{sj}). \quad (1.7)$$

Here $\hat{r}_{sj}$ is the coordinate system with the z-axis along the vector r_{sj}.

Coefficients (1.7) correspond to the spherical wave propagating from the sth particle to the jth one. In other words, if the near-field components are ignored in (1.6), the coefficients $H^{(q)}_{LMlm}(\hat{\mathbf{n}}_0,\ \hat{\mathbf{r}}_{js})$ describe spherical waves propagating between particles regardless of the distances between them. This means that the sizes of particles are negligible as compared to the distance between them. Coefficients (1.7) can be obtained formally from (1.6) under conditions $k_0 r_{js} \gg 1$, $r_{js} \gg a_j + a_s$, and $k_0 r_{js} \gg \max(L + l)$, where a_i is the radius of the ith particle. Nevertheless, we will use them for any distance between particles to compare two models: the model with the near field and the model ignoring the near field.

In the next sections, the influence of the near field on the intensity and the degree of linear polarization of light scattered by clusters of spherical particles will be considered. We will restrict ourselves only to these characteristics, since they are precisely the quantities most frequently measured in the investigations of different laboratory and naturally occurring objects, particularly the atmosphereless celestial bodies.

The scattering characteristics are determined by the scattering matrix $\mathbf{F}(\mathbf{k}_0,\ \mathbf{k}_{sc})$ which describes the transformation of the Stokes vector of the incident light $\mathbf{I}_0$ into that of the scattered light $\mathbf{I}$

$$\mathbf{I} = \frac{1}{(k_0 r)^2}\mathbf{F}(\mathbf{k}_0,\ \mathbf{k}_{sc})\mathbf{I}_0\,.$$

If the incident light is unpolarized, the relative intensity I (i.e., normalized to the intensity of the incident radiation) and the degree of linear polarization P of scattered light can be determined as follows (see, for example, Mishchenko et al., 2002, 2010):

$$I = F_{11} = \frac{1}{2}\sum_{pn}\left\langle |S_{pn}|^2\right\rangle, \qquad P = -\frac{F_{21}}{F_{11}} = \frac{\frac{1}{2}\sum\limits_{pn}\left\langle S_{pn}S^{*}_{-pn}\right\rangle}{I}\,. \tag{1.8}$$

Here S_{pn} is the amplitude scattering matrix of a cluster in the CP-representation, $p, n = \pm 1$, the angular brackets denote averaging over the cluster orientation, if required. The amplitude scattering matrix S_{pn} of a cluster can be represented similar to Eq. (1.2) as a sum of the amplitude scattering matrices for all particles

$$S_{pn} = \sum_{j=1}^{N} t^{(j)}_{pn}(\mathbf{k}_0,\ \mathbf{k}_{sc}) \tag{1.9}$$

where $t^{(j)}_{pn}(\mathbf{k}_0,\ \mathbf{k}_{sc})$ is the amplitude matrix of jth particle of the cluster.

In the circular-polarization basis (the CP-representation) used here, the basis unit vectors in the meridional planes $(z,\ \mathbf{k}_0)$ and $(z,\ \mathbf{k}_{sc})$ are the covariant helical basis vectors (Varshalovich et al., 1988) $\mathbf{e}^{(h)}_n(\hat{\mathbf{k}}_0)$ and $\mathbf{e}^{(h)}_p(\hat{\mathbf{k}}_{sc})$, respectively. The helical orts $\mathbf{e}^{(h)}_n(\hat{\mathbf{k}}_0)$ and $\mathbf{e}^{(h)}_n(\hat{\mathbf{k}}_{sc})$ are formed by the unit vectors $\mathbf{e}_\vartheta$ and $\mathbf{e}_\varphi$ of the spherical coordinate system (Varshalovich et al., 1988) in the given meridional planes. The rotation from the vector $\mathbf{e}^{(h)}_n(\hat{\mathbf{k}}_0)$ to the vector $\mathbf{e}_n(\hat{\mathbf{k}}_0)$ and from the vector $\mathbf{e}_p^{(h)}(\hat{\mathbf{k}}_{sc})$ to the vector $\mathbf{e}_p(\hat{\mathbf{k}}_{sc})$ is described by the rotation matrices depending on the Euler angles ψ_0 and ψ, respectively (Fig. 1.1). Then, after accounting for Eq. (1.4), we have

$$t_{pn}^{(j)}(\mathbf{k}_0,\ \mathbf{k}_{sc}) = \exp\left(-i\mathbf{k}_{sc}\mathbf{R}_j + in\psi_0 - ip\psi\right) \sum_{LM} \frac{2L+1}{2} A_{LM}^{(j)(pn)} D_{Mp}^{*L}(\hat{\mathbf{n}}_0, \hat{\mathbf{k}}_{sc}) \tag{1.10}$$

Matrix (1.10) describes transformation of components of the electric vector of the incident wave determined in one meridional plane into those of the scattered wave determined in another meridional plane. Transformation of matrix (1.10) to the scattering plane can be made with the help of the addition theorem for the Wigner functions (Varshalovich et al., 1988).

1.3 Shielding of particles by each other in the near field

1.3.1 Mutual shielding in simple systems of particles

The idea of shielding (shadowing) is usually associated with the case when the sizes of scatterers are comparable to or larger than the wavelength (see, e.g., Hapke, 1993). However, for the scatterers that are substantially smaller than the wavelength, the influence of shielding can be also noticeable at distances comparable to the wavelength, i.e. in the near field of the scatterers. Though the near field can manifest itself in all of the elements of the scattering matrix, the mutual shielding mostly affects the intensity. Because of this, while analyzing the near-field effect in the mutual shielding for the clusters of spherical particles, we will mainly consider the intensity of the scattered radiation (the manifestations of the near field in polarization will be considered in the next sections).

In the external field, the scatterers, the sizes of which are smaller than the wavelength, are polarized as dipoles, and the coefficients (1.6) should describe the distribution of the field of charges induced in them. Let us consider qualitatively the peculiarities of light scattering by a pair of closely located scatterers of small sizes as compared to the wavelength, which are polarized as dipoles in the external field. Figure 1.2 shows two pairs of such scatterers illuminated by the external field $\mathbf{E}^{(0)}$ and the configuration of charges induced in them. The scatterers are located in the scattering plane (in the picture plane), and the incident radiation is polarized in the scattering plane too. In Fig. 1.2(a), the configuration of charges is shown for the case when the particles do not interact in the near (electrostatic) fields. In such configuration of charges, the intensity of light scattered by particles along the line AB, passing through their centers, differs from zero. Figure 1.2(b) shows the configuration of charges in the scatterers interacting in the electrostatic fields. In this case, the intensity of light scattered along the line AB is equal to zero. In other words, implication of the near field leads to "shielding" of a scatterer by another one in the direction passing through their centers.

Of course, a contribution of the near field in Fig. 1.2(b) is exaggerated. Angular dependencies of the intensity of scattered light by a pair of identical small scatterers in contact (bisphere) are shown in Fig. 1.3. The size parameter of particles is

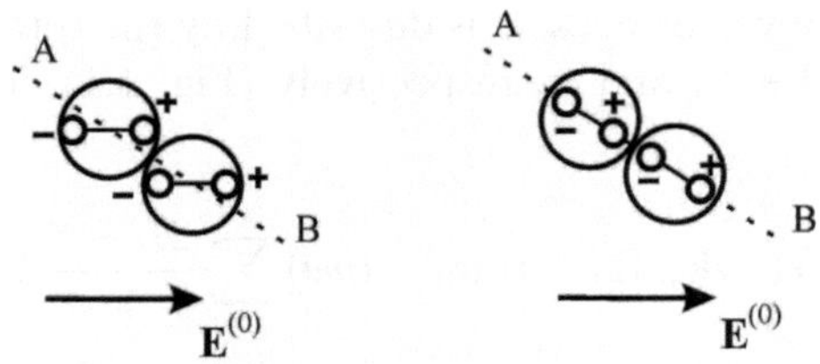

Fig. 1.2. The scheme explaining 'shielding' of small scatterers (dipoles) in the near field. The line AB passes through the centers of scatterers. If the interaction of the induced charges is ignored, the intensity of light scattered by particles along this line differs from zero (a). The intensity of light scattered along the line AB by particles interacting in the near field is equal to zero (b).

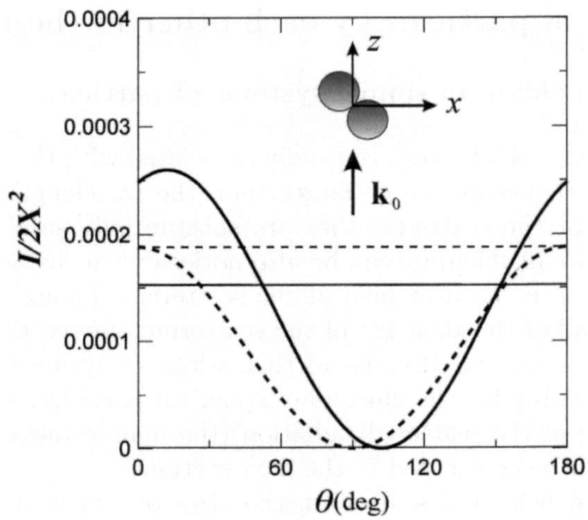

Fig. 1.3. The relative intensity of light scattered by a bisphere consisting of Rayleigh scatterers in contact versus a scattering angle. The size parameter of the bisphere components is 0.01, and the refractive index is $\tilde{m} = 10.0 + i0.0$. Solid and dashed curves show the models accounting for and ignoring the near field, respectively. Thick and thin curves correspond to the case when the incident radiation polarized in the scattering plane and perpendicular to it, respectively.

$X = 0.01$, and the refractive index is $\tilde{m} = 10.0 + i0.0$. Scatterers are located in the xz plane, and the angle between the z-axis and the symmetry axis of the bisphere is 45°. The intensity of the scattered light is divided by a quantity $2X$. The scattering angle ϑ is measured in the xz plane from the z-axis in the direction of the positive values of x (clockwise in the picture).

As is seen from the plot, in the model ignoring the near field, the angular dependence of intensity is identical to that for a spherical particle in the Rayleigh limit. In the model accounting for the near field, the intensity of light scattered in directions $\vartheta \approx 0$ and $\vartheta \approx 180°$ strongly depends on the polarization of the incident light. If the incident light is polarized in the scattering plane, intensity of the scattered light in these directions is much higher than in the previous model. This phenomenon is caused by substantial increase of the dipole moments because of electrostatic interaction of scatterers (Fig. 1.2(b)). The minimum of the intensity is located at $\vartheta \approx 105°$ while the schematic representation in Fig. 1.2(b) predicts it to be in the direction along the line AB passing through the centers of dipoles. (In a given example, this direction corresponds to 135°.) This means that the dipole moments of scatterers in Fig. 1.3 are oriented in the direction $\vartheta \approx 105°$. At the interval of scattering angles of $100° < \vartheta < 160°$, the intensity in the model considering the near field is appreciably less than that in the model ignoring the near field. This diminution of the intensity is caused by "shielding" of scatterers illustrated qualitatively by Fig. 1.2.

In the previous example the near field is the electrostatic field. Shielding of scatterers arising in the electromagnetic interaction of the wavelength-sized scatterers will be considered in the following examples. Figures 1.4(a) and 1.4(b) depict the intensity of light scattered by bispheres as a function of the scattering angle. The axis of the bispheres is perpendicular to the direction of propagation of the incident unpolarized light indicated by the wave vector $\mathbf{k}_0$. The size parameter of the bisphere components is $X = 4.0$, and their refractive index is $\tilde{m} = 1.32 + i0.05$.

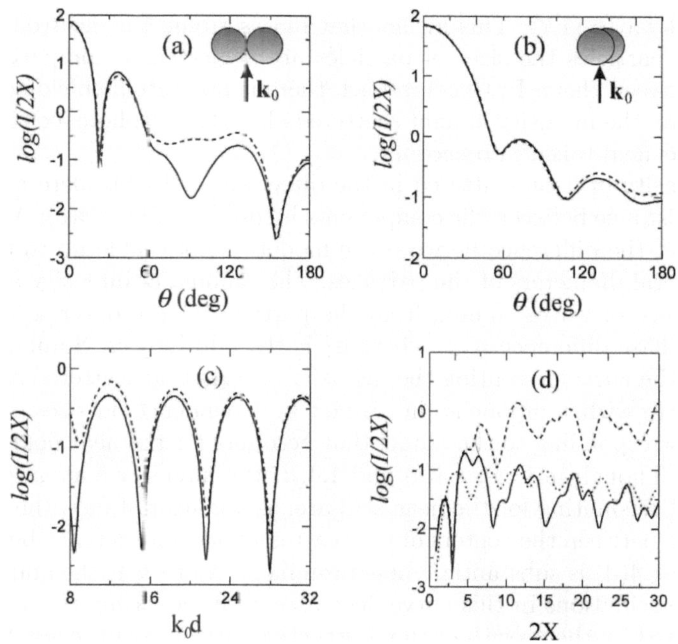

Fig. 1.4. The relative intensity of radiation (in logarithmic scale) scattered by bispheres. Plots (a) and (b) depict the dependence on a scattering angle for two orientations of the bisphere; panel (c) presents the relative intensity of radiation scattered along the bisphere axis versus a distance k_0d between the bisphere components; and panel (d) shows the same quantity versus a size parameters of the components being in contact. The models accounting for (coefficients (1.6)) and ignoring (coefficients (1.7)) the near field are shown with solid and dashed curves, respectively. The dotted curve in panel (d) presents the model for the individual component, and these data are divided by a quantity X, not $2X$.

A relatively small real part of the refractive index and a rather large imaginary part were chosen in order to avoid the appearance of sharp "splashes" peculiar to nonabsorbing scatterers with high values of the real part of the refractive index. For the above refractive index, the curves for intensity of the scattered radiation are relatively smooth, which makes the comparison of the considered models to be much easier.

The scattering plane corresponds to the picture plane. Two cases of orientation of the bispheres with respect to the scattering plane are considered. Orientation of bispheres is shown in the right upper corners of Figs. 1.4(a) and 1.4(b). The intensity of scattered light (the element F_{11} Eq. (1.8)) is divided by a quantity $2X$.

As is seen from a comparison of the curves in Fig. 1.4(a), the intensity of light scattered along the axis of the bisphere (in the direction $\vartheta = 90°$) calculated with the near field taken into account, is significantly lower than that calculated without the near field components. The intensity of light scattered in the direction perpendicular to the bisphere axis (Fig. 1.4(b)) is virtually the same in both models. As was shown in the previous section, coefficients (1.6) completely describe all the peculiarities of the electric field between the scattering particles. These peculiarities manifest themselves, in particular, in mutual shielding of particles, since the sizes of particles and distances between them are comparable. If the field components decreasing more rapidly than r_{js}^{-1} are ignored in the coefficients (1.6), they describe the spherical waves (1.7). This means that in describing the electromagnetic field between the particles the sizes of particles are neglected in comparison with the distances between them. In this approach there is no mutual shielding of particles and, therefore, the intensity of light scattered along the bisphere axis is larger than with the near field taken into account.

The intensity of light scattered in the direction of the bisphere axis in dependence on a distance between the components is shown in Fig. 1.4(c). As can be seen from the plot, the difference between the models is noticeable up to the distances of about several diameters of the particles. The minima of intensity are caused by the interference of waves coming from the particles to the observation point and having the phase difference $n_1\pi$, where n_1 is the odd integer. Similar minima are also seen in the curve presenting the intensity of radiation scattered along the axis of the bisphere with components in contact in dependence on sizes of the components and corresponding to the model that accounts for the near field (Fig. 1.4(d), solid curve). Though in Figs. 1.4(c) and 1.4(d) the intensity curves corresponding to the model accounting for the near field are, as a matter of fact, the dependences on a distance between the centers of the scatterers, the interference behavior of the curve in Fig. 1.4(d) is substantially less prominent. Moreover, the amplitude of the interference oscillations in this curve decreases with increasing the scatterer sizes, which is caused by the growth of the destructive influence of the scatterer sizes on the wave interference due to the shielding strengthening.

For comparison, the curve corresponding to the individual scatter is also shown in Fig. 1.4(d) (dotted curve). In contrast to the other curves normalized to $2X$, this function is normalized to X. It is seen that the shielding tends to the limit of geometric optics, when the scatterer sizes increases. At this limit, the intensity of radiation scattered by the bisphere should be halved and become equal to that for the individual particle. The substantial difference in intensity between the models accounting for and ignoring the near field (approximately by an order of magnitude) is caused by a high contribution of multiple scattering in the case, when the near filed is ignored. Specifically, a substantial contribution of multiple scattering, which decreases with increasing distance between the bisphere components, is well noticeable in Fig. 1.4(c) in the minima of intensity at $k_0 d \approx 9$ and $k_0 d \approx 15$.

In order to gain insight into the mechanism of mutual shielding of scatterers, let us expand the coefficients (1.5) to a series in terms of iterations. In other words, we write the solution of the system of equations (1.5) as

$$\begin{aligned} A_{LM}^{(j)(pn)} &= a_L^{(j)(pn)} \exp\left(i\mathbf{k}_0\mathbf{R}_j\right) D_{Mn}^{L}(\hat{\mathbf{n}}_0,\ \hat{\mathbf{k}}_0) + \\ &+ \sum_q a_L^{(j)(pq)} \sum_{slm} a_l^{(s)(qn)} H_{LMlm}^{(q)}(\hat{\mathbf{n}}_0\hat{\mathbf{r}}_{js}) \exp(i\mathbf{k}_0\mathbf{R}_s) D_{mn}^{l}(\hat{\mathbf{n}}_0,\ \hat{\mathbf{k}}_0) \\ &+ \ldots . \end{aligned} \tag{1.11}$$

This series is convenient to be interpreted as an order-of-scattering expansion of the solution of the system (1.5). Here, the first term in the right-hand side corresponds to the single scattering by a particle j located at the point $\mathbf{R}_j$, the second one corresponds to the double scattering, first by particle s and then by particle j. It is worth noting that such an interpretation should be considered as a mathematical construction (Mishchenko et al., 2011). The expressions for the scattered field (1.2)–(1.4) and the systems of equations (1.5) are time-independent. These relations describe the radiation scattered by a cluster of particles as if it is one object. Therefore, such interpretation does not correspond to the real multiple scattering process requiring the lag of waves to be accounted for, when considering their propagation between the particles. However, this presentation is a convenient basis for considering the forming mechanisms of different scattering characteristics.

Let us use the double scattering approximation to consider how the shielding arises. Figure 1.5 gives us an idea of the contribution of each scattering order to the intensity of radiation scattered by the cluster shown in Fig. 1.4(a).

In Fig. 1.5, curve *1* corresponds to the single scattering contribution including the interference of singly scattered waves, model *2* additionally takes into account the double scattering contribution including the interference of doubly scattered waves, and the points present the model, which additionally includes the contribution from the interference of waves scattered once and twice. The latter practically coincides with the exact solution considering all of the scattering orders (solid curve). As is seen from Fig. 1.5, under the current approximation, the mutual shielding is described by the interference of singly and doubly scattered waves. In the next section, we will show that this interference is also responsible for the mechanism inducing negative polarization in the densely packed systems of particles. It is worth stressing that the interference of this type (the interference of waves from different scattering orders) is ignored in the currently used theories of multiple scattering by discrete systems.

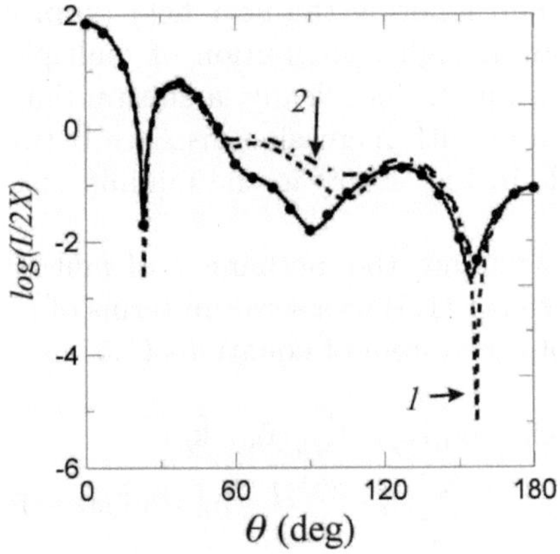

Fig. 1.5. As in Fig. 1.4(a), but for contributions of different scattering orders. Model *1* (dashed) takes into account the singly scattered radiation and its interference; model *2* (dash-dotted) is model *1* plus the contribution of the doubly scattered radiation and its interference; the dots present the contribution of model *2* plus that from the interference of radiation scattered once and twice. The exact solution (see also Fig. 1.4(a)) is shown by a solid curve.

1.3.2 Mutual shielding of particles in chaotically oriented large clusters

Let us consider now the shielding phenomenon in more complex randomly oriented clusters of spherical particles. The characteristics of light scattered by randomly oriented clusters can be calculated with the very efficient computer codes, developed by Mackowski and available on the Internet (Mackowski, ftp). These codes naturally take into account the near field between the particles composing the clusters. For our specific purpose, the codes were adapted to neglect the near-field components.

Clusters of identical particles (monomers) (see Fig. 1.6) were generated according to the procedure described by Mackowski (1995). Their shape is close to spherical, and the monomers are randomly positioned. The size parameter of the constituting monomers is assumed to be $X = 1.5$, and the size parameters of the clusters $X_0 = k_0 a_m$ (where a_m is the radius of the minimal sphere circumscribing the cluster) are approximately 9.25, 11.9, and 14.67, respectively, which means that their diameter is approximately 3, 4, and 5 wavelengths, respectively. The packing density of monomers in the clusters is $\xi = N(X/X_0)^3 \approx 0.2$ (where N is the number of monomers in the cluster).

Figures 1.7 and 1.8 depict the relative intensity I and the degree of linear polarization P of radiation scattered by randomly oriented clusters in dependence on a scattering angle. The considered values of the refractive index are $\tilde{m} = 1.5 + i0.001$ and $\tilde{m} = 1.5 + i0.1$ for Figs. 1.7 and 1.8, respectively. The thick and thin curves correspond to the models calculated with the near field accounted for (with coefficients (1.6)) and ignored (with coefficients (1.7)), respectively. For all cases, the values of intensity are divided by X_0^2. As is seen from these plots, if the near field is ignored, the intensity substantially increases in the whole range of scattering angles.

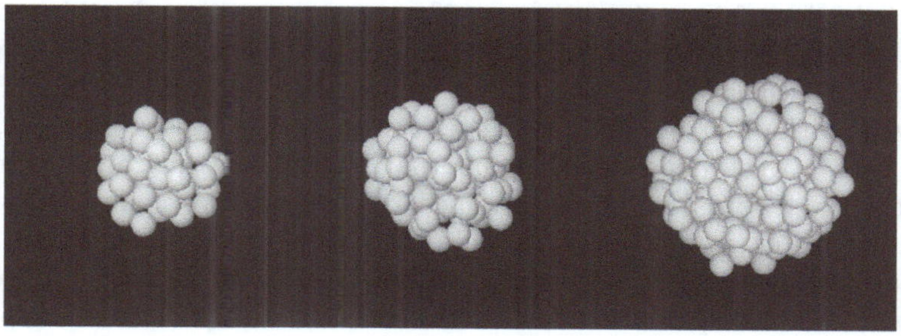

Fig. 1.6. The clusters consisting of 50, 100, and 200 spherical monomers, which are considered in this section.

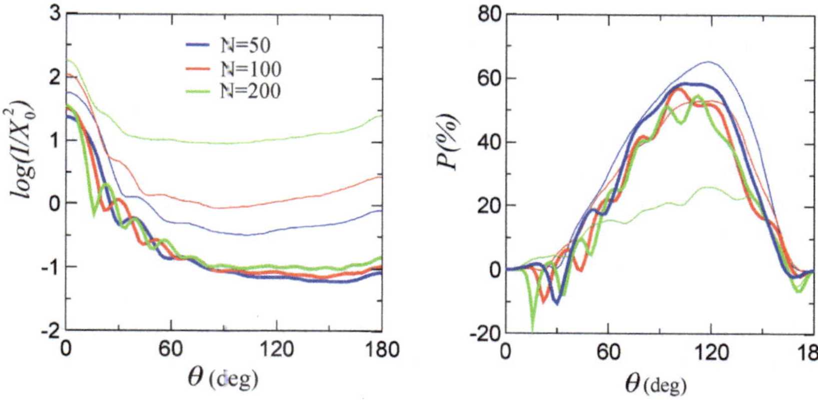

Fig. 1.7. The relative intensity I (in logarithmic scale) and the degree of linear polarization P of radiation scattered by randomly oriented clusters of spherical particles ($X = 1.5$) calculated with the near field taken into account (thick curves) and ignored (thin curves). The numbers of particles corresponding to different types of the curves are shown in the legend. The refractive index is $\tilde{m} = 1.5 + i0.001$.

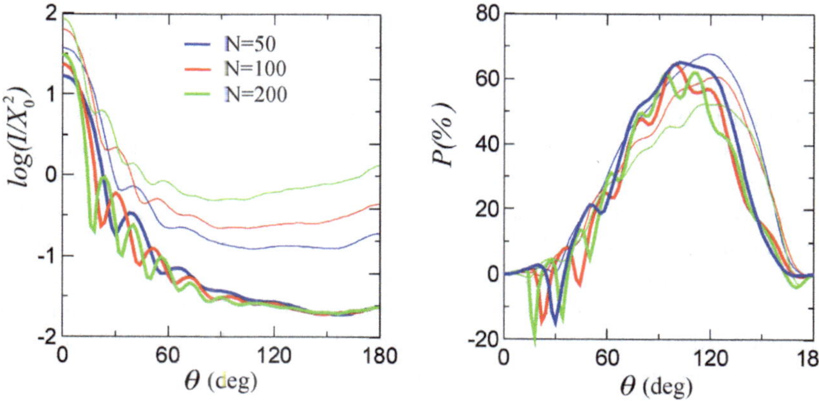

Fig. 1.8. As in Fig. 1.7, but for the refractive index $\tilde{m} = 1.5 + i0.1$.

The behavior of intensity at $\vartheta > 60°$ attracts attention. In this region, the intensity calculated with accounting for the near field weakly depends on a number of particles N (especially for strongly absorbing scatterers). Since the intensity is normalized to the unit area of the cluster's cross-section (more precisely, divided by X_0^2), such a weak dependence means that, at least in this range of scattering angles, the intensity is mainly determined by the particles composing a surface layer of the cluster. The other particles in the cluster are shielded by the particles of the surface layer. When the near field is ignored, the particles do not shield each other, and more particles are involved into the multiple scattering processes. This increases the contribution of the multiple scattering and, consequently, the intensity of the scattered radiation in comparison to that in the model accounting for the interaction in the near field. Due to the same cause, the intensity more strongly depends on a number of particles in the cluster, if the near field is ignored, especially for weakly absorbing scatterers.

At the scattering directions close to $\vartheta \approx 0°$, the interference of waves scattered once gives the main contribution to the intensity of the scattered radiation. If the near field is taken into account, the mutual shielding of particles and the shifts of waves inside the cluster make the particles located near the equator (limb) of the cluster to be the main contributors to this interference (Tishkovets et al., 2004a). The number of these particles $\tilde{N}$ is approximately proportional to the radius of the cluster. Since the contribution of the interference is proportional to $\tilde{N}(\tilde{N}-1)$, the intensity of the scattered radiation in the direction $\vartheta \approx 0°$ is proportional to the area of the cluster's cross-section. Because of this, under the specified normalization, the intensity in this scattering direction relatively weakly depends on a number of particles in the cluster. If the near field is ignored, all of the particles take part in the interference, which makes the intensity of the scattered radiation to be higher and more strongly dependent on a number of particles in the cluster.

Let us consider the behavior of linear polarization in more detail. As is seen from the above plots, the ignoring of the near field generally induces smaller changes in this characteristic relative to those in intensity. Only for $N = 200$ and $\tilde{m} = 1.5 + i0.001$ (Fig. 1.7), the polarization of radiation scattered in side directions is substantially smaller in the model ignoring the near field, which is caused by a considerable contribution of multiple scattering. At the same time, almost all polarization curves demonstrate the feature in the backscattering domain, which is not typical for individual constituents of this size. Direct calculations show that the degree of linear polarization for particles with $X = 1.5$ and the refractive indeces specified here is positive in the whole range of scattering angles. However, the clusters of such particles do demonstrate the branch of negative values of polarization at scattering angles close to $\vartheta = 180°$. Since this behavior of linear polarization is similar to that characteristic of the most atmosphereless bodies of the Solar system and the cometary dust (see, e.g., Mishchenko et al., 2010), to understand the mechanisms inducing the branch of negative polarization is enormously important for correct interpretation of the groundbased observations of these celestial objects.

As is seen from Figs. 1.7 and 1.8, the branch of negative polarization may also appear in the case, when the near field is ignored. Because of this, the supposition on the effect of weak localization of waves as a cause of this branch (Muinonen, 1990, 2004; Muinonen et al., 2002; Shkuratov, 1989; Shkuratov et al., 1994, 2002) seems to be well grounded. The WL effect appears due to the constructive interference of multiply scattered waves propagating along the same string of particles in the medium but in opposite directions. It is supposed that this interference is also responsible for the opposition effect observed for atmosphereless celestial bodies as a nonlinear enhancement of brightness, when the scattering angle approaches opposition (see, e.g., Mishchenko et al., 2010). The weakly pronounced opposition effect can be seen in the intensity curves that present the models ignoring the near field (Figs. 1.7 and 1.8). Recall that the contribution of multiple scattering from all particles of the cluster is significant in this case, since the mutual shielding is absent. When the near field is taken into account, the scattering characteristics of the clusters are determined by the particles composing a surface layer of the cluster. As can be seen from the plots, the opposition effect in intensity is less prominent (if at all), and the branch of negative polarization is better developed. This suggests that in the scattering by systems of closely packed particles, the branch of negative polarization is not only formed due to the WL effect, but can be also induced by the other mechanism that works effectively at small distances between the scatterers. Let us consider in more detail the field structure in the close vicinity of a scattering particle and the manifestations of its peculiarities in the intensity and polarization of radiation scattered by simple ensembles of closely positioned spherical particles.

1.4 Interaction of particles in the near field and the opposition phenomena

1.4.1 The field inhomogeneity near the scatterers

Everywhere in space, only the total field, i.e. a sum of the incident (1.1) and scattered components (1.2), is a real physical field (Mishchenko et al., 2011)), i.e. only the field

$$\mathbf{E}_{tot} = \mathbf{E}^{(0)} + \mathbf{E} \tag{1.12}$$

has a physical meaning.

At the distances significantly exceeding the wavelength, the scattered wave is a spherical wave going away from the particle. Since the directions of the incident and scattered waves do not coincide in this case, they can be considered to be independent (except the direction $\vartheta = 0$, where these waves are connected). At the wavelength-order distances from the scatterer, the incident and scattered fields cannot be treated as being independent.

Let us analyze the behavior of the field (1.12) near a spherical particle. We assume that the incident wave is linearly polarized in the plane $x_0 y_0$ (see Fig. 1.9). To calculate the field distribution in the particle's vicinity, the formulas of the Mie theory (Mishchenko et al., 2002) can be used. Figure 1.9 depicts the results of calculations of the max(Re $|\mathbf{E}_{tot}|$) surfaces and the directions of the field vector near the particle (Tishkovets, 1998; Tishkovets and Litvinov, 1999; Tishkovets et al. 1999; 2004). The particle's characteristics are: $X = 4$ and $\tilde{m} = 1.32 + i0.05$. (If the field is homogeneous, the max(Re $|\mathbf{E}_{tot}|$) surfaces are the constant phase surfaces.)

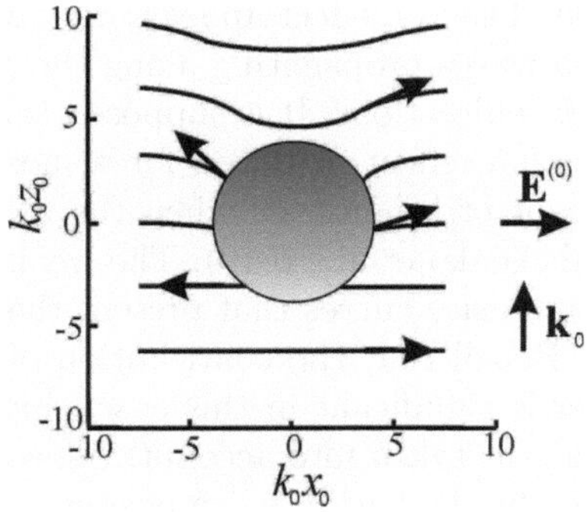

Fig. 1.9. The field structure in the close vicinity of a spherical particle with a size parameter $X = 4$ and refractive index $\tilde{m} = 1.32 + i0.05$. $\mathbf{k}_0$ is the wave vector, and $\mathbf{E}^{(0)}$ is the field vector of the incident wave. The vector signs near the particle indicate the direction of the strength vectors of the total field.

Figure 1.9 shows the cross-sections of the max(Re $|\mathbf{E}_{tot}|$) surfaces near the particle by the plane x_0, z_0. Such a shape of the surfaces is caused by the lag (retardation) of the induced field within a particle with respect to the incident field.

Under a specified polarization of the incident wave, the vector (1.12) in the x_0, z_0 plane is parallel to this plane, and its components E_ϑ, E_r are nonzero, while the component $E_\varphi = 0$. In the y_0, z_0 plane, the field vector $\mathbf{E}_{tot}$ (1.12) is parallel to the vector $\mathbf{E}^{(0)}$, and its component E_φ is nonzero, while the components $E_\vartheta = E_r = 0$ (see, e.g., Tishkovets et al., 1999; 2004). Generally speaking, the field vector in the x_0, z_0 plane is not tangent to the max(Re $|\mathbf{E}_{tot}|$) surfaces. The direction of the vector (1.12) in this plane can be reliably determined only in the close vicinity of the particle's surface up to the wavelength-order distances, if the particle is highly absorbing and not very large. In the close vicinity of the particle, on the max(Re $|\mathbf{E}_{tot}|$) surfaces, the real parts of the components E_x, E_z of the vector (1.12) substantially exceed the imaginary parts. When the distance to the particle increases, the imaginary part of the component E_z becomes comparable to the real one, and it is difficult to determine the direction of the vector (1.12) there. It is worth noting that the larger the real part of the refractive index $Im(\tilde{m})$ and smaller the quantity $Re(\tilde{m}) - 1$, the simpler the field structure near the scatterer.

When explaining the characteristics of radiation scattered by closely packed systems of particles, we will consider the following properties of the field (1.12) near the particle.

(1) Near the scattering particle, the fields of the incident and scattered waves are connected. This means that the contribution of the interference of the radiation scattered once by particle i and the radiation scattered first by particle j, a neighbor of particle i, and then by particle i is nonzero, i.e. the contribution of the interference of singly and doubly scattered waves is nonzero. Note that, if the secondary waves propagating between the scatterers are assumed spherical (except the case, when the particles are in the straight line along the wave vector of the incident wave), this contribution is zero, since the directions of propagation of the incident and scattered waves do not coincide and, therefore, these waves are not connected. This connection is provided by the near field.

(2) The turn of the field vector (1.12) relative to the field vector of the incident wave leads to the appearance of the component E_z, which is always located in the scattering plane. Consequently, in the radiation scattered by the particle being in such a field, the portion of radiation polarized in the scattering plane, $I_{||}$, increases. Finally, this can result in the negative branch of the degree of linear polarization, which resembles that observed in some atmosphereless celestial bodies and cometary dust.

(3) The scale of the inhomogeneous zone of the wave is of an order of the wavelength λ. This means that the considered effects are essential only for the particles with sizes comparable to or less than the wavelength and are of no importance for large particles with sizes substantially exceeding λ. These effects are also insignificant for random media composed of the scatterers that are much smaller than the wavelength, since they practically induce no distortion in the exciting wave.

Let us consider the scattering of the field shown in Fig. 1.9 by the neighbor particles placed in the inhomogeneous zone. For simplicity, we will assume that the sizes of these 'test' particles are small relative to the wavelength (i.e., they are Rayleigh scatterers). In this case, the inhomogeneity of the field exciting each of these particles can be ignored.

In Fig. 1.10, the scattering schemes for small particles (dipoles) in the homogeneous (Fig. 1.10(a)) and inhomogeneous (Fig. 1.10(b)) fields are compared. If the field is homogeneous, the dipole moments $\mathbf{p}$ induced in each of the dipoles are parallel to the field vector of the incident wave, and their p_z components are zero. If the field is inhomogeneous, the p_z components for two dipoles (1 and 3 in Fig. 1.10(b)) are nonzero.

Let us compare the behavior of the intensity and the degree of linear polarization of radiation scattered by the dipoles placed in homogeneous (Fig. 1.10(a)) and inhomogeneous (Fig. 1.10(b)) field. To simplify the analysis, we will assume that the modules of all of the dipoles are the same and, at the specified polarization of the incident radiation, the dipole moment of dipole 3 in Fig. 1.10(b) is parallel to the axis z_0, while the dipole moment of dipole 1 is antiparallel to it.

The plane x_0, z_0 is chosen to be the scattering one. Then, the intensity of radiation scattered by the dipoles in Fig. 1.10(a) is determined by expressions

$$\begin{aligned} I_{||} &= 4p^2 \cos^2 \vartheta \,, \\ I_{\perp} &= 0 \,. \end{aligned}$$

Here, $I_{||}$ and $I_{\perp}$ are the intensities of the components of the scattered radiation polarized parallel and perpendicularly to the scattering plane, respectively, and p is the module of the induced dipole moment of one of the dipoles.

For the case shown in Fig. 1.10(b), the corresponding quantities are

$$\begin{aligned} I_{||} &= 2p^2 \cos^2 \vartheta + 2p^2 \sin^2 \vartheta \,, \\ I_{\perp} &= 0 \,. \end{aligned}$$

Let us now consider the intensities for the case of the incident radiation polarized perpendicularly to the scattering plane, i.e. the field vector of the incident wave is

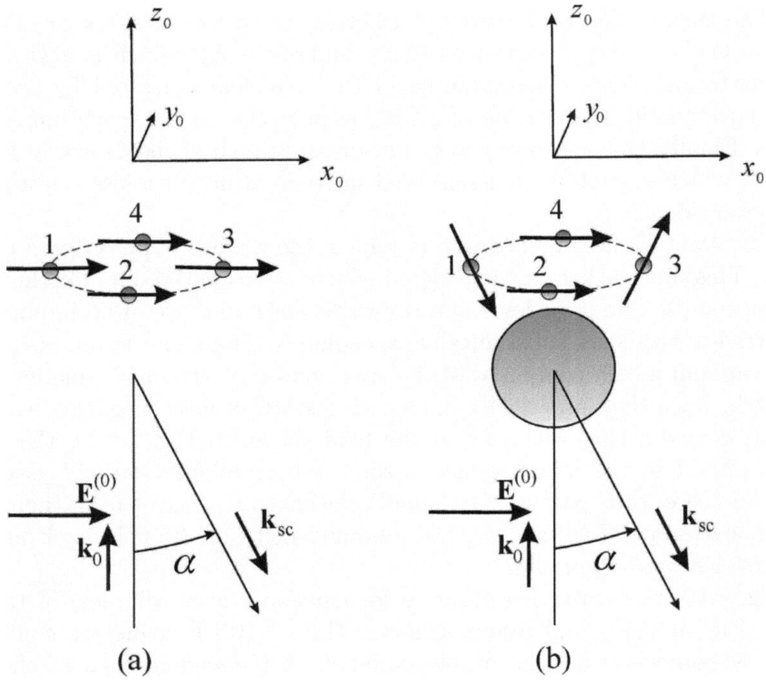

Fig. 1.10. Schemes illustrating the peculiarities of scattering of homogeneous and inhomogeneous waves by four dipoles. $\mathbf{k}_0$ is the wave vector, and $\mathbf{E}^{(0)}$ is the field vector of the incident wave. The vector signs near the dipoles indicate the direction of the induced dipole moments. The phase angle $\alpha = \pi - \vartheta$.

parallel the axis y_0. Then, for the homogeneous field (Fig. 1.10(a)),

$$\begin{aligned} I_{||} &= 0\,, \\ I_{\perp} &= 4p^2\,, \end{aligned}$$

while for the inhomogeneous field (Fig. 1.10(b)),

$$\begin{aligned} I_{||} &= 2p^2 \sin^2\vartheta\,, \\ I_{\perp} &= 2p^2\,. \end{aligned}$$

If the incident radiation is nonpolarized, the intensity I and the degree of linear polarization P of the scattered radiation are

$$\begin{aligned} I &= I_{||} + I_{\perp} = 4p^2(1+\cos^2\vartheta)\,, \\ P &= \frac{I_{\perp} - I_{||}}{I} = \frac{1-\cos^2\vartheta}{1+\cos^2\vartheta}\,, \end{aligned}$$

and

$$\begin{aligned} I &= 2p^2(2+\sin^2\vartheta)\,, \\ P &= -\frac{\sin^2\vartheta}{2+\sin^2\vartheta}\,. \end{aligned}$$

for the cases depicted in Fig. 1.10(a) and 1.10(b), respectively. The corresponding angular dependences of the intensity and the degree of linear polarization of the scattered radiation are shown in Fig. 1.11. It is worth noting that the analysis of the scattering characteristics of dipoles shown in Fig. 1.10(b) takes into account the double scattering, while that for Fig. 1.10(a), the single scattering. However, the accounting for the double scattering for Fig. 1.10(a) will cause no essential changes in the curves in Fig. 1.11: due to the random positioning of dipoles in the medium, the curves will be only smoothed in part. The real key peculiarity of the field shown in Fig. 1.10(b) is the presence of the component E_z in the close vicinity of the exiting particle. This component is parallel to the scattering plane, which leads to the described behavior of the scattering characteristics.

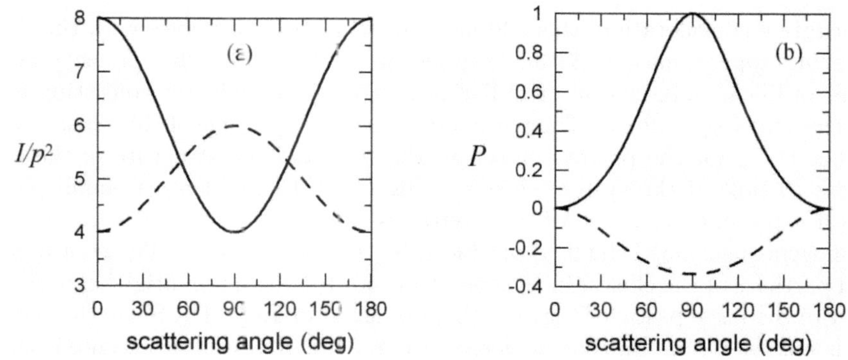

Fig. 1.11. The relative intensity I/p^2 (a) and the degree of linear polarization P (b) of the radiation scattered by the dipoles (shown in Fig. 1.10) versus the scattering angle. The solid and dashed curves correspond to Figs. 1.10(a) and 1.10(b), respectively. (From Petrova et al., 2007.)

As is seen from the plots in Fig. 1.11, the field inhomogeneity affecting the dipoles manifests itself in both intensity and polarization of the scattered radiation. First, the intensity substantially decreases in the directions $\vartheta = 0$ and $\vartheta = \pi$ and increases in side directions. Second, the degree of linear polarization becomes negative.

Though the dipole moments of dipoles *3* and *1* (Fig. 1.10(b)) were assumed to be directed parallel and antiparallel to the axis z, respectively, the problem can be also considered for the directions of the dipole moments shown in Fig. 1.10(b). For this case, the formulas are more complicated, but the above conclusions remain valid. Moreover, due to the screening of dipole *1* by the large particle, the branch of negative polarization will be more pronounced, and the polarization minimum will be close to the direction perpendicular to the dipole moment of dipole *3*.

The above quantitative analysis of characteristics of the radiation scattered by dipoles placed in the inhomogeneous field near the scatterer shows that the scattering of radiation by densely packed media may result in the forming of the branch of negative polarization due to the near-field effects. At the same time, the backscattering peak in brightness is weakened. This mechanism is mostly efficient in the media composed of the wavelength-sized scatterers, since the scales of the field inhomogeneities near the scatterers are comparable to the wavelength. This mechanism is ineffective for large particles (because the extension of the wave inhomogeneity zone is small relative to the particle sizes) and for particles that are much smaller than the wavelength (because the field inhomogeneities themselves are small).

1.4.2 Different scattering mechanisms: comparison of contributions to the scattering characteristics of simple clusters

Let us illustrate the formation of the scattering characteristics, which was considered above in the model of Rayleigh particles, with the examples of simple clusters of spherical particles. The strongly absorbing particles will be analyzed, because their scattering characteristics are caused by several first orders of scattering. In our quantitative consideration, this will allow us to restrict ourselves with the double-scattering approximation. As an example, we will consider the intensity and the degree of linear polarization of radiation scattered by a cluster consisting of nine particles (see Fig. 1.12(a)). The parameters of a large particle of this cluster are the same as those for the particle shown in Fig. 1.9, and the structure of the cluster is close to that of the system shown in Fig. 1.10. Eight identical small particles with size parameters $X = 1.5$ and refractive indices $\tilde{m} = 1.5 + i0.1$ are in the inhomogeneous zone of the field produced by the large particle. The coordinates of small particles of the cluster in the spherical coordinate system originating from the center of the large particle (Fig. 1.12(b)) are listed in Table 1.1. Since the structure of this cluster resembles that depicted in Fig. 1.10, the results presented below of calculations of intensity and polarization of the radiation scattered by this cluster allow us to estimate qualitatively the near-field effects considered quantitatively in the previous section.

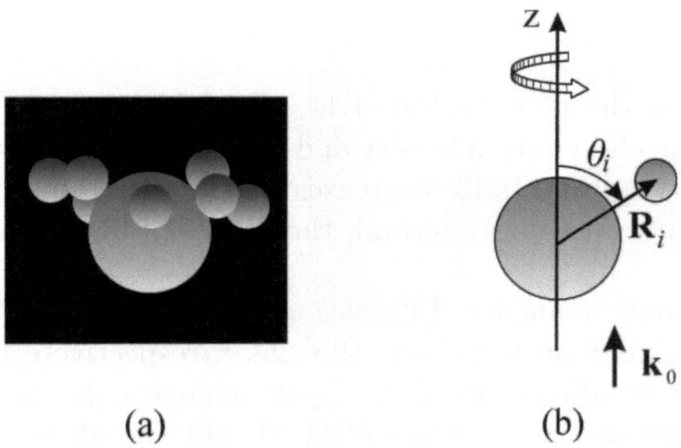

Fig. 1.12. The structure of the cluster (a) and the scheme explaining the averaging of the scattering characteristics of this cluster (b). (The azimuth of particles φ_i is not denoted in the scheme.)

To reduce the effects connected with the regular mutual position of particles in the cluster, which can 'mask' the near-field effects, the distances R_i and angles ϑ_i of small particles of the cluster were selected different. The results of calculations of the scattering matrix element F_{11} and the degree of linear polarization P are depicted in Figs. 1.13 and 1.14, respectively.

Table 1.1. The initial coordinates and the parameters of particles composing the cluster shown in Fig. 1.12.

i	X_i	$\tilde{m}_i$	$k_0 R_i$	ϑ_i(deg)	φ_i(deg)
1	4.0	$1.32 + i0.05$	0.0	0	0
2	1.5	$1.5 + i0.1$	5.5	85	0
3	1.5	$1.5 + i0.1$	6.5	80	45
4	1.5	$1.5 + i0.1$	5.8	75	90
5	1.5	$1.5 + i0.1$	6.7	70	135
6	1.5	$1.5 + i0.1$	6.2	65	180
7	1.5	$1.5 + i0.1$	5.9	75	225
8	1.5	$1.5 + i0.1$	7.1	85	270
9	1.5	$1.5 + i0.1$	6.8	65	315

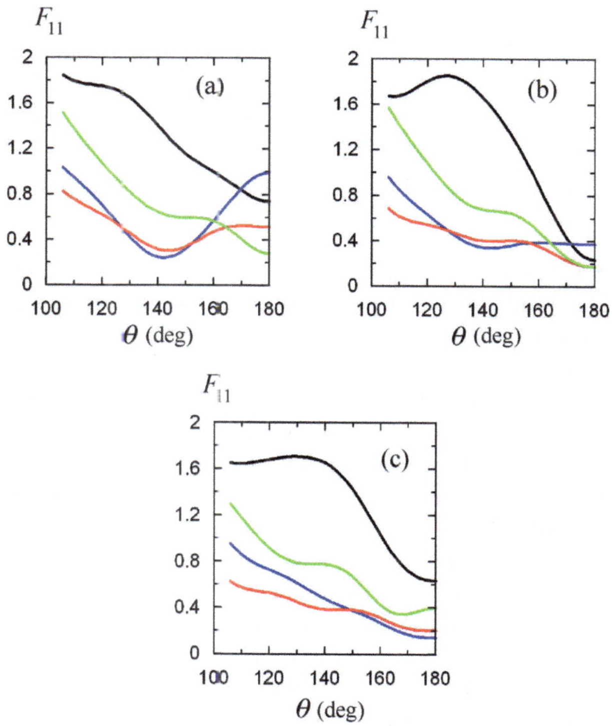

Fig. 1.13. The scattering matrix element F_{11} versus the scattering angle for different models: the cluster presented in Table 1.1 (a); the cluster with angles ϑ_i decreased by 10° (b); the cluster with angles ϑ_i decreased by 20° (c). The black curves correspond to the double-scattering approximation minus the interference of singly and doubly scattered waves. The blue curves correspond to the previous model (shown with black curves) plus the interference of singly and doubly scattered waves. The red curves present the model accounting for all of the scattering orders. The green curves present the model accounting for all of the scattering orders, but ignoring the near field.

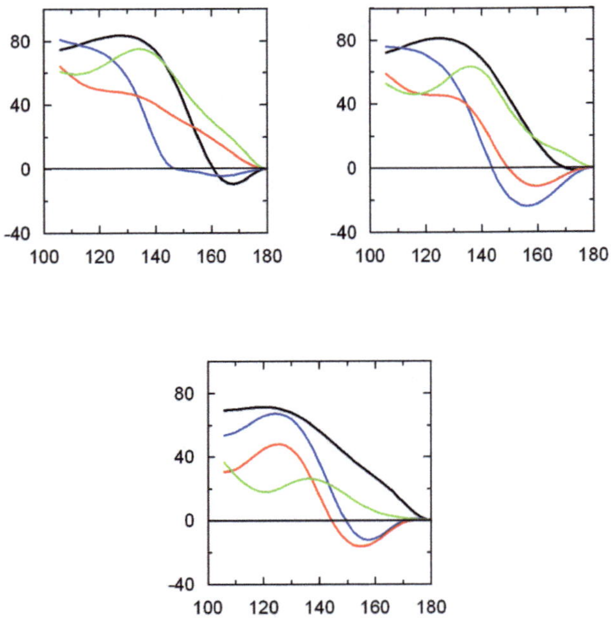

Fig. 1.14. As in Fig. 1.13, but for the degree of linear polarization.

The incident radiation propagates along the z -axis of the laboratory coordinate system. (The propagation direction is shown with vector $\mathbf{k}_0$ in Fig. 1.12(b). The incident radiation is assumed to be nonpolarized.) Since the characteristics of the scattered radiation strongly depend on the cluster orientation relative to the propagation direction of the incident radiation, which also 'masks' the near-field effects, the results of calculations were averaged by rotation of the cluster over the z-axis. To diminish further the influence of the orientation effects, the results of calculations were additionally averaged by varying simultaneously all angles ϑ_i within the interval of $\pm 5°$. While averaging by both the cluster rotation and the angles ϑ_i, the distribution function was assumed to be constant (the equiprobable distribution function over angles is assumed).

Figures 1.13 and 1.14 present the results of model calculations for the clusters of three types. The models depicted in Figs. 1.13(a) and 1.14(a) correspond to the cluster described in Table 1.1. The models in Figs. 1.13(b) and 1.14(b) are for the cluster with angles ϑ_i decreased by 10°, and the models in Figs. 1.13(c) and 1.14(c) are for the cluster with angles ϑ_i decreased by 20°. Thus, practically all of small particles in the cluster of the first type are in the almost homogeneous field (see Fig. 1.9), and they all are not shielded by the large particle in the scattering direction of $\vartheta = 180°$. In the cluster of the second type, almost all of small particles are in a stronger inhomogeneous field, and some of them are 'shadowed' by the large particle in the scattering direction of $\vartheta = 180°$. All small particles of the cluster of the third type are in a strong inhomogeneous field, and most of them are 'shadowed' by the large particle in the scattering direction of $\vartheta = 180°$. In all of the plots of Figs. 1.13 and 1.14, the models corresponding to the double-scattering approximation minus the interference of singly and doubly scattered waves (black curves) take into account the near-field effect only partly – in the diffuse (noncoherent) scattering and in the coherent backscattering. In the models considering the interference of singly and doubly scattered waves (blue curves) (compare with the contribution of this interference in Fig. 1.5), the near-field effect is completely taken into account in the double-scattering approximation. The exact solutions and the models ignoring the near-field effect, but accounting for all scattering orders, are also shown (red and green curves, respectively).

Let us consider in short the contribution of different mechanisms to the scattering matrix element F_{11} (Fig. 1.13). The ignoring of the interference of singly and doubly scattered radiation in the double-scattering approximation (black curves) increases the intensity to the values substantially exceeding the exact solution. The accounting for this interference (blue curves) yields the values of intensity that are much closer to the exact solution. It can be seen that this component describes the effect of mutual shielding considered in the previous section. This effect is well expressed for $\vartheta \leq 160°$ in Figs. 1.13(a) and 1.13(b) and for all ϑ in Fig. 1.13(c). One can also observe the decrease of the intensity of the scattered radiation in the direction of $\vartheta = 180°$, when the number of small particles being in the 'shadow' of the large particle increases. This phenomenon can be considered as a 'shadow effect' for the cluster, and it does not appear in the model ignoring the near field.

Now consider the manifestation of the near field in the degree of linear polarization (Fig. 1.14). This characteristic is believed to be mostly informative for interpretation of the optical observations of the solar system bodies (Mishchenko et al., 2010). Therefore, to understand the mechanisms of its formation is of especial importance. Note that nowadays, the interpretation of observations of the surfaces of atmosphereless celestial bodies covered with regolith (a relatively closely packed medium) ignores the near field. Figure 1.14 shows that the near field should be taken into account. It is seen that the model ignoring the near field produces no branch of negative polarization that are observed in the most atmosphereless bodies (Mishchenko et al., 2010; Belskaya and Shevchenko, 2000; Belskaya et al., 2002, 2003, 2005). If the near field is accounted for, the negative branch appears, and it is induced by the field inhomogeneity near the large particle of the cluster. In the double-scattering approximation, the polarization in this branch in Fig. 1.14(a) is relatively weak, since most small particles of the cluster are in almost homogeneous field (see Fig. 1.9). In this configuration of the cluster, the contribution of higher scattering orders is significant (Fig. 1.13(a)), which leads to the inversion of the polarization sign. In Fig. 1.14(b), the polarization is higher, since small particles of the cluster are in the field, which is strongly inhomogeneous. For this cluster, the contribution of high scattering orders is also noticeable, which is seen from some depolarization of the radiation. In Fig. 1.14(c), the polarization in the double scattering approximation is weaker than that in Fig. 1.14(b), since most of small particles of the cluster are shielded by the large particle.

Though the near-field effects have been considered here for oriented or partly oriented clusters, this analysis remains valid for randomly oriented ensembles (with the equiprobable distribution function over the Euler angles). Specifically, the branch of negative values of linear polarization also arises from the interaction between the particles of the cluster in the inhomogeneous near field (Tishkovets et al., 2004b; Tishkovets and Jockers, 2006; Tishkovets, 2008). This conclusion is important for the interpretation of the optical observations of the cometary dust (Mishchenko et al., 2010).

1.4.3 Near-field effects in the large clusters

Let us now consider the manifestations of the near-field effects in the scattering characteristics in more detail using the clusters with different structure of the surface layer. These clusters were generated in the following way. The compact cluster with a tetrahedron lattice composed of 50 particles served as a core (see the upper insert in Fig. 1.15). The overall shape of the cluster is close to spherical. First, 10 monomers were added to its surface in a random way. Then 40 more monomers were added to the surface in the same way. The intensity and polarization of the original cluster and the two modified ones are depicted as functions of scattering angle in Fig. 1.15 and compared with those of a compact cluster of irregular structure consisting of 100 monomers (see the lower insert in Fig. 1.15). Even a few monomers added to the surface of the regular cluster significantly suppress the interference oscillations in the curves of the original cluster and make the curves to look similar to those of the cluster with a more random structure. This can be explained as follows.

A layer of random particles added to the regular cluster works as an amplitude–phase inhomogeneity for the incident wave. After having passed through this layer, the wave becomes strongly inhomogeneous; the variations of its amplitude, phase,

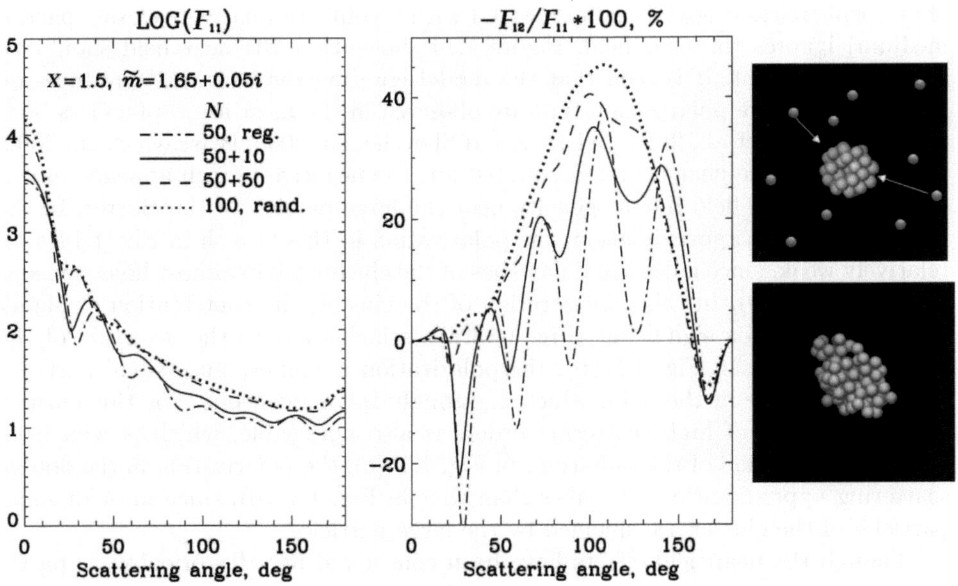

Fig. 1.15. The intensity and polarization of light scattered by clusters with different structure of the surface layer. Additional monomers are added in a random way to the outside of the completely regular compact cluster. The model for the random cluster is also shown for comparison. The number of monomers is shown for each model. (Adapted from Petrova et al. (2004).)

and propagation direction become chaotic. If the number of particles placed randomly in the outer layer is large enough, then there is almost no correlation between the phases of the radiation produced in the scattering of such a wave by the individual underlying particles. Consequently, the location of particles deeper inside the cluster with respect to each other is unimportant for the scattering process. For the angular dependence of the intensity, the inner structure of the cluster is even less important. The intensity level increases with the number of monomers, albeit this growth also depends on the imaginary part of the refractive index. Since the scattering properties of the inner particles are not sensitive to their location, the polarization of the cluster depends only weakly on their number in the cluster and on the regularity of its inner structure, but more strongly on its packing density.

In the backscattering region, the enhancement of intensity, which is caused by the interference of multiply scattered waves, is formed by the particles in the outer layer of the cluster, where the radiation field is practically homogeneous. At the same time, the negative polarization, which is to a great extent determined by the interaction of particles in the near field, is mainly generated by the particles below the surface layer of the cluster, where the radiation field is inhomogeneous, and the amplitude, phase, and propagation direction of the wave change randomly. Note that the situation is the same in a powder-like layer, which makes conclusions also relevant to regolith surfaces. Thus, the field inhomogeneity below the surface layer of the cluster (or regolith) reduces the dependence of the negative polarization on the location of particles in the deeper layers, but not on the compactness. Depending on the structure of the aggregate, the interference of multiply scattered waves or the near-field interactions are more efficient for a given cluster. This means that the opposition effects in intensity and polarization do not always go in parallel.

Since the relative contribution of different scattering mechanisms, determining the behavior of the branch of negative polarization, strongly depends on the properties of the particles composing the scattering ensemble, let us now consider how the appearance of the branch of negative polarization changes versus the cluster structure, the number of constituents, their size and refractive index. Figure 1.16 displays the negative branch of polarization calculated for several clusters shown in the inset. The structure of the core of cluster *1* is regular, and several monomers are added randomly (as in the case presented in Fig. 1.15). The other clusters are irregular in structure, and their porosity grows from cluster *2* to cluster *4*. The size parameter of the monomers is 1.5 for all of the clusters from *1* to *4*. The last cluster consists of particles of different sizes, and the curves for the ensembles of this kind are presented in Fig. 1.16(g) for different scaling coefficients applied to the size of the constituents.

The upper plot in the left column (Fig. 1.16(a)) shows the effect of the cluster structure, and the numbers next to the curves correspond to the cluster number. To eliminate the interference oscillations (see Fig. 1.15) and to show the effect of structure more clearly, the intensity and polarization of four realizations (differing by the starting set of random numbers) of each cluster were averaged. While the difference in the inversion angle of polarization for various clusters is about 5–6°, the models averaged over realizations show a difference in the inversion angle of 2–4° between clusters of different types. The negative branch is somewhat stronger

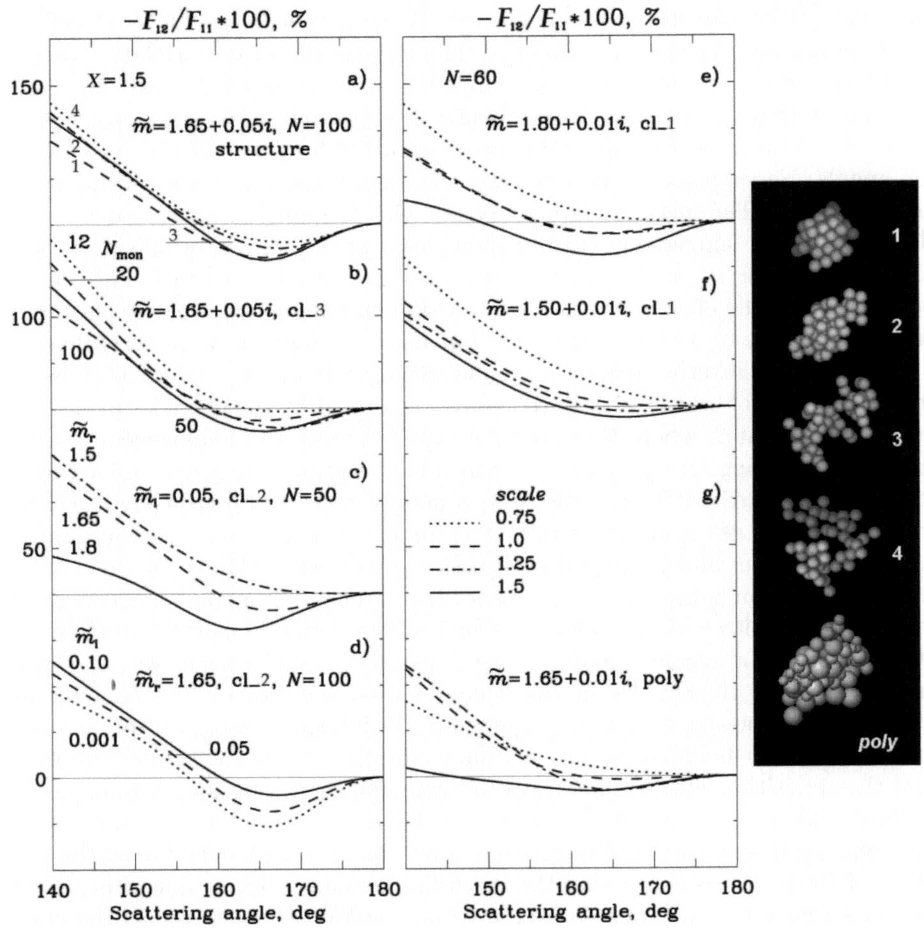

Fig. 1.16. The negative branch of polarization for the clusters shown in the inset. The parameters of the clusters used for each set of models are shown in the corresponding plots. Each set of curves is shifted upward from the previous one by 40%. (a) The dependence on cluster structure. (b) The dependence on the number of monomers. (c) The dependence on the real part of the refractive index. (d) The dependence on the imaginary part of the refractive index. The right column of the figure (plots (e)–(g)) presents the dependence on a size parameter of the constituent particles. The scale applied to the monomer size in these plots is shown in plot (g). (e) and (f) Two values of the refractive index (the negative branch deepens with increasing monomer sizes in both cases). (g) A case of constituent particles of different sizes (from 0.75 to 1.5). The influence of the largest particles is noticeable, probably because they are not completely covered by the smaller ones. (Adapted from Tishkovets et al. (2004a).)

for more compact clusters than for sparse ones. The minimum is deeper, and the inversion angle is shifted somewhat to smaller scattering angles.

Figure 1.16(b) shows the effect of a number of constituents. When N increases, there is a tendency for the inversion angle to decrease, but the decrease is not monotonic. Similarly, there is a tendency for the minimum of polarization to become deeper. The non-monotonic behavior may be caused by interaction between the fragments of wavelength size in the comparatively sparse cluster *3*, which leads to additional averaging of the scattering contributions over the fragments. It is worth noting that the polarization minimum is closer to the inversion angle for smaller N and moves to $\vartheta = 180°$ for larger N; i.e., for clusters consisting of a larger number of particles, the negative branch becomes more symmetric, and its shape much more resembles the parabolic one characteristic to the negative branch observed for cometary dust. This is caused by the increase of the contribution of constituents located far enough from each other to the wave interference.

Many parametric studies of the scattering properties of aggregate particles revealed that the negative branch of polarization strongly depends on the refractive index of the constituents. For completeness, we show this dependence for the backscattering region in Figs. 1.16(c) and 1.16(d). In general, the effect of the real part of the refractive index is more important than that of the imaginary one. When $\mathrm{Re}(\tilde{m})$ grows and/or $\mathrm{Im}(\tilde{m})$ decreases, the negative branch becomes more prominent.

The right column of Fig. 1.16 presents the influence of the size parameters of constituting particles on the negative branch. Larger monomers produce a stronger and wider negative branch (Figs. 1.16(e) and 1.16(f)). This is valid for a wide range of refractive indices. For even larger monomers ($X > 1.5$–1.75), the negative branch becomes very prominent. When the cluster consists of particles of various sizes (Fig. 1.16(g)), larger particles (probably, only when they are not buried deep in the cluster) manifest themselves in strengthening the negative branch and depressing the polarization maximum.

1.4.4 The near-field and weak-localization effects: the ranges of influence

As has been shown in subsections 1.3.2, 1.4.1, and 1.4.2, the phenomena existing in the near-field zone of the scattering particles, i.e. the mutual shielding and the field inhomogeneity, may substantially influence the phase dependences of intensity and polarization of light scattered by densely packed ensembles of particles. From the standpoint of interpretation of the results of laboratory measurements (and astronomical observations), the accounting for this influence can be a key point in the explanation of the facts, when the observed behavior of the scattering characteristics deviates from that predicted by the classical radiative-transfer and weak-localization (coherent-backscattering) theories strictly valid only for the sparse media. The backscattering domain is of particular interest, since the most astronomical objects are observed at phase angles, $\alpha = \pi - \vartheta$, smaller than $\approx 50°$. Because of this, it is important to determine the area of applicability of the low-density theories and to estimate the influence of the interaction of particles in the near field on the scattering characteristics of relatively large ensembles of particles.

For analysis of packing density effects, the most robust method has been that proposed by Mishchenko et al. (2007) and later adopted by Mishchenko and Liu (2009), Mishchenko et al. (2009a, 2009b), Okada and Kokhanovsky (2009), Mackowski and Mishchenko (2011a, 2011b), Dlugach et al. (2011), and Petrova and Tishkovets (2011). In these studies, the random particulate medium is modeled by a spherical volume randomly filled with N identical non-overlapping spherical particles. Although the mutual positions of the particles with respect to each other remain the same, they are sufficiently 'random' from the outset. Therefore, averaging over all orientations of this configuration yields an infinite continuous set of random realizations of the scattering volume while enabling one to apply the orientation averaging technique afforded by the superposition T-matrix method (Mackowski and Mishchenko, 1996). Of course, this model of aggregate particles of varying porosity cannot be expected to reproduce exactly the great diversity of morphologies of discrete random media encountered in natural and laboratory settings. However, it is sufficiently representative for considering the influence of the packing density on the manifestation of the coherent backscattering and near-field effects.

Recall that according to the low-density theory the intensity of light scattered by the fixed-volume ensembles of particles exhibits a backscattering peak, which rapidly grows with a number of constituents N, but has almost the same angular width (see Fig. 1.17 depicting the example of the models for the volume filled with non-absorbing particles with relatively low packing density). At scattering angles larger than $\approx 150°$, linear polarization rapidly develops a pronounced minimum of negative values. It is worth noting that the intensity of light scattered by the individual constituent particles of the specified properties shows no increase to opposition, and the corresponding polarization is positive in the whole range of angles. The scattering angle of minimal polarization is essentially independent of

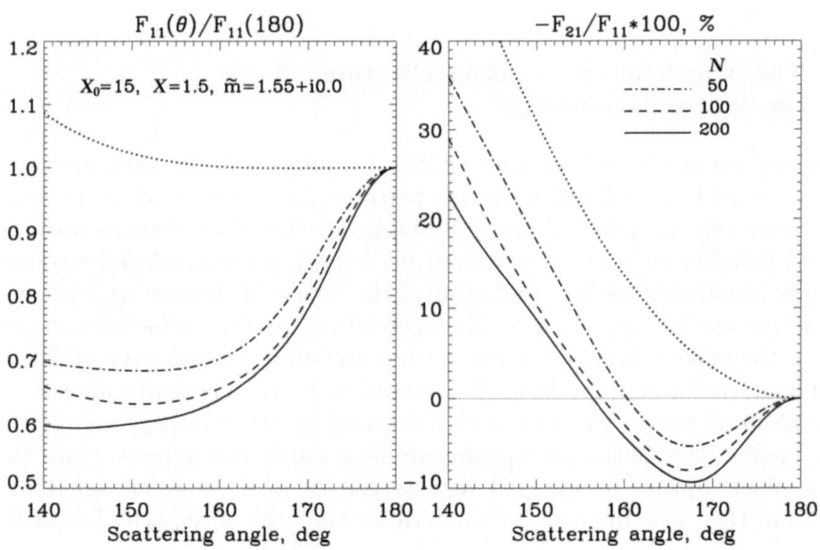

Fig. 1.17. The phase dependencies of the relative intensity (left) and the degree of linear polarization (right) of light scattered by the spherical volumes of different packing densities. (The packing density is defined as NX^3/X_0^3 and changes from 5% to 20% in the present examples.) The backscattering domain is shown. The size parameter of the monomers X and the volume X_0, the refractive index $\tilde{m}$, and the number of particles N in the volume are indicated in the plot. The scattering characteristics of the individual monomer are shown with dotted curves.

N and is comparable to the angular semi-width of the brightness peak. Extensive computations (Mackowski and Mishchenko, 1996; Mishchenko et al., 2007, 2009a, 2009b; Mackowski and Mishchenko, 2011a, 2011b; Dlugach et al., 2011) performed for the particulate spherical volumes (with size parameters ranging from 20 up to 60) confirmed that the phase dependences of intensity and polarization for volumes containing nonabsorbing particles with packing densities less than 20–30% are in perfect qualitative agreement with the predictions of the low-density theory of coherent backscattering. The modeling showed that the polarization minimum moves to opposition, when the size parameter of the volume increases. Furthermore, if the size of the scattering volume (i.e., the interference base) is large enough (that is not the case shown in Fig. 1.17), the profile of the negative branch is asymmetric, so that its minimum is closer to the opposition than to the inversion point. This means that the polarization minima observed in the relatively sparse systems of non-absorbing particles have the same basic morphology as that predicted by the classical theory of coherent backscattering (Mishchenko et al., 2000, 2006).

Starting from packing densities $\approx 30\%$, the phase curves develop high-frequency interference ripples typical of a single spherical particle with a size greater than the wavelength (Mishchenko et al., 2002) This behavior is obviously inconsistent with the radiative transfer and coherent backscattering predictions. To better distinguish packing-density effects in closely packed aggregates, the interference features can be smoothed by distorting the regular spherical shape of the particulate volume (Petrova and Tishkovets, 2011).

The examples of the results obtained in the computations of scattering characteristics of the distorted spherical volume are shown in Fig. 1.18. The case of absorbing constituents is presented here. In general, the extensive simulations have revealed a complex behavior of the opposition features of absorbing particles (Dlugach et al., 2011; Petrova and Tishkovets, 2011). The changes in the amplitude and width of the backscattering intensity peak as well as in the depth and width of the negative-polarization branch with increasing absorption can be nonmonotonic. However, a key difference in the behavior of the backscattering peak and the negative branch of polarization observed for non-absorbing and absorbing particles with increasing packing density can be noticed. As one could expect merely from the weakening of multiple scattering in absorbing systems of particles, the opposition surge of intensity is smoothed with increasing packing density, which is opposite to the trend observed in non-absorbing particulate volumes. For the same reason, the branch of negative polarization should become shallower, when the packing density grows. However, the extensive model calculations showed that its behavior is much more complex, though the polarization minimum always moves from opposition. Depending on the size parameters of the constituents and the real part of the refractive index, the negative branch of polarization may become more or less pronounced, while the opposition effect in intensity is always weakened in more closely packed absorbing aggregates (see Fig. 1.18) (Petrova and Tishkovets, 2011). Evidently, this behavior of the phase curves does not follow from the low-density theory of coherent backscattering, and it can be explained by the influence of the interaction of particles in the near field.

The scattering mechanism accounting for the near-field effects allows one to explain the branch of negative polarization observed in some dark asteroids, though

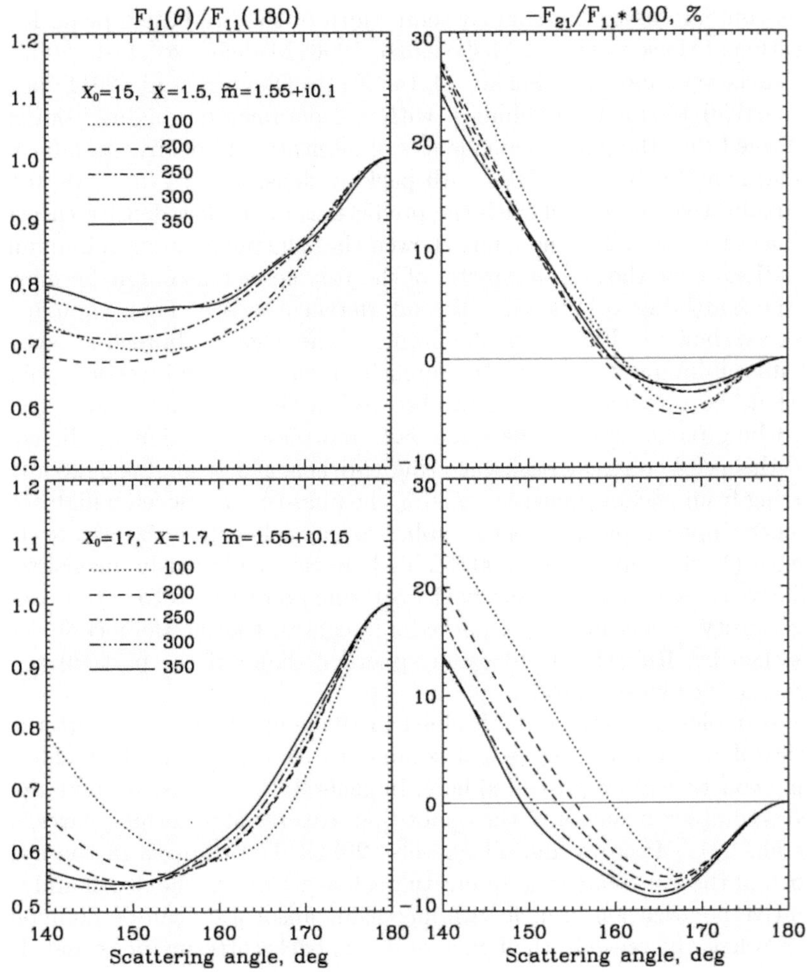

Fig. 1.18. The phase dependencies of the relative intensity (left) and the degree of linear polarization (right) of light scattered by the 'distorted' spherical volumes of different packing densities (varying from 10% to 35%). The backscattering domain is shown. Two cases of the cluster parameters are presented in the upper and lower panels, respectively. The size parameters of the monomers X and the volume X_0, the refractive index $\tilde{m}$, and the number of particles in the volume are indicated in the plot.

they demonstrate no noticeable opposition peak in brightness (Belskaya et al., 2002). In these cases, the interaction of scattering particles in the near-field zones promotes the negative polarization and depresses the interference peak due to the decrease of intensity in the backscattering domain (Fig. 1.11).

The model T-matrix calculations performed for the ensembles of particles with varying porosity showed that the optical effect of increasing the number of particles in a volume can be expected to be twofold. At the outset, it facilitates multiple scattering and thus enhances the classical radiative transfer and coherent backscattering features. Eventually, however, it causes scattering patterns not predicted by the low-packing-density theories, and their interpretation requires the mechanism(s) accounting for the interaction in the near field to be invoked.

1.5 Concluding remarks

The existing theories of multiple scattering of radiation, specifically the radiative transfer theory, are based on the assumption that the secondary waves propagating from one particle of the medium to another are spherical, which is equivalent to the statement that the scatterers are in the far-field zones of each other. This supposition completely ignores the near field and, therefore, is true only for sparse media. However, in closely packed media, the effects caused by the near field can be observed. One of these effects is the mutual shielding. It is worth noting that the mutual shielding may manifest itself in the systems of particles smaller than the wavelength rather than only for the particles comparable to or larger than the wavelength. The other manifestation of the near field is some damping of the opposition effect in intensity of the scattered radiation and the strengthening of the negative values of the degree of linear polarization in the backscattering domain. The latter is observed as the branch of negative polarization in many atmosphereless celestial bodies. Such a behavior of the scattering characteristics allows the phase dependences of intensity and polarization observed in low-albedo asteroids to be explained. Their peculiarity is that the opposition effect in intensity is practically unnoticeable, while the branch of negative polarization is rather well pronounced. Unfortunately, the description of multiple scattering by discrete media presently lacks methods correctly accounting for the near field, which substantially restricts the potentialities of interpretation of the remote sensing data of different objects.

Acknowledgments

We gratefully acknowledge Klaus Jockers, Pavel Litvinov, Anatoli Minakov, and Michael Mishchenko for many constructive discussions of problems touched upon in this chapter.

References

Akkermans E., P. E. Wolf, R. Maynard, and G. Maret, 1988: Theoretical study of the coherent backscattering of light by disordered media, *J. Phys. France*, **49**, 77–98.

Barabanenkov Yu. N., Yu. A. Kravtsov, V. D. Ozrin, and A. I. Saichev, 1991: Enhanced backscattering in optics, Progress in Optics, **29**, 65–197.

Bass F. G., and I. M. Fuks, 1979: *Wave Scattering from Statistically Rough Surfaces*, Pergamon Press, Oxford, New York.

Belskaya I. N., and V. G. Shevchenko, 2000: Opposition effect of asteroids, *Icarus*, **147**, 94–105.

Belskaya I. N., V. G. Shevchenko, Yu. S. Efimov et al., 2002: Opposition polarimetry of the low albedo asteroid 419 Aurella, *Proceedings of Asteroids, Comets, Meteors (ACM 2002), 2002, 29 July–2 August, Technical University Berlin*, Berlin, Germany, 489–491.

Belskaya I. N., V. G. Shevchenko, N. N. Kiselev et al., 2003: Opposition polarimetry and photometry of S- and E-type asteroids, *Icarus*, **166**, 276–284.

Belskaya I. N., Yu. G. Shkuratov, Yu. S. Efimov et al., 2005: The F-type asteroids with small inversion angle of polarization, *Icarus*, **178**, 213–221.

Borghese F., P. Denti, and R. Saija, 2002: *Scattering from Model Nonspherical Particles. Theory and Applications to Environmental Physics*, Springer-Verlag, New York.

Dlugach J. M., M. I. Mishchenko, L. Liu, and D. W. Mackowski, 2011: Numerically exact computer simulations of light scattering by densely packed random particulate media, *J. Quant. Spectrosc. Radiat. Transfer*, doi:10.1016/j.jqsrt.2011.02.009

Felderhof B. U., R. B. Jones, 1987: Addition theorems for spherical wave solutions of the vector Helmholtz equation, *J. Math. Phys.*, **28**, 836–839.

Fuller K. A., and D. A. Mackowski, 2000: Electromagnetic scattering by compounded spherical particles, *Light Scattering by Nonspherical Particles: Theory, Measurements, and Applications* (eds. Mishchenko M. I., Hovenier J. W., Travis L. D.), Academic Press, San Diego, 225–272.

Grandshtein I. S., and I. M. Ryzhik, 1980: *Table of Integrals, Series, and Products*, Academic Press, New York.

Greffet J.-J., and R. Carminati, 1998: *Image formation in near-field optics*, Progr. Surface Sci., 56, 133-237.

Hapke B., 1993: *Theory of Reflectance and Emittance Spectroscopy*, University Press, Cambridge.

Hoekstra A. G., and M. A. Sloot, 2000: Biophysical and biomedical applications of nonspherical scattering, in *Light Scattering by Nonspherical Particles: Theory, Measurements, and Applications* (eds. Mishchenko M. I., Hovenier J. W., Travis L.D.), Academic Press, San Diego.

Horan P. K., and L. L. Wheeles, 1977: Quantitative single cell analysis and sorting, *Science*, **198**, 149–157.

Ishimaru A., 1978: *Wave Propagation and Scattering in Random Media*, Vols. 1, 2, Academic Press, New York.

Mackowski D. W., 1995: Electrostatics analysis of sphere clusters in the Rayleigh limit: application to soot particles, *Appl. Opt.*, **34**, 3535–3545.

Mackowski, D.W., /ftp://ftp.eng.auburn.edu/pub/dmckwski/scat codes/index.html

Mackowski D. W., and M. I. Mishchenko, 1996: Calculation of the T matrix and the scattering matrix for ensembles of spheres, *J. Opt. Soc. Am. A*, **13**, 2266–2278.

Mackowski D. W., and M. I. Mishchenko, 2011a: Direct simulation of multiple scattering by discrete random media illuminated by Gaussian beams, *Phys. Rev. A*, **83**, 013804.

Mackowski D. W., and M. I. Mishchenko, 2011b: A multiple sphere T-matrix Fortran code for use on parallel computer clusters. *J. Quant. Spectrosc. Radiat. Transfer*, doi:10.1016/j.jqsrt.2011.02.019.

McCartney Earl J., 1976: *Optics of the Atmosphere: Scattering by Molecules and Particles*, John Wiley & Sons, New York.

Mishchenko M. I. and L. Liu, 2009: Electromagnetic scattering by densely packed particulate ice at radar wavelengths: exact theoretical results and remote-sensing implications, *Appl. Opt.*, **48**, 2421–2426.

Mishchenko M. I., and D. W. Mackowski, 2011: Coherent backscattering in the cross-polarized channel, *Phys. Rev. A*, **83**, 013829.

Mishchenko M. I., J.-M. Luck, and T. Nieuwenhuizen, 2000: Full angular profile of the coherent polarization opposition effect, *J. Opt. Soc. Am. A*, **17**, 888–891.

Mishchenko M. I., L. D. Travis, and A. A. Lacis, 2002: *Scattering, Absorption, and Emission of Light by Small Particles*, Cambridge University Press, Cambridge.

Mishchenko M. I., L. D. Travis, and A. A. Lacis, 2006: *Multiple Scattering of Light by Particles. Radiative Transfer and Coherent Backscattering*, Cambridge University Press, Cambridge.

Mishchenko M. I., L. Liu, D. W. Mackowski et al., 2007: Multiple scattering by random particulate media: exact 3D results, *Opt. Express*, **15**, 2822–2836.

Mishchenko M. I., J. M. Dlugach, L. Liu et al., 2009a: Direct solutions of the Maxwell equations explain opposition phenomena observed for high albedo solar system objects, *Astrophys. J.*, **705**, L118–L122.

Mishchenko M. I., J. M. Dlugach, L. Liu et al., 2009b: Azimuthal asymmetry of the coherent backscattering cone: theoretical results, *Phys. Rev. A*, **80**, 053824.

Mishchenko M. I., V. K. Rosenbush, N. N. Kiselev, D. F. Lupishko, V. P. Tishkovets, V. G. Kaydash, I. N. Belskaya, Y. S. Efimov, and N. M. Shakhovskoy, 2010: *Polarimetric Remote Sensing of Solar System Objects*, Akademperiodyka, Kyiv.

Mishchenko M. I., V. P. Tishkovets, L. D. Travis, B. Cairns, G. M. Dlugach, L. Liu, V. K. Rosenbush, and N. N. Kiselev, 2011: Electromagnetic scattering by a morphologically complex object: Fundamental concepts and common misconceptions, *J. Quant. Spectrosc. Radiat. Transfer*, **112**, 671–692.

Muinonen K., 1990: *Light scattering by inhomogeneous media: backward enhancement and reversal of linear polarization*, PhD thesis, University of Helsinki.

Muinonen K., 2004: Coherent backscattering of light by complex random media of spherical scatterers: numerical solution, *Waves Random Media*, **14**, 365–388.

Muinonen K., J. Piironen, Yu. G. Shkuratov et al., 2002: Asteroid photometric and polarimetric phase effects, *Asteroids III* (eds. Bottke, Jr., W. F., Cellino, A., Paolicchi, P., Binzel, R. P.), University of Arizona Press, Tucson, 123–138.

Muinonen K., J. Tyynela, E. Zubko, and G. Videen, 2010: Coherent backscattering in planetary regoliths, in *Light Scattering Reviews* (ed. Kokhanovsky A.A.), Springer-Praxis, Chichester, Vol. 5, 477–518.

Okada Y., and A. A. Kokhanovsky, 2009: Light scattering and absorption by densely packed groups of spherical particles, *J. Quant. Spectrosc. Radiat. Transfer*, **110**, 902–917.

Petrova E. V. and V. P. Tishkovets, 2011: Light scattering by aggregates of varying porosity and the opposition phenomena observed in the low-albedo particulate media, *J. Quant. Spectrosc. Radiat. Transfer*, doi:10.1016/j.jqsrt.2011.04.010.

Petrova, E.V., Tishkovets, V.P., and Jockers, K., 2004: Polarization of light scattered by Solar system bodies and the aggregate model of dust particles, *Astron. Vestn.* **38**(4), 309–324 [Sol. Syst. Res. (Engl. Transl.) 38 (4), 354–371].

Petrova E. V., V. P. Tishkovets, and K. Jockers, 2007: Modeling of opposition effects with ensembles of clusters: Interplay of various scattering mechanisms, *Icarus*, **188**, 233–245.

Quinby-Hunt M. S., P. G. Hull, and A. J. Hunt, 2000: *Polarized light scattering in the marine environment, in Light Scattering by Nonspherical Particles: Theory, Measurements, and Applications* (eds. Mishchenko M. I., Hovenier J. W., Travis L. D.), Academic Press, San Diego.

Shkuratov Yu. G., 1989: New mechanism of the negative polarization of light scattered by atmosphereless cosmic bodies, *Astron Vestnik*, **23**, 176–180 [in Russian].

Shkuratov Yu. G., K. Muinonen, E. Bowell, et al., 1994: A critical review of theoretical models of negatively polarized light scattered by atmosphereless solar system bodies, *Earth Moon Planets*, **65**, 210–246.

Shkuratov Yu., A. Ovcharenko, E. Zubko, O. Miloslavskaya, K. Muinonen, J. Piironen, R. Nelson, W. Smythe, V. Rosenbush, and P. Helfenstein, 2002: The opposition effect and negative polarization of structural analogs for planetary regoliths, *Icarus*, **159**, 396–416.

Tishkovets V. P., 1998: Backscattering of light by close-packed system of particles, *Optics and Spectroscopy*, **85**, 212–17.

Tishkovets V. P., 2008: Light scattering by closely packed clusters: Shielding of particles by each other in the near field, *J. Quant. Spectrosc. Radiat. Transfer*, **109**, 2665–2672.

Tishkovets V. P., and K. Jockers, 2006: Multiple scattering of light by densely packed random media of spherical particles: Dense media vector radiative transfer equation, *J. Quant. Spectrosc. Radiat. Transfer*, **101**, 54–72.

Tishkovets V. P., and P. V. Litvinov, 1996: Coefficients of light extinction by randomly oriented clusters of spherical particles in the double scattering approximation, *Optics and Spectroscopy*, **81**, 288–291.

Tishkovets V. P., and P. V. Litvinov, 1999: Opposition effects in light scattering by regolith-type media, *Sol. Syst. Res.*, **33**, 2, 162–168.

Tishkovets V. P., and M. I. Mishchenko, 2004: Coherent backscattering of light by a layer of discrete random medium, *J. Quant. Spectrosc. Radiat. Transfer*, **86**, 161–180.

Tishkovets V. P., Yu. G. Shkuratov, and P. V. Litvinov, 1999: Comparison of collective effects at scattering by randomly oriented cluster of spherical particles, *J. Quant. Spectrosc. Rad. Transfer*, **61**, 767–773.

Tishkovets V. P., E. V. Petrova, and K. Jockers, 2004a: Optical properties of aggregate particles comparable in size to the wavelength, *J. Quant. Spectrosc. Radiat. Transfer*, **86**, 241–265.

Tishkovets V. P., P. V. Litvinov, E. V. Petrova, K. Jockers, and M. I. Mishchenko, 2004b: Backscattering effects for discrete random media: theoretical results, in *Photopolarimetry in Remote Sensing: NATO Science Series* (eds. Videen G., Yatskiv Ya., Mishchenko M. I.), Kluwer Academic Publishers, Dordrech, 221–242.

Tsang L., J. A. Kong, and K.-H. Ding, 2000: *Scattering of Electromagnetic Waves. Theories and Applications*, John Wiley & Sons, New York.

Tsang L., and J. A. Kong, 2001a: *Scattering of Electromagnetic Waves. Advanced Topics*, Wiley-Interscience, New York.

Tsang L., J. A. Kong, K.-H. Ding, and C.O. Ao, 2001: *Scattering of Electromagnetic Waves. Numerical Simulations*, John Wiley & Sons, New York.

Varshalovich D. A., A. N. Moskalev, and V. K. Khersonskii, 1988: *Quantum Theory of Angular Momentum*, World Scientific, Singapore.

van Rossum M. C. W., and Th. M. Nieuwenhuizen, 1999: Multiple scattering of classical waves: microscopy, mesoscopy, and diffusion, *Reviews of Modern Physics*, **71**, No. 1, 313–371.

Watson K. M., 1969: Multiple scattering of electromagnetic waves in an underdense plasma, *J. Math. Phys.*, **10**, 688–702.

Wolf P. E., and G Maret, 1985: Weak localization and coherent backscattering of photons in disordered media, *Phys. Rev. Lett.*, **55**, 2696–2699.

2 Multi-spectral luminescence tomography with the simplified spherical harmonics equations

Alexander D. Klose

2.1 Introduction

In the past, biomedical imaging methods, such as X-ray computed tomography (CT), magnetic resonance imaging (MRI), and ultrasound (US) imaging, only visualized macroscopic tissue structures. Although these imaging modalities may allow differentiating pathological from normal tissue on a macroscopic scale, they failed to elucidate the molecular mechanisms of a pathological dysfunction. Therefore, functional MRI was introduced in the early 1990s, which enabled measuring changes in blood flow and blood oxygenation levels in response to neural activity [1]. In addition to MRI, magnetic resonance spectroscopy (MRS) allowed for studying metabolic changes in the brain involving specific nuclei such as protons, phosphorus, carbon, and sodium [2–4].

Molecular imaging, which has emerged in recent years, goes far beyond the visualization of tissue morphology or a few endogenous molecules [5–10]. In fact, it aims to directly monitor complex biochemical processes on a microscopic scale, i.e. on a cellular or sub-cellular level. Molecular imaging methods use reporter probes, which interact with specific molecular targets inside the cell, the cell-surface, or within the intra-cellular space. Such targets can be, for example, cell surface receptors, macrophages, or enzymes, which relate to specific biological processes or pathological dysfunctions. Because molecular imaging visualizes changes in the disease's molecular make-up, it promises to detect disease development much earlier and more differentiated than anatomical imaging at later stages of disease progression. Thus, it will not only lead to a deeper understanding of patho-physiological processes, but therapy and pharmacological intervention could also be initiated at an earlier stage of disease progression.

Nuclear and luminescence imaging methods have become the pillars in preclinical molecular imaging of small animals, which are utilized as models for studying human disease and for drug development [11–14]. While the former approach requires ionizing radionuclides as reporter probes, the latter uses optical sources as signaling component. Single photon emission computed tomography (SPECT) and positron emission tomography (PET) generate three-dimensional (3D) volume maps of the *in vivo* radionuclide bio-distribution. Relatively high image resolution as compared to luminescence imaging can be obtained because scattering effects of high-energy photons inside tissue are negligible.

Luminescence imaging has several advantages over nuclear imaging modalities. First of all, optical sources emit low-energy photons (2–3 eV), which are less harmful than more energetic gamma radiation (>100 keV) emitted from radioactive probes. Second, optical imaging typically offers a higher sensitivity (as compared to SPECT) and its imaging technology is relatively inexpensive (as compared to PET and SPECT). Third, luminescence imaging has the advantage that multiple light sources with different spectral characteristics can be used for multiplexed imaging in the same animal. And last, an optical reporter system itself is comparatively inexpensive, as the costs associated with imaging of radionuclides are generally higher due to technological implications associated with their production, storage, and disposal. Despite these advantages, strong scattering and partial absorption of light limits the ability for deep tissue imaging. Thus, the main application of luminescence imaging remains only to be found in pre-clinical imaging of small animals including mice and rats.

Luminescence imaging can be subdivided into different categories according to its physical or chemical mechanisms of light emission by optical sources. These sources can either directly be expressed in transgenic mice or be administered from outside and include, for example, fluorescent proteins [15–18], organic dyes [19], enzyme-substrate systems [20, 21], and nanoparticles [22]. Imaging of fluorescence sources, also termed *fluorescence imaging*, uses a halogen lamp or a laser diode that illuminates the tissue surface, and a fluorescent ligand or protein is stimulated for light emission. The fluorescence light is collected at the tissue surface and displayed in planar images for further analysis [23–26]. Imaging of bioluminescence sources, also termed *bioluminescence imaging*, does not require an external light source for emission stimulation. In fact, target cells are transfected with a *luc* reporter gene that acts as an optical tag. The reporter gene expresses the enzyme luciferase, which catalyses the emission of visible light by chemically reacting with an administered substrate luciferin [27]. The emitted light is detected on the tissue surface by an optical camera [28, 29]. Bioluminescence imaging is a very sensitive technique because of relatively low ambient light levels [30, 31]. The last category, also termed as *Cerenkov light imaging*, can be considered as a hybrid approach of optical and nuclear imaging, where a radionuclide is used as source for light emission and an optical detector collects the luminescence light at the tissue boundary [32–34]. This modality is based on the *Cerenkov* effect where light is emitted by a charged particle propagating faster than the speed of light in the same medium [35].

While PET and SPECT are the most understood technology for providing 3D image datasets, luminescence imaging only yields two-dimensional (2D) surface images, which limit the ability for quantitative imaging. Therefore, methods for luminescence tomography have been developed that build 3D spatial maps of light-emitting sources inside highly scattering tissue from light intensity measurements taken on the tissue surface. In the next few sections, we will describe the underlying physical and mathematical aspects of multi-spectral luminescence tomography that capitalizes on the multi-spectral light–tissue interaction for the purpose of 3D image formation. We will focus on a light propagation model based on the simplified spherical harmonics (SP_N) equations, and on its analytical and numerical solutions, briefly describe different methods for solving the inverse source problem, and show some imaging examples.

2.2 Challenges in tissue optics

2.2.1 Tissue scattering and absorption

Direct imaging of light sources inside biological tissue is difficult, because several physical and technological challenges need to be overcome. A major barrier for forming 3D volume images of light sources is (i) multiple light scattering and (ii) accurate modeling of light propagation beyond the diffusion limit. Unlike high-energy photons, as they are used in PET or SPECT, visible and near-infrared (NIR) light is strongly scattered in biological tissue [36–39]. The scattering coefficienti, $\mu_s(\lambda)$, in units of cm^{-1}, is wavelength-dependent and is typically in the range 100 to 200 cm^{-1}. The reduced scattering coefficienti, $\mu_s' = (1-g)\mu_s$, is given by the scattering coefficienti and the anisotropy factor, $0.5 < g < 1$, which describes highly forward-peaked scattered photons.

Because of multiple scattering, a point source buried deep inside tissue will only yield diffuse light intensity distributions on the tissue surface. A direct image of the light-emitting source according to the laws of geometrical optics cannot be obtained. Although the measured light intensity is a function of the source emission strength, the actual spatial source location inside tissue cannot be retrieved from a single 2D surface image. Figure 2.1 illustrates the problem of planar luminescence imaging. Here, a mouse had been engineered such that its kidneys constitutively expressed luciferase, which becomes part of a photochemical reaction and will emit visible light [40]. Figure 2.1(A) shows a luminescence image taken on the tissue surface of the mouse, whereas Figure 2.1(B) is an image of the light-emitting kidneys directly exposed to the camera after tissue dissection. Although the actual expression level of luciferase in both kidneys is the same, the planar surface image of the still living animal does not show the correct expression levels.

Another challenge in imaging of light sources constitutes the accurate modeling of photon propagation beyond the diffusion limit [41–43]. The light absorption coefficient of tissue, $\mu_a(\lambda) = C\epsilon(\lambda)$ in units of cm^{-1}, is wavelength-dependent and is a function of the tissue deoxy-hemoglobin (Hb) and oxy-hemoglobin (HbO_2) concentrations, C_{Hb} and C_{HbO_2}, in units of mol cm^{-3}. Hb and HbO_2 are chromophores with strongly varying extinction coefficients, $\epsilon_{Hb}(\lambda)$ and $\epsilon_{Hb0_2}(\lambda)$, between 550 nm and 650 nm. As seen in Figure 2.2, visible light at wavelengths smaller than 650 nm will be absorbed much more strongly than red or NIR light with wavelengths larger than 650 nm.

Combining the scattering and absorption parameters, the transport albedo, $\mu_s'/(\mu_s' + \mu_a)$, characterizes the physical properties of light propagation in tissue. For example, Figure 2.3 shows the transport albedo as a function of wavelength for three different tissue components. The transport albedo at wavelengths smaller than 650 nm is $\mu_s'/(\mu_s' + \mu_a) < 1$. The diffusion model, a low-order approximation to the radiative transfer equation (RTE), has limited validity at these wavelengths

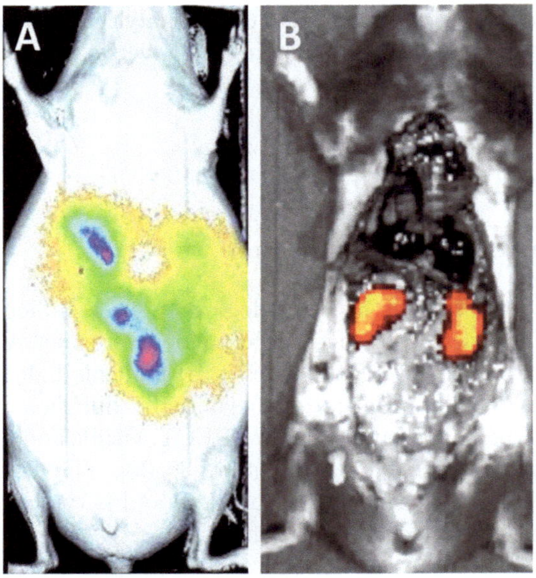

Fig. 2.1. Optical images of a mouse (ventral side) with luminescent kidneys [40]. Both images show different light distributions caused by different levels of light scattering of tissue. (A) Image of the luminescence light intensity on the tissue surface. The light is strongly scattered and only diffuse intensity distributions can be measured. (B) Direct image of the bioluminescence light intensity of the dissected animal. Both kidneys are directly exposed to the camera and no light scattering occurs.

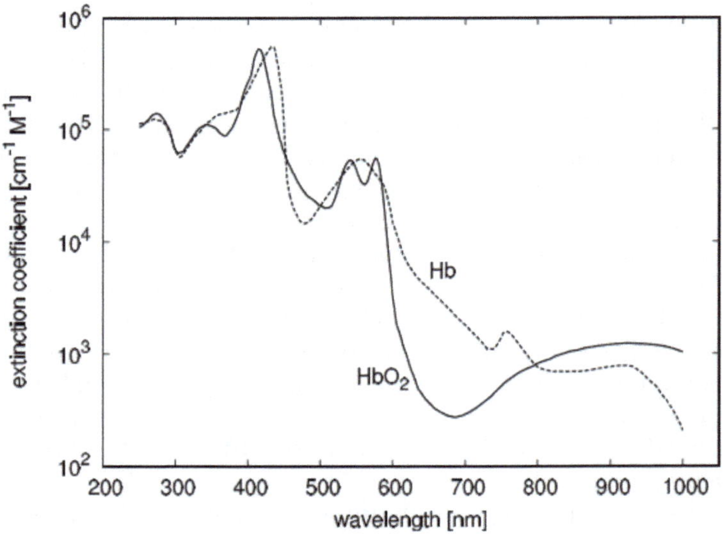

Fig. 2.2. Extinction spectrum of Hb and HbO_2. $\epsilon(\lambda)$ is given in units of $cm^2\,mol^{-1} = cm^{-1}M^{-1}$. Strong light absorption can be found below 650 nm which lends blood its red color [47].

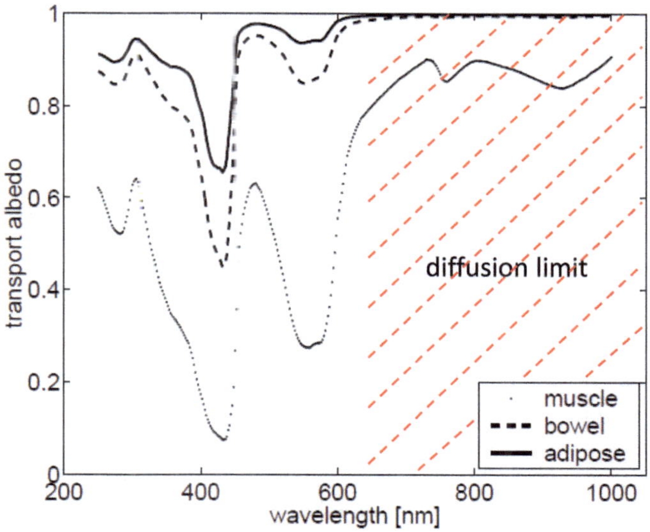

Fig. 2.3. Transport albedo, $\mu_s'/(\mu_s' + \mu_a)$, of different tissue components (μ_s' and μ_a are based on data taken from http://omlc.ogi.edu/spectra/hemoglobin/index.html, Dr Scott Prahl, Oregon Medical Laser Center, Portland). The diffusion limit mostly applies to wavelengths > 650 nm.

because its solutions exceed the diffusion limit ($\mu_s'/(\mu_s' + \mu_a) \approx 1$) [41–43]. Figure 2.4 demonstrates on an example the impact of the transport albedo on the light diffusion. Given a constant light intensity at a detector point with distance d to the light source, solutions of the diffusion equation overestimate the source emission strength with respect to solutions of a high-order radiative transfer model. Modeling of radiative transfer beyond the diffusion limit constitutes a challenge, because a high-order approximation to the RTE instead of the mathematically simple diffusion model needs to be employed [44–46]. One of those high-order approximations is the SP_N model that will be discussed in Section 2.3.1. A detailed review of different approximations to the RTE in luminescence imaging is given in [47].

2.2.2 Tomography and light source reconstruction

Considering the physical constraints for direct imaging of light sources, tomographic methods are needed that retrieve the actual source emission strength and location from light intensity measurements taken on the tissue surface. A major difficulty in determining the 3D source distribution with well-established mathematical techniques from transmission or emission tomography [48] is imposed, however, by multiple scattering of photons. Figure 2.5 illustrates the physical differences between luminescence tomography of visible light and emission and transmission tomography of high-energy radiation in nuclear medicine. For example, X-ray CT deals only with non-scattered or single-scattered photons and photon transport within the tissue can be described by a solution

$$\psi = \psi_0 \, e^{-\int_s \sigma_t(x,y)\, ds} \tag{2.1}$$

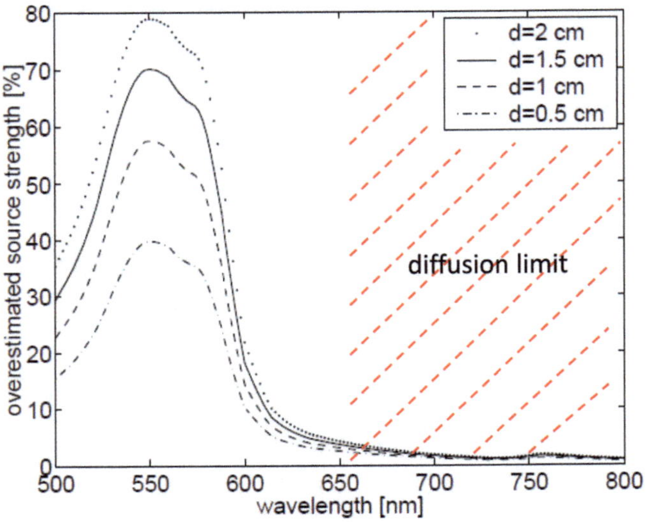

Fig. 2.4. Overestimated source strength based on the diffusion model w.r.t. an radiative transfer solution for bowel tissue beyond the diffusion limit. An isotropic point source was placed in different distances, *d*, from a detector and the light intensity was calculated with the diffusion model and a radiative transfer model. Assuming the same light intensity for both models at the detector location, the source emission strength estimated by the diffusion model deviates from the true source strength by up to 80%.

of a first-order partial differential equation (PDE)

$$\boldsymbol{s} \cdot \nabla \psi + \sigma_t \psi = 0 \tag{2.2}$$

$\sigma_t(x, y)$ is the spatially varying attenuation coefficient for X-ray photons, whereas ψ and ψ_0 are the photon flux along direction $\boldsymbol{s}$ for a given source ψ_0. The attenuation of photons along $\boldsymbol{s}$ is

$$a = \int_{\boldsymbol{s}} \sigma_t(x, y)\, d\boldsymbol{s} \tag{2.3}$$

With (2.1) and (2.3), we can write for the projections of $\sigma_t(x, y)$:

$$a = -\ln \frac{\psi}{\psi_0} = \int_{\boldsymbol{s}} \sigma_t(x, y)\, d\boldsymbol{s} \tag{2.4}$$

The spatial distribution of $\sigma_t(x, y)$ can be obtained from the projection a in (2.4) by using a *filtered back-projection method* [9].

Equation (2.4) can also be applied to high-energy photons emitted by a radionuclide. Its gamma radiation propagates along straight lines inside tissue and, therefore, similar projection and inversion methods can be found for SPECT and PET. For example, the projections a of the spatial activity distribution $\rho(x, y)$ of a radionuclide is given by a line integral expressed by equation (2.4). In turn, an inversion algorithm calculates the spatial maps $\rho(x, y)$ for given measurements of a. Figure 2.5 summarizes the differences and similarities of X-ray CT, SPECT, PET, and luminescence tomography.

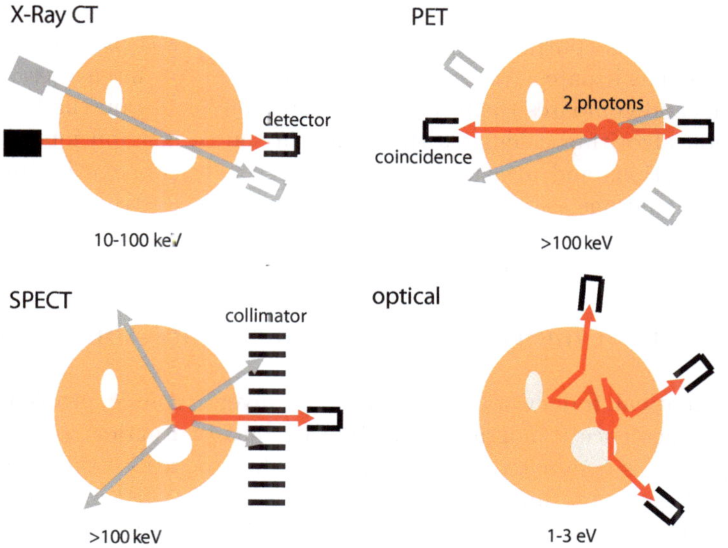

Fig. 2.5. Comparison of X-ray CT, PET, SPECT, and luminescence tomography. The former techniques use a hardware discriminator (collimator, pinhole camera, incident circuit) for eliminating multiple scattering events. Mostly non-scattered photons are detected. Such photons travel along a straight line and the detected radiation intensity is described by an integral equation. In contrast, luminescence tomography requires a light propagation model, which takes multiple scattering into account and calculates the diffuse light intensities on the tissue surface. Since no *straight* line path is present, inversion methods such as the back-projection method have limited success.

Because luminescence tomography needs to take multiple scattering of light into account, an integro-differential equation instead of a first-order PDE for $\psi(\boldsymbol{r}, \boldsymbol{\Omega})$ needs to be solved. This integro-differential equation is known as the RTE:

$$\boldsymbol{\Omega} \cdot \nabla\psi + \mu_t\psi = \mu_s \int_{4\pi} p(\Omega, \Omega')\psi(\Omega')\, d\Omega' + q \tag{2.5}$$

The RTE has, in addition to equation (2.2), an in-scatter and source term on the right-hand side and its photon flux is a function of the direction $\boldsymbol{\Omega}$. The in-scatter term describes the gain of photons at $\boldsymbol{r} = (x, y, z)$ along direction $\boldsymbol{\Omega}$ from scattered photons of all directions $\boldsymbol{\Omega}'$. The source term $q(\boldsymbol{r}, \boldsymbol{\Omega}) = Q(\boldsymbol{r})/4\pi$ constitutes the source power density in units of photons $\mathrm{s}^{-1}\ \mathrm{cm}^{-3}\ \mathrm{sr}^{-1}$, which is proportional to the optical reporter probe uptake or expression at location $\boldsymbol{r}$ inside tissue. The attenuation coefficient for light, μ_t, is the sum of the scattering, μ_s, and absorption, μ_a, coefficients. The phase function $p(\boldsymbol{\Omega}, \boldsymbol{\Omega}')$ describes the probability that a photon coming from direction $\boldsymbol{\Omega}$ is scattered into the solid angle $d\Omega'$ along direction $\boldsymbol{\Omega}'$. Solutions to the RTE cannot be formulated as in (2.1) and, thus, no inversion approach similar to the back-projection method can be applied.

Furthermore, the inverse problem for retrieving the spatial distribution of $Q(\boldsymbol{r})$ is not only highly ill-posed but also under-determined. Additional linearly-independent light intensity measurements are required for making the problem less under-determined. Therefore, luminescence tomography tries to capitalize on the strongly varying tissue absorption coefficient within the spectral window of 560–660 nm for the purpose of constructing spatial maps of $Q(\boldsymbol{r})$. A radiative transfer model calculates the spectrally dependent light intensities at different wavelengths and compares them to the measured intensities on the tissue surface. An inversion algorithm calculates the unknown source distribution based on the spectral predictions and measurements. Both the light propagation model based on the SP_N equations and the multi-spectral source reconstruction approach will be explained next.

2.3 Methods of multi-spectral luminescence tomography

The mathematical framework for solving the inverse source problem in multi-spectral luminescence tomography can be cast into two major components. The first component is a radiative transfer model, which describes the propagation of light inside tissue as a function of wavelength λ. This model predicts the partial boundary current or light intensities, J^+ in units of photons $\mathrm{s}^{-1}\,\mathrm{cm}^{-2}$, at the tissue surface for an assumed distribution of light sources, Q, inside the tissue. Q is the emission density in units of photons $\mathrm{s}^{-1}\,\mathrm{cm}^{-3}$ of an isotropic source.

The second component is a solution method for solving the functional relationship between the unknown source distribution inside tissue and the measured boundary current. Such relationship can be either defined as an optimization problem of an error function or cast into an algebraic system of equations. In both cases, the radiative transfer model is utilized for aiding the solution process by calculating the photon flux as a function of wavelength. Light source reconstruction by means of multi-spectral light intensity measurements and radiative transfer modeling has also found applications in ocean optics, where the surface depth and emission strength of luminescent algae needs to be determined [49–53].

2.3.1 Radiative transfer model

We will provide a brief overview of recent achievements in solving the radiative transfer model. The light propagation model, F, is generally described by PDEs for the photon flux inside the tissue. Most multi-spectral luminescence source reconstruction methods use the diffusion equation, which is a second-order PDE and, thus, a vast amount of analytical and numerical solution techniques are available. Furthermore, numerical solutions to the diffusion equation can be obtained with relative little computational effort. Despite its attractiveness in luminescence tomography, the diffusion equation is only a low-order approximation to the RTE. Therefore, diffusion solutions can lead to substantial errors beyond the diffusion limit at wavelengths smaller than 650 nm where light absorption dominates scattering [41–43]. In such cases, high-order approximations to the RTE need to be employed. These approximations include, for example, the discrete-ordinates (S_N) and the spherical harmonics (P_N) methods. Although both methods are computationally very expensive, they can be applied to problems beyond the diffusion limit. In 2006, Klose and Larsen [54] developed a light propagation model based on the SP_N equations that combines the advantages of both the diffusion model and the

RTE. Originally, the SP_N model was devised in the 1960s and had been successfully applied in nuclear engineering [55–64], but had not been applied in tissue optics due to missing boundary conditions and a mathematical description for modeling anisotropic scattering of photons.

SP_3 equations

The propagation of light in tissue, which originates from a light source $Q(\boldsymbol{r})$ inside the tissue, is described by a set of $(N+1)/2$ coupled diffusion equations, termed as the SP_N equations of Nth order. The 3D-SP_N equations are derived from the one-dimensional (1D) P_N equations and by replacing the 1D spatial derivative d/dx with the 3D operator $\nabla = (d/dx, d/dy, d/dz)$. Moreover, the Legendre moments, ϕ_n, of the 1D-P_N approximation are used to define $(N+1)/2$ *composite moments*, φ_n. For example, the two composite moments $\varphi_1(\boldsymbol{r})$ and $\varphi_2(\boldsymbol{r})$ for the 3D-SP_3 $(N=3)$ equations are given by the Legendre moments $\phi_0(\boldsymbol{r})$ and $\phi_2(\boldsymbol{r})$:

$$\varphi_1 = \phi_0 + 2\phi_2 \,, \tag{2.6a}$$

$$\varphi_2 = 3\phi_2 \,. \tag{2.6b}$$

Thus, the photon flux or fluence $\phi(\boldsymbol{r}) = \phi_0(\boldsymbol{r}) = \int_{4\pi} \psi((\boldsymbol{r}), \boldsymbol{\Omega})\, d\boldsymbol{\Omega}$ inside tissue is defined as:

$$\phi_0 = \varphi_1 - \frac{2}{3}\varphi_2 \,. \tag{2.7}$$

Let us assume we only have light-emitting sources inside the tissue, which do not require an external light source for emission stimulation. Then we will obtain the following SP_3 model for a single spectral band $\Delta\lambda$ centered at λ_l:

$$-\boldsymbol{\nabla} \cdot \frac{1}{3\mu_{a1}} \boldsymbol{\nabla}\varphi_1 + \mu_a\varphi_1 = Q(\lambda_l) + \left(\frac{2}{3}\mu_a\right)\varphi_2 \tag{2.8a}$$

$$-\boldsymbol{\nabla} \cdot \frac{1}{7\mu_{a3}} \boldsymbol{\nabla}\varphi_2 + \left(\frac{4}{9}\mu_a + \frac{5}{9}\mu_{a2}\right)\varphi_2 = -\frac{2}{3}Q(\lambda_l) + \left(\frac{2}{3}\mu_a\right)\varphi_1 \,. \tag{2.8b}$$

The spectrally-dependent nth-order absorption coefficients are defined as:

$$\mu_{an} = \mu_t - \mu_s g^n \tag{2.9}$$

with $\mu_{a0} = \mu_a$ and $\mu_{a1} = \mu_a + \mu_s'$. The SP_3 boundary conditions for a partially reflective air–tissue interface are given by:

$$\left(\frac{1}{2} + A_1\right)\varphi_1 + \left(\frac{1+B_1}{3\mu_{a1}}\right)\boldsymbol{n}\cdot\boldsymbol{\nabla}\varphi_1 = \left(\frac{1}{8} + C_1\right)\varphi_2 + \left(\frac{D_1}{\mu_{a3}}\right)\boldsymbol{n}\cdot\boldsymbol{\nabla}\varphi_2 \tag{2.10a}$$

$$\left(\frac{7}{24} + A_2\right)\varphi_2 + \left(\frac{1+B_2}{7\mu_{a3}}\right)\boldsymbol{n}\cdot\boldsymbol{\nabla}\varphi_2 = \left(\frac{1}{8} + C_2\right)\varphi_1 + \left(\frac{D_2}{\mu_{a1}}\right)\boldsymbol{n}\cdot\boldsymbol{\nabla}\varphi_1 \,. \tag{2.10b}$$

The remaining coefficients $A_1, \ldots, D_1, \ldots, A_2, \ldots, D_2$ are defined in [54]. The solution of equations (2.8) and (2.10) yields the partial current $J^+(\lambda_l)$ at the boundary with surface normal $\boldsymbol{n}$:

$$J^+(\lambda_l) = \left(\frac{1}{4} - \frac{1}{2}R_1\right)\left(\varphi_1 - \frac{2}{3}\varphi_2\right) + \frac{1}{3}\left(\frac{5}{16} + \frac{5}{4}R_1 - \frac{15}{4}R_3\right)\varphi_2 - \left(\frac{0.5 - 1.5R_2}{3\mu_{a1}}\right)\boldsymbol{n}\cdot\nabla\varphi_1 - \frac{\frac{21}{4}R_2 - \frac{35}{4}R_4}{7\mu_{a3}}\boldsymbol{n}\cdot\nabla\varphi_2 \,. \quad (2.11)$$

The moments R_1, R_2, R_3, R_4 of the reflectivity R can be found again in [54].

In the case of emission stimulation by an external light source, we need a two-level SP_3 model and equations (2.8) and (2.10) are slightly modified for each level. Moreover, a fluorescence light source can either be stimulated with a spectrally varying excitation field $\phi(\lambda^x)$ or the fluorescence light, $J^+(\lambda^m)$, emerging at the tissue surface, can be measured at different spectral bands. In Section 2.4.3, we will only consider the former case in more detail, which is also termed *excitation-resolved fluorescence tomography.*

At the emission stimulation level, an external light source $S(\lambda^x, \boldsymbol{\Omega})$ is placed at the tissue boundary, which emits light within the spectral band $\Delta\lambda$ centered at λ^x. It illuminates along $\boldsymbol{\Omega}$ the tissue boundary with surface normal $\boldsymbol{n}$, and the excitation light, represented by the flux $\phi^x = \phi(\lambda^x)$, stimulates sources $Q(\lambda^m)$ for fluorescence emission at wavelengths λ^m. We obtain the following SP_3 equations:

$$-\nabla\cdot\frac{1}{3\mu_{a1}}\nabla\varphi_1^x + \mu_a\varphi_1^x = \frac{2}{3}\mu_a\varphi_2^x \quad (2.12a)$$

$$-\nabla\cdot\frac{1}{7\mu_{a3}}\nabla\varphi_2^x + \left(\frac{4}{9}\mu_a\frac{5}{9}\mu_{a2}\right)\varphi_2^x = \frac{2}{3}\mu_a\varphi_1^x \,. \quad (2.12b)$$

The partial-reflective boundary conditions are given with:

$$\left(\frac{1}{2} + A_1\right)\varphi_1^x + \left(\frac{1+B_1}{3\mu_{a1}}\right)\boldsymbol{n}\cdot\boldsymbol{\nabla}\varphi_1^x = \left(\frac{1}{8} + C_1\right)\varphi_2^x + \left(\frac{D_1}{\mu_{a3}}\right)\boldsymbol{n}\cdot\boldsymbol{\nabla}\varphi_2^x + \int_{\boldsymbol{\Omega}\cdot\boldsymbol{n}<0} S(\lambda^x, \boldsymbol{\Omega})2\left|\boldsymbol{\Omega}\cdot\boldsymbol{n}\right| d\Omega \quad (2.13a)$$

$$\left(\frac{7}{24} + A_2\right)\varphi_2^x + \left(\frac{1+B_2}{7\mu_{a3}}\right)\boldsymbol{n}\cdot\boldsymbol{\nabla}\varphi_2^x = \left(\frac{1}{8} + C_2\right)\varphi_1^x + \left(\frac{D_2}{\mu_{a1}}\right)\boldsymbol{n}\cdot\boldsymbol{\nabla}\varphi_1^x + \int_{\boldsymbol{\Omega}\cdot\boldsymbol{n}<0} S(\lambda^x, \boldsymbol{\Omega})\left(5\left|\boldsymbol{\Omega}\cdot\boldsymbol{n}\right|^3 - 3\left|\boldsymbol{\Omega}\cdot\boldsymbol{n}\right|\right) d\Omega \,. \quad (2.13b)$$

Last, the excitation field $\Phi^x(\boldsymbol{r}, \lambda^x)$ is given by :

$$\Phi^x = \varphi_1^x - \frac{2}{3}\varphi_2^x \,. \quad (2.14)$$

Once the excitation field has been calculated, we will solve the SP_3 equations for the fluorescence light originating from light emitting sources $Q(\lambda^m) = \Phi^x\eta\epsilon c_0$:

$$-\nabla\cdot\frac{1}{3\mu_{a1}}\nabla\varphi_1^m + \mu_a\varphi_1^m = \Phi^x\eta\epsilon c_0 + \frac{2}{3}\mu_a\varphi_2^m \quad (2.15a)$$

$$-\nabla\cdot\frac{1}{7\mu_{a3}}\nabla\varphi_2^m + \left(\frac{4}{9}\mu_a\frac{5}{9}\mu_{a2}\right)\varphi_2^m = -\frac{2}{3}\Phi^x\eta\epsilon c_0 + \frac{2}{3}\mu_a\varphi_1^m \,. \quad (2.15b)$$

The fluorescence source is defined by its quantum yield, η, concentration, c, and extinction coefficient, ϵ. The partial-reflective boundary conditions for fluorescence light are given with:

$$\left(\frac{1}{2}+A_1\right)\varphi_1^m+\left(\frac{1+B_1}{3\mu_{a1}}\right)\boldsymbol{n}\cdot\boldsymbol{\nabla}\varphi_1^m=\left(\frac{1}{8}+C_1\right)\varphi_2^m+\left(\frac{D_1}{\mu_{a3}}\right)\boldsymbol{n}\cdot\boldsymbol{\nabla}\varphi_2^m \tag{2.16a}$$

$$\left(\frac{7}{24}+A_2\right)\varphi_2^m+\left(\frac{1+B_2}{7\mu_{a3}}\right)\boldsymbol{n}\cdot\boldsymbol{\nabla}\varphi_2^m=\left(\frac{1}{8}+C_2\right)\varphi_1^m+\left(\frac{D_2}{\mu_{a1}}\right)\boldsymbol{n}\cdot\boldsymbol{\nabla}\varphi_1^m\,. \tag{2.16b}$$

The partial current, $J^+(\lambda^m)$, of the fluorescence light at the boundary with surface normal $\boldsymbol{n}$ is obtained from φ_1^m and φ_2^m:

$$\begin{aligned}J^+(\lambda^m)=&\left(\frac{1}{4}-\frac{1}{2}R_1\right)\left(\varphi_1^m-\frac{2}{3}\varphi_2^m\right)+\\&\frac{1}{3}\left(\frac{5}{16}+\frac{5}{4}R_1-\frac{15}{4}R_3\right)\varphi_2^m-\\\left(\frac{0.5-1.5R_2}{3\mu_{a1}}\right)\boldsymbol{n}\cdot\nabla\varphi_1^m-&\frac{\frac{21}{4}R_2-\frac{35}{4}R_4}{7\mu_{a3}}\boldsymbol{n}\cdot\nabla\varphi_2^m\,.\end{aligned} \tag{2.17}$$

The SP_3 equations can be solved with analytical or numerical methods as described next.

Analytical solutions

Analytical solutions for the SP_3 equations (2.8) have been found for (i) optically uniform tissue with spatially non-varying scattering and absorption coefficients and (ii) non-bounded tissue geometries [65,66]. The Green's function for equation (2.8) is obtained by using the plane wave expansion of the composite moments $\varphi_1(\boldsymbol{r})$, $\varphi_2(\boldsymbol{r})$, and an isotropic point source $Q(\boldsymbol{r})=\delta(r)/(4\pi r^2)$ for an infinite medium with spherical symmetry and ($r=|\boldsymbol{r}|$):

$$\varphi_{1,2}(r) = \frac{1}{2\pi^2 r}\int_0^\infty k\varphi_{1,2}(k)\sin(kr)\,dk \tag{2.18}$$

$$Q(r) = \frac{1}{2\pi^2 r}\int_0^\infty k\sin(kr)\,dk\,. \tag{2.19}$$

We obtain a system of linear equations for the plane wave expansion coefficients $\varphi_{1,2}(k)$ by substituting the composite moments (2.18) and the source (2.19) into equations (2.8):

$$\left(\frac{k^2}{3\mu_{a1}}+\mu_a\right)\varphi_1(k)-\frac{2}{3}\mu_a\varphi_2(k)=1 \tag{2.20a}$$

$$-\frac{2}{3}\mu_a\varphi_1(k)+\left(\frac{k^2}{7\mu_{a3}}+\frac{4}{9}\mu_a+\frac{5}{9}\mu_{a2}\right)\varphi_2(k)=-\frac{2}{3}\,. \tag{2.20b}$$

The solutions for the expansion coefficients are given by

$$\varphi_1(k) = \frac{f_1(k^2)}{k^4 + c_1 k^2 + c_2} \tag{2.21a}$$

$$\varphi_2(k) = \frac{f_2(k^2)}{k^4 + c_1 k^2 + c_2} \tag{2.21b}$$

with functions

$$f_1(k^2) = \frac{35}{3}\mu_{a1}\mu_{a2}\mu_{a3} + 3\mu_{a1}k^2 \tag{2.22a}$$

$$f_2(k^2) = -\frac{14}{3}\mu_{a3}k^2 \tag{2.22b}$$

and constants

$$c_1 = 3\mu_a\mu_{a1} + \frac{28}{9}\mu_a\mu_{a3} + \frac{35}{9}\mu_{a2}\mu_{a3} \tag{2.23a}$$

$$c_2 = \frac{35}{3}\mu_a\mu_{a1}\mu_{a2}\mu_{a3}\,. \tag{2.23b}$$

Next, we find the partial functions of (2.21) by calculating the zeros k_1 and k_2 of the denominator $k^4 + c_1k^2 + c_2$. With the substitution $x = k^2$, we arrive at the zeros $x_{1,2} = -\frac{1}{2}c_1 \pm \frac{1}{2}\sqrt{c_1^2 - 4c_2}$ of the modified denominator $x^2 + c_1x + c_2$, yielding the partial functions:

$$\frac{f_1(x)}{x^2 + c_1x + c_2} = \frac{f_1(x)}{(x - x_1)(x - x_2)} = \frac{\frac{f_1(x_1)}{x_1 - x_2}}{x - x_1} + \frac{\frac{f_1(x_2)}{x_2 - x_1}}{x - x_2} \tag{2.24a}$$

$$\frac{f_2(x)}{x^2 + c_1x + c_2} = \frac{f_2(x)}{(x - x_1)(x - x_2)} = \frac{\frac{f_2(x_1)}{x_1 - x_2}}{x - x_1} + \frac{\frac{f_2(x_2)}{x_2 - x_1}}{x - x_2}\,. \tag{2.24b}$$

Hence, the partial functions for the expansion coefficients are given by:

$$\varphi_1(k) = \frac{f_1(k^2)}{(k^4 + c_1k^2 + c_2)} = \frac{f_1(k^2)}{(k^2 + k_1^2)(k^2 + k_2^2)} = \frac{\frac{f_1(x_1)}{k_2^2 - k_1^2}}{k^2 + k_1^2} + \frac{\frac{f_1(x_2)}{k_1^2 - k_2^2}}{k^2 + k_2^2} \tag{2.25a}$$

$$\varphi_2(k) = \frac{f_2(k^2)}{(k^4 + c_1k^2 + c_2)} = \frac{f_2(k^2)}{(k^2 + k_1^2)(k^2 + k_2^2)} = \frac{\frac{f_2(x_1)}{k_2^2 - k_1^2}}{k^2 + k_1^2} + \frac{\frac{f_2(x_2)}{k_1^2 - k_2^2}}{k^2 + k_2^2}\,, \tag{2.25b}$$

where $-x_1 = k_1^2$ and $-x_2 = k_2^2$. Last, we obtain solutions for the composite moments $\varphi_{1,2}(r)$ by comparing the solutions (2.25) of $\varphi_{1,2}(k)$ and the plane wave expansion (2.18) of $\varphi_{1,2}(r)$ to the Green's function $G(r)$ for the infinite space diffusion equation:

$$G(r) = \frac{3\mu_{a1}}{4\pi r}e^{-\mu r} = \frac{3\mu_{a1}}{2\pi^2 r}\int_0^\infty \frac{k\sin(kr)}{k^2 + \mu^2}\,dk \tag{2.26}$$

with $\mu = \sqrt{3\mu_a\mu_{a1}}$. Thus, the composite moments are a superposition of two Green's functions:

$$\varphi_1(r) = \frac{f_1(x_1)}{k_2^2 - k_1^2}\frac{1}{4\pi r}e^{-k_1 r} + \frac{f_1(x_2)}{k_1^2 - k_2^2}\frac{1}{4\pi r}e^{-k_2 r} \tag{2.27a}$$

$$\varphi_2(r) = \frac{f_2(x_1)}{k_2^2 - k_1^2}\frac{1}{4\pi r}e^{-k_1 r} + \frac{f_2(x_2)}{k_1^2 - k_2^2}\frac{1}{4\pi r}e^{-k_2 r}\,. \tag{2.27b}$$

Finally, the photon flux inside tissue as a function of distance to the source is obtained with (2.7) and (2.27). Figure 2.6 shows the fluence inside tissue as a function of the distance to an isotropic light source for different absorption coefficients.

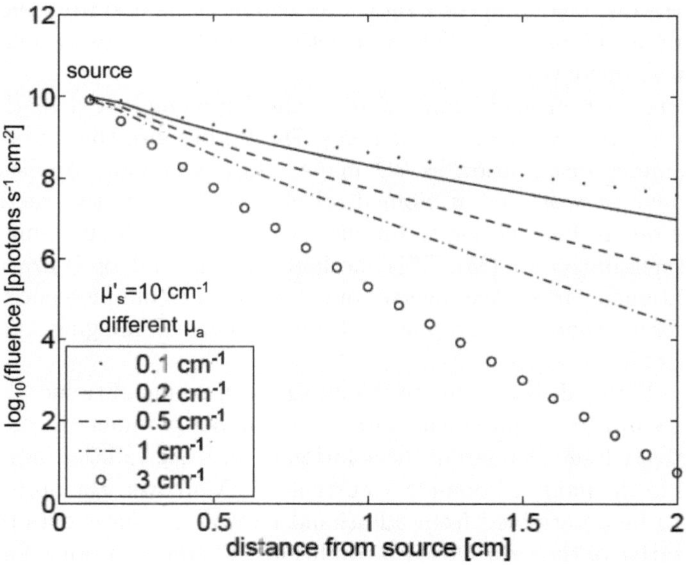

Fig. 2.6. Fluence, ϕ, inside tissue as a function of distance, r, to the light source, $Q(r_0 = 0)$, for different absorption coefficients. The solution of the fluence has been obtained by solving the SP_3 equations and is given by (2.7) and (2.27).

Numerical solutions

In the case of (i) optically non-uniform tissue with spatially varying scattering and absorption coefficients and (ii) complex tissue geometries, we will need to use numerical methods for solving the SP_N equations. Klose and Larsen [54] have solved the SP_N equations up to order $N = 7$ with a finite difference method on structured Cartesian grids. Here, a second-order spatial differencing scheme for discretizing the Laplacian operator was employed, which allowed for solving the SP_N equations on coarse grids while achieving solutions with relatively small numerical discretization error. The SP_3 equations could be solved approximately 100–200 times faster than the S_{16} discrete-ordinates equations of the RTE, which required a first-order spatial differencing scheme (*step method*) and, hence, fine grids for obtaining unconditionally stable solutions. Since then, the SP_3 equations for photons have been solved, for example, with finite-difference methods [40,67,68] on structured grids and with finite-element methods [69–75, 129] on unstructured grids.

2.3.2 Source reconstruction methods

The second component for solving the linear inverse source problem is a multi-spectral source reconstruction algorithm. It determines the spatial distribution and emission strength of the light sources $Q(\boldsymbol{r})$ inside the tissue domain given the planar light intensity images at the boundary, represented by $J^+(\boldsymbol{r})$. As seen in Figure 2.6, the distance from the light source can be estimated from the photon flux by measuring a unique set of light intensities at different absorption coefficients, i.e. different wavelengths.

The reconstruction algorithms in biomedical imaging are sometimes also referred to as *image reconstruction methods* [78, 79], because they visualize the unknown parameter distribution as 3D maps. Most commonly applied techniques for solving inverse problems in biomedical imaging are iterative methods including the algebraic reconstruction technique (ART) [80, 81], the maximum-likelihood expectation-maximization (ML-EM) method [82, 83], and optimization methods [84]. As mentioned earlier, the inverse source problem in luminescence tomography is quite different than tomographic methods in nuclear imaging and it has two challenges.

First, it is highly ill-posed due to strong light scattering of tissue. It implies that small changes in the measurement data, Y, or model parameters, μ_a and μ_s, will lead to relatively large changes in the solution of Q. Uncertainties in μ_a and μ_s will cause errors in the modeled boundary current J^+. Moreover, the optical properties often need to be determined from additional *ex vivo* measurements that may not be representative of the actual *in vivo* conditions. Therefore, prior information may be included into the image reconstruction process, which will restrict the space of possible solutions.

Second, the reconstruction problem is also largely under-determined due to the relatively small amount of boundary data when compared to the large number of unknown source points inside tissue. Therefore, multi-spectral light intensity data will be utilized that increase the total amount of linearly independent boundary measurements and, subsequently, decrease the solution space making the inverse problem less under-determined.

Algebraic reconstruction technique and maximum-likelihood expectation-maximization

The ART and ML-EM method are image reconstruction methods used in X-ray CT, SPECT, and PET [79]. They solve large systems of linear equations for the unknown image variable. Similar to these applications, the system of linear equations for the unknown source distribution is derived from the forward model

$$Y = F(Q)\,. \tag{2.28}$$

We define the continuous variable $Q(\boldsymbol{r})$ on a grid supported by M grid points. Thus, the unknown source distribution inside tissue is given by the M-dimensional source vector $\boldsymbol{Q}$ representing a stack of tomographic images with a total of M voxels. The $N = D \cdot L$-dimensional measurement vector $\boldsymbol{Y}$ is given by the noise-corrupted partial current J^+. Each element Y_n of $\boldsymbol{Y}$ depicts one of D detectors on the tissue surface that measures the light intensity within L spectral bands $\Delta\lambda$ centered around λ_l. The elements Y_n with doublet $n = (d, l)$ are ordered as follows:

$$\boldsymbol{Y} = \left(Y_{(1,1)}, \ldots, Y_{(D,1)}, \ldots, Y_{(d,l)}, \ldots, Y_{(D,L)}\right) . \qquad (2.29)$$

The light propagation model, which links Q with its partial current at the boundary, is represented by the $N \times M$ dimensional matrix $\boldsymbol{F}$, and we obtain the following system of equations for the source vector $\boldsymbol{Q}$:

$$\boldsymbol{Y} = \boldsymbol{F} \cdot \boldsymbol{Q} . \qquad (2.30)$$

Each matrix element F_{nm} is given by the calculated partial boundary current $J^+_{n=(d,l),m}$ for a detector $d \in D$ with wavelength interval $l \in L$, and source point $m \in M$. J^+ at the boundary point d is calculated by solving the SP_3 equations for a given unit source $\tilde{Q}_m$ defined at grid point m and absorption coefficient $\mu_a(\lambda_l)$. The formation of $\boldsymbol{F}$ would be computationally demanding if the SP_3 equations had been solved for all M unit sources $\tilde{Q}_m$. However, the construction of $\boldsymbol{F}$ can be sped up by utilizing the *reciprocity theorem* and the matrix elements F_{nm} are determined by placing a virtual unit source $\tilde{Q}_d$ at each detector point position. The photon flux, $\phi_{(m,l)}$, obtained at interior grid point m and wavelength interval λ_l is assigned to F_{nm}. Since $D < M$, an approximate speed up of M/D is obtained.

The matrix equation (2.30) is solved by an iterative scheme, either the ART or ML-EM method, for the mth element of $\boldsymbol{Q}$. The ART is based on *Kaczmarz's method* [81,85] and solves linear systems of equations with large number of unknown variables. Its updating scheme is given by:

$$Q_m^{k+1} = Q_m^k + F_{nm} \frac{Y_n - \sum_{m'} F_{nm'} Q_{m'}^k}{\sum_{m'} F_{nm'}^2} . \qquad (2.31)$$

The ML-EM algorithm is based on the Richardson–Lucy algorithm [86,87] and its iterative scheme is given by:

$$Q_m^{k+1} = \frac{Q_m^k}{\sum_{n'} F_{n'm}} \sum_n F_{nm} \frac{Y_n}{\sum_{m'} F_{nm'} Q_{m'}^k} . \qquad (2.32)$$

We will demonstrate the multi-spectral ART source reconstruction on an example. A gaseous tritium light source (GTLS), made from a glass capillary tube, has been placed within a rectal catheter and positioned directly beneath the spine of a mouse. The GTLS tube emits light over a range of 100 nm with a peak at 531 nm. The position of the GTLS tube can be seen within the X-ray CT image in Figure 2.7(B). Luminescence images were taken with an IVIS 200 system (Caliper Life Sciences Inc.) at $L = 6$ different spectral bands ranging from 560 nm to 660 nm with $\Delta\lambda = 20$ nm (Figure 2.7(A)). The source reconstruction algorithm calculated the distribution of $Q(\boldsymbol{r})$. As seen in Figure 2.7(C), the reconstructed tube location is in accordance with the X-ray CT. More details can be found in [40].

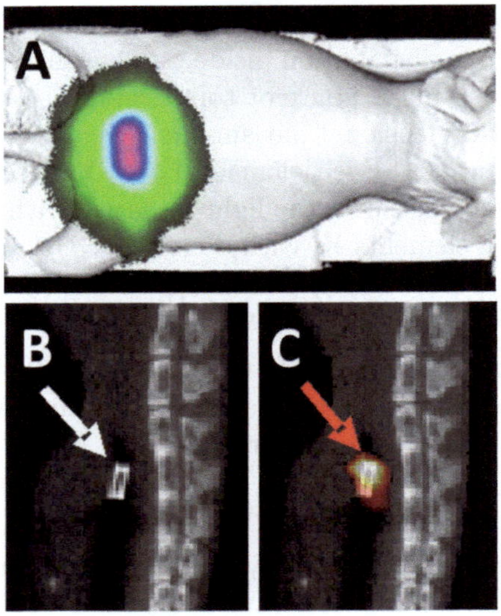

Fig. 2.7. Luminescence source reconstruction with a ART [40]. (A) Luminescence image (640 nm) of GTLS implanted into the rectum of a mouse. (B) Slice of X-ray CT scan with GTLS positioned in rectum. (C) Luminescence image reconstruction of GTLS.

Gradient-based source reconstruction

The spatial source distribution can also be reconstructed with a gradient-based optimization method that minimizes a χ^2-error function. The error function describes the discrepancy between the experimental, $\boldsymbol{Y}$, and predicted, $J^+(\boldsymbol{Q})$, light intensity data for all D detector points on the tissue surface:

$$\chi^2(\boldsymbol{Q}) = \sum_{d,\lambda} \left(\frac{J^+_{d,\lambda}(\boldsymbol{Q}) - Y_{d,\lambda}}{\sigma_{d,\lambda}} \right)^2 . \tag{2.33}$$

$\sigma_{d,\lambda}$ exhibits the confidence we have in the accuracy of our measurement data. It is mainly influenced by the system noise (e.g. shot noise) of the instrument that measures the light intensities. Gradient-based reconstruction techniques are local optimization methods including, for example, the steepest descent (SD) method, the conjugate gradient (CG) method, or different quasi-Newton (QN) methods [84, 89]. These techniques iteratively update an initial guess, $\boldsymbol{Q}_{k=0}$, of the unknown source vector $\boldsymbol{Q}$ along a search direction $\boldsymbol{u}$. Once the minimum is found, the final result is the sought source distribution. For example, the updating procedure of a *limited-memory Broyder-Fletcher-Goldfarb-Shanno* (lmBFGS) method [84, 88, 89], a particular implementation of a QN method, at iteration step k is formulated as:

$$\boldsymbol{Q}^{k+1} = \boldsymbol{Q}^k + \alpha^k \boldsymbol{u}^k . \tag{2.34}$$

The parameter α^k is the step length along the direction $\boldsymbol{u}^k$. Once an update has been performed, a new search direction $\boldsymbol{u}^{k+1}$ will be determined with

$$\boldsymbol{u}^{k+1} = -\frac{d\chi^2}{d\boldsymbol{Q}}^{k+1} + \gamma_1 \boldsymbol{s} + \gamma_2 \boldsymbol{y} \tag{2.35}$$

with the vectors

$$s = c^{k+1} - Q^k \tag{2.36}$$

and

$$y = \frac{d\chi^2}{dQ}^{k+1} - \frac{d\chi^2}{dQ}^{k} . \tag{2.37}$$

The scalars $\gamma_{1,2}$ are defined by

$$\gamma_1 = -\left(1 + \frac{y^T y}{s^T y}\right) \frac{s^T \frac{d\chi^2}{dQ}^{k+1}}{s^T y} + \frac{y^T \frac{d\chi^2}{dQ}^{k+1}}{s^T y} \tag{2.38}$$

and

$$\gamma_2 = \frac{s^T \frac{d\chi^2}{dQ}^{k+1}}{s^T y} . \tag{2.39}$$

At each iteration step, k, a new derivative $d\chi^2/dQ$ needs to be calculated in order to determine the search direction u. More information regarding its application in fluorescence tomography can be found in [90, 91].

The non-uniqueness of the inverse source problem implies that multiple solutions for Q could exist [92]. That may also indicate the existence of multiple local minima of an error function. Gradient-based optimization methods, that explore the parameter space only in the vicinity of a single local minimum, may consequently lead to premature convergence without finding the global or a nearby smaller local minimum. Moreover, a local optimization method depends on the starting condition Q_0 and always leads to the same solution for that given starting point.

Stochastic source reconstruction

Instead of using a local optimization technique as shown in the previous section, a quite different approach for solving the inverse problem is stochastically sampling the global search space of all possible source distributions [93]. Stochastic source reconstruction approaches utilize a *random* process for minimizing the error function and they are not limited to a local parameter space close to the starting point Q_0. These methods comprise simulated annealing [94], genetic algorithms [95], and evolution strategies [96, 97].

In general, a stochastic source reconstruction method consists of three basic elements: (i) a mechanism for randomly generating and altering source distributions $Q(\theta) = \sum_{m=1}^{M} \theta_m b_m$ represented by the data vector θ and M source basis functions b_m, (ii) a light-propagation model based on the SP_N equations for determining $J^+(\theta)$ for each sampled source distribution, and (iii) a mechanism for controlling the sampling process by selecting a new set of source distributions for the next iteration step. The error function, which will be minimized, is defined as the $\chi^2(\theta)$-error norm as a function of θ:

$$\chi^2(\theta) = \sum_{d,\lambda} \left(\frac{J^+_{d,\lambda}(\theta) - Y_{d,\lambda}}{\sigma_{d,\lambda}} \right)^2 . \tag{2.40}$$

A multi-membered evolution strategy has been developed for luminescence tomography, where the alteration mechanism is based on a mutation and recombination operator, and the unknown data vector θ is randomly altered for consecutive generations of populations. A parent population consisting of μ data vectors θ^{μ} at generation k produces a population of λ offspring vectors θ^{λ} by randomly recombining vector elements $\theta^{\mu}_{I,II}$ of two parent individuals:

$$\theta^{\lambda}_{m} = \frac{\theta^{\mu}_{I} + \theta^{\mu}_{II}}{2} . \tag{2.41}$$

The new offspring generation $(k+1)$ is altered by pointwise mutation and each vector element θ^{λ}_{m} of all members λ is randomly modified:

$$\hat{\theta}^{\lambda}_{m} = \theta^{\lambda}_{m} + N(0, \hat{\sigma}^{\lambda}_{m}) . \tag{2.42}$$

Normally distributed random numbers with expectation zero and standard deviations σ are added to the variables θ. Similar to natural evolution, this selection process favours individuals best adapted to their environment. Only the fittest individuals of both populations, θ^{μ} and $\hat{\theta}^{\lambda}$, with smallest error function become the new parent population θ^{μ} of the next generation $(k+1)$. The iterative process is terminated after the relative change of two generations is smaller than a defined threshold.

The strategy parameters σ^{λ}_{m} in (2.42) are modified in order to facilitate the search progress. This process is also known as *self-adaptation*, where the strategy parameters σ^{λ}_{m} will be subject to mutation and recombination operators as well. A new strategy parameter $\hat{\sigma}^{\lambda}_{m}$ of the next generation is obtained by:

$$\hat{\sigma}^{\lambda}_{m} = \sigma^{\lambda}_{m}\, e^{\tau N(0,1)} . \tag{2.43}$$

The self-adaptation parameter τ is empirically determined and is bound by $0 < \tau < 1$.

The stochastic source reconstruction is demonstrated on a numerical example mimicking a 2D tissue slice of a mouse with two interior sources and different distances: (i) 0.75 cm, (ii) 0.5 cm, and (iii) 0.25 cm. First, J^{+} at $N = 180$ detector points was calculated for $M = 675$ source basis functions b_m. Each source function b_m had a cell size of 0.0675 cm and a power density of 10^9 photons s^{-1} cm^{-3}. Next, synthetic measurement data Y were generated for two light sources with size of 0.125 cm × 0.125 cm. The measurement data were corrupted with 1% Gaussian noise. Last, the parent and offspring population had μ = 3,000 and λ = 18,000 members and the global search process was finished after 900 generations yielding a total of 16,200,000 sampled source distributions. Figure 2.8 shows the reconstructed source distribution. Two sources could still be separated when they were more than 0.25 cm apart. The spatial location of single sources could be recovered with a deviation of less than 0.1 cm from its original location.

2.4 Applications

2.4.1 Multi-spectral bioluminescence tomography

Multi-spectral bioluminescence tomography has seen a rapid growth of different applications in molecular imaging and particularly in cancer research [98–112].

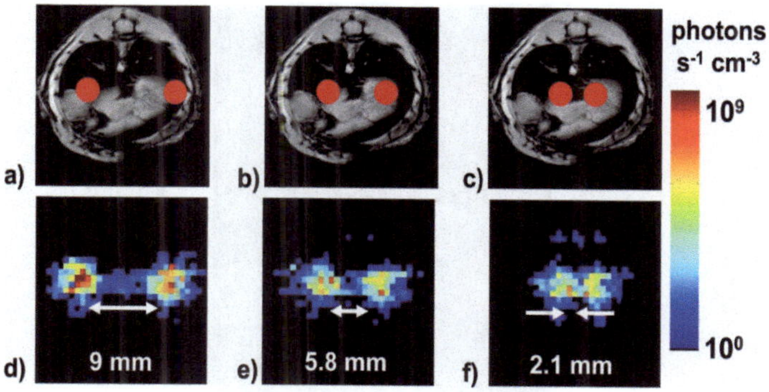

Fig. 2.8. Reconstruction of source distribution Q based on synthetic boundary measurement data. (a)–(c) MRI of mouse with two luminescent sources. Source separation is: (a) 7.5 mm, (b) 5 mm, and (c) 2.5 mm. (d)–(f) reconstructed source distribution Q.

The very first application of multi-spectral luminescence tomography with the SP_3 equations has been in 3D bioluminescence imaging of tumor development in mice [40, 112, 113]. The optical reporter system for monitoring tumor growth is based on the luciferase-luciferin reporter system. The cancer cells of the tumor have been genetically modified with the *luc* gene isolated from either the firefly or the click-beetle. The luciferase-luciferin reporter system has a broad emission spectrum with wavelengths smaller than 650 nm. Light at wavelengths < 650 nm, however, is strongly absorbed inside tissue and, thus, yields only small light intensities at the tissue surface, which are detected with a highly sensitive and often cooled CCD camera.

The impact of multi-spectral light intensity measurements is shown on an example. First, it demonstrates that an increase in spectral bands improves the source reconstruction; and second, it shows that including spectral bands at wavelengths smaller than 620 nm lead to a significant improvement in source reconstruction accuracy. Click-beetle green (CBG) luciferase-expressing U87 cells were injected into a small animal and were expected to accumulate in the lung. A set of luminescence surface images filtered at $L = 6$ different spectral bands between 560 nm and 660 nm ($\Delta\lambda = 20$ nm) were acquired from the dorsal and ventral view. Unfiltered images were acquired between each spectrally filtered image and were used to correct for the time-course of the changing luciferin uptake. The spatial distribution of the CBG expressing tumor cells were reconstructed using the full spectrum ($L = 6$) between 560 nm and 660 nm an the ML-EM method. Figure 2.9(A) shows a cross-sectional image where both lung lobes are clearly visible. Figure 2.9(B) shows a reconstruction where only $L = 2$ spectral bands centered at 560 and 580 nm were used. Again, both lung lobes could be identified, but they were not as well filled as in the full spectrum case. In contrast, Figure 2.9(C) shows a reconstruction based on $L = 2$ spectral bands centered at 640 nm and 660 nm. Here, only one lung lobe was shown to contain significant signal and its distribution tended to be overly shallow.

Efforts towards *quantitative* bioluminescence tomography are also undertaken that address the challenges for overcoming the uncertainty in the optical tissue pa-

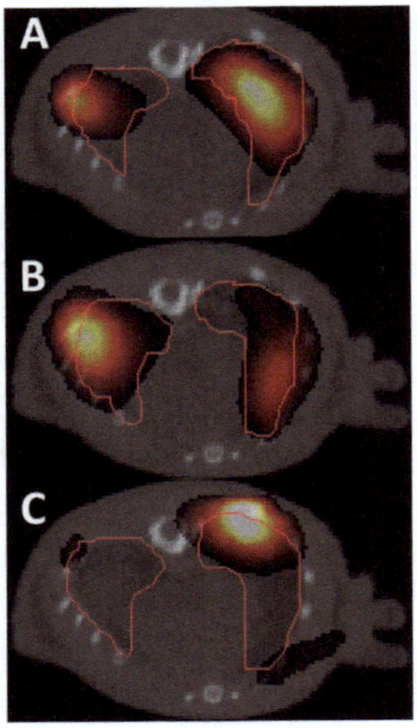

Fig. 2.9. *In vivo* source reconstruction of CBG expressing U87 cells in the lung (tumor border circled by red line) overlaid on top of an X-ray CT scan. (A) Full set of six wavelengths, (B) set of only two short wavelengths, (C) set of only two long wavelengths.

rameters. The optical tissue properties have a direct impact on the light intensities on the tissue surface. These properties are, however, partially unknown leading to erroneous image reconstructions. Therefore, the absorption and scattering parameters need to be determined prior to the source reconstruction in order to yield accurate and quantitative results.

2.4.2 Multi-spectral Cerenkov light tomography

Cerenkov luminescence tomography is very similar to bioluminescence tomography in terms of the source reconstruction process. In both cases, a source inside tissue emits light that is measured at different spectral bands on the tissue surface. A reconstruction algorithm recovers the spatial source distribution given the boundary data. The only difference between both methods is the different mechanism for light-emission generation and different light-emission spectra.

The Cerenkov effect is the emission of visible light caused by the discharge of an electron or positron during β^- or β^+ decay of a radionuclide [35]. The emitted particle, while traveling faster than light inside the medium, loses kinetic energy due to the polarization of electrons of the medium. The polarized electrons relax back to their native state while giving off visible Cerenkov light. A sensitive optical camera detects the light on the tissue surface. Therefore, the Cerenkov effect permits imaging of already existing and clinically-approved PET or SPECT radiotracers with conventional optical imaging systems [114–126]. The major advantage of Cerenkov light imaging is the use of low-cost optical instrumentation instead of expensive PET or SPECT instrumentation. Moreover, optical imaging could potentially also be used in back-reflectance geometry as a handheld device or in an intra-operative setting, which is not feasible with PET or SPECT imaging technology. Last, Cerenkov light imaging, similar to bioluminescence imaging, is not interfered by ambient tissue autofluorescence and, hence, promises a higher sensitivity than fluorescence imaging with relatively large signal to background levels. First attempts for Cerenkov light tomography have also been made [127–132], but limitations concerning a multi-spectral light propagation model beyond the diffusion limit and unknown absorption and scattering properties of tissue have not been sufficiently addressed yet.

2.4.3 Multi-spectral fluorescence tomography

Historically, fluorescence tomography has been the first application for source reconstruction methods in luminescence imaging [133–137]. These fluorescence tomography methods use source–detector multiplexing at a single spectral band of excitation and emission wavelengths for the purpose of tomographic source reconstruction. The tissue surface is illuminated by placing a series of light sources at different locations $\boldsymbol{r}_s$ on the tissue surface and a fluorescence source at location $\boldsymbol{r}$ inside the tissue is excited. In turn, the fluorescence source emits luminescence light and the partial current is detected at $\boldsymbol{r}_d$ on the surface. A source reconstruction algorithm calculates the unknown source distribution inside the tissue given the boundary measurements $\boldsymbol{Y}(\boldsymbol{r}_s, \boldsymbol{r}_d)$ for multiple spatial configurations of source–detector pairs (s,d) on the tissue surface.

Multi-spectral fluorescence tomography, however, does not rely on multiplexing of source–detector positions (s,d) for the purpose of image reconstruction. Instead, it utilizes, similar to bioluminescence and Cerenkov light tomography, the strongly varying extinction spectrum of Hb and $Hb0_2$ for solving the inverse source problem [67, 68, 138] and multiple wavelength–detector pairs (λ,d) are used as linearly independent data points. Moreover, multi-spectral fluorescence tomography can also operate in two different modes. Regarding the first mode of operation, a fluorescence source is stimulated for light emission using only a single spectral band for the excitation light. The emission light is collected on the tissue surface for multiple spectral bands centered at λ_l. This mode of operation is very similar to bioluminescence and Cerenkov light tomography, where luminescence light is detected at multiple spectral bands. The second mode of operation is reversed and the fluorescence source is stimulated for light emission using excitation light with different λ_l, whereas the fluorescence light is measured only at a single spectral window. In comparison to methods with source–detector multiplexing where boundary measurements $\boldsymbol{Y}(\boldsymbol{r}_s, \boldsymbol{r}_d)$ are taken for multiple source and detector positions, the measurement data $\boldsymbol{Y}(\boldsymbol{r}_d, \lambda)$ of the multi-spectral inverse source problem is now given by a set of multiple wavelengths and detector positions. Nevertheless, both multi-spectral modes could also be combined forming a multi-spectral excitation and emission collection approach.

Hyper-spectral excitation-resolved fluorescence tomography

Hyper-spectral excitation-resolved fluorescence tomography (HEFT) of quantum dots is an implementation of the second mode of operation [67, 68]. It uses partially overlapping instead of discrete spectral bands $\Delta\lambda$ around λ_l and, therefore, it conforms to the term *hyper-spectral*. HEFT capitalizes on the broad absorption spectrum, $\mu_{af}(\lambda) = C_f \epsilon_f(\lambda)$, of quantum dots and the spectral characteristics of tissue hemoglobin. The extinction spectrum $\epsilon_f(\lambda)$ of quantum dots with concentration C_f overlaps with the strongly varying absorption spectrum of Hb and HbO_2. Since the fluorescence emission directly depends on the strength of the excitation field $\Phi^x(\boldsymbol{r}, \lambda)$, the emission strength, $\Phi^x \eta C_f \epsilon_f$, of the fluorescence source will encode for its location $\boldsymbol{r}$ inside tissue. Therefore, the fluorescence light is measured for different excitation wavelengths and the partial current on the tissue surface is given by $J^+(\boldsymbol{r}_d) = F(\Phi^x(\lambda, \boldsymbol{r})\eta C_f(\boldsymbol{r})\epsilon_f)$.

Quantum dots make an excellent tool for excitation-resolved fluorescence tomography because their relatively broad extinction spectrum extends over a few hundreds of nanometers. Quantum dots consist of a nanometer-sized semiconductor core (3–20 nm) made of CdSe, CdT, or CdTe. The core is coated by a shell typically consisting of ZnS. The emission peak of quantum dots can be *tuned* to any chosen wavelength by simply changing the size of the Cd core. Thus, the spectral distance between emission and excitation wavelengths of quantum dots is not limited to the small Stokes shift of organic dyes and proteins. Quantum dots emitting in the red and NIR spectrum also circumvent the limitations of strong light absorption and auto-fluorescence at shorter wavelengths, while facilitating deep tissue imaging.

The functionality of HEFT is demonstrated on an *ex vivo* tissue experiment (Figure 2.10(A)) [68]. As seen in Figure 2.10(B), an optically transparent vessel of fluorescent quantum dots (Qtracker705, Invitrogen, USA) was submerged into a slab of chicken breast tissue, which was slightly compressed between two glass plates. The confined tissue slice had dimensions of 4 cm $\times$ 3 cm $\times$ 1.2 cm and the depth of the vessel, measured from the top plane (at $z = 12$ mm), was 5 mm. The tissue slab was imaged with a pre-clinical small animal imaging system (Photon Imager, Biospace Lab, France), which is a light-tight box containing a white light source, an automatic filter system, and a cooled CCD camera. A halogen lamp with a tunable bandpass filter ($\Delta\lambda = 10$ nm spectral width) provided the light for fluorescence stimulation at λ_l. The top surface of the tissue slice was illuminated with light of nine different λ_l ranging from 580 nm to 660 nm. Figure 2.10(C) shows a single fluorescence image in back-reflectance that was taken at $\lambda = 700$ nm for each illumination band. A filter blocked off the excitation light from entering the detection unit. Next, the tissue slab geometry was discretized on a structured grid with $48 \times 60 \times 13 = 37{,}440$ grid points and with grid point separation of 0.1 cm. The top plane of the phantom is indicated by $z = 12$, whereas $z = 0$ depicts the bottom plane. The partial current of fluorescence light was calculated for 529 boundary grid points on the glass plates at $z = 0$ and $z = 12$. Therefore, the matrix $\boldsymbol{F}$ in equation (2.30) consisted of 37,440 columns and 529×9 rows and a maximum of nine wavelengths. Figure 2.10(D) show horizontal (x-y) and transversal (x-z) cross-sections of source reconstructions as a function of excitation wavelength intervals ($L = 1, 3, 5, 9$) between 580 and 660 nm. The reconstructions indicate that an increase of the number of wavelengths improve the localization ability of the reconstruction. A precise and concisely confined source localization in the horizontal plane could only be obtained for $L = 9$. The lower the number of wavelengths involved in the reconstruction, the less is the capability of accurately resolving the source depth. In the case $L = 1$, almost all of the reconstructed source

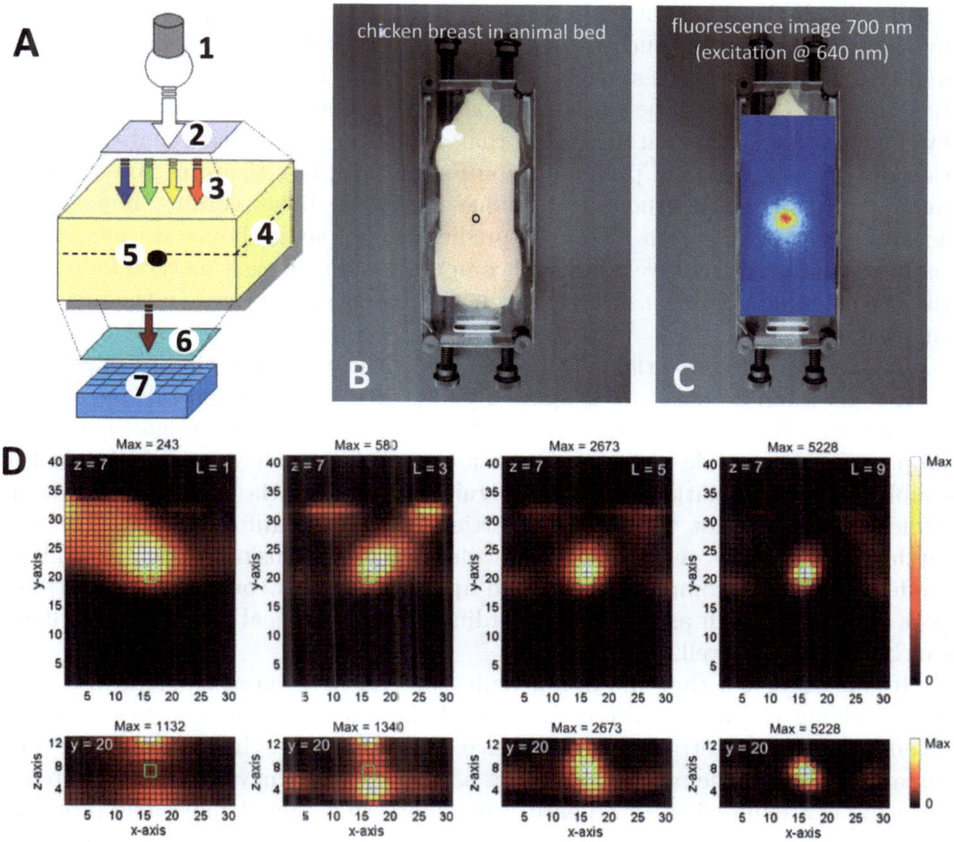

Fig. 2.10. HEFT of quantum dots submerged into a slab of chicken breast tissue [68]. (A) Experimental set-up with white (1) light source, (2) tunable bandpass filter, (3) spectrally filtered excitation light, (4) tissue slab, (5) quantum dot light source, (6) emission band filter, and (7) CCD camera. (B) Imaging bed with chicken breast tissue. (C) Fluorescence light intensity image taken on the imaging bed surface ($z = 12$). (D) Quantum dot source reconstructions (top row: horizontal slice ($z = 7$); bottom row: transversal slice ($y = 20$)) for different sets ($L = 1$, 3, 5, 9) of used excitation bands.

density is concentrated right beneath the tissue boundary close to the detector plane (as shown in the transversal slice). Only a low amount of source density can be observed at the true depth of the quantum dot source. Increasing the number of wavelengths $L > 1$, increases the spatial reconstruction accuracy, and the source localization starts to grow into the volume. Best results are obtained for $L = 9$.

2.5 Summary and Outlook

In summary, multi-spectral luminescence tomography reconstructs the 3D biodistribution of light-emitting sources in tissue. These sources report on biological processes and molecular pathways of disease development, which are studied in small animals. Visible light is, however, strongly scattered and partially absorbed by tissue chromophores such as hemoglobin. This leads to diffuse light intensity distributions on the tissue surface, which prohibit the direct observation of the optical sources. Therefore, the measured light at the tissue surface is further processed by a source reconstruction algorithm that calculates the location and concentration of the sought optical probe distribution and displays them in cross-sectional images. The 3D reconstruction technique for visible light requires, however, an accurate light propagation model for predicting the partial boundary flux on the tissue surface and a relatively fast and computationally efficient method for solving the inverse source problem. Both components have been discussed in the presented chapter.

The SP_3 equations, originally devised for nuclear reactor physics, have been derived for luminescence tomography of biological tissue with anisotropic scattering and refractive index mismatch. The SP_3 equations accurately model the light propagation for a wide range of absorption coefficients. They consist of only two coupled diffusion equations and, thus, significantly reduce the computational effort while preserving most transport properties beyond the diffusion limit. The SP_3 method significantly outperforms the widely applied diffusion model in terms of model accuracy. A computational speed-up of over 100 as compared to full transport techniques, such as the discrete ordinates and spherical harmonics methods, can be achieved as well.

It was observed, that the use of multi-spectral light intensity data also facilitates solving the highly ill-posed and underdetermined inverse source problem. Therefore, multi-spectral bioluminescence tomography was introduced by several research groups in the past years. It uses a light-emitting enzymatic reaction based on a luc gen expression and, hence, can report on targeted molecular processes *in vivo*. During the past two years, excitation-resolved fluorescence tomography of light-emitting quantum dots has been developed. It capitalizes on the broad absorption spectrum of quantum dots and, thus, 3D spatial source distributions can be retrieved from spectral data alone. Last, multi-spectral Cerenkov luminescence tomography based on the SP_3 equations has been introduced. This modality uses radionuclides as source for Cerenkov light emission in tissue. Based on multi-spectral datasets, the radionuclide distribution can be retrieved from diffuse light emission measurements taken at the tissue surface.

While macroscopic imaging modalities, such as X-ray CT and MRI, visualize tissue structures only on a macroscopic level, luminescence tomography employs optical reporter probes on a cellular level to determine the expression of indicative molecular markers of disease development. Therefore, multi-spectral luminescence tomography may enable non-invasive, cost-effective, and high throughput imaging in ambulatory or bedside settings in the near future.

Acknowledgments

This work was supported in part by grants (UL1RR024156 and 1R44RR030701-01) of the National Center for Research Resources (NCRR) and by a grant (1R21EB011772-01A1) of the National Institute of Biomedical Imaging and Bioengineering (NIBIB).

References

1. S. Ogawa, T.M. Lee, A.S. Nayak, P. Glynn: Oxygenation-sensitive contrast in magnetic resonance image of rodent brain at high magnetic fields. *Magnetic Resonance in Medicine*, **14(1)**, 68–78 (1990).
2. J. Frahm, H. Bruhn, M.L. Gyngell, K.D. Merboldt, W. Hnicke, R. Sauter: Localized high-resolution proton NMR spectroscopy using stimulated echoes: initial applications to human brain in vivo. *Magn. Reson. Med.*, **9(1)**, 79–93 (1989).
3. M.W. Weiner, H.P. Hetherington: The power of the proton. *Radiology*, **172(2)**, 318–320 (1989).
4. M.S. van der Knaap, J. van der Grond, P.C. van Rijen, J.A. Faber, J. Valk, K. Willemse: Age-dependent changes in localized proton and phosphorus MR spectroscopy of the brain. *Radiology*, **176(2)**, 509–15 (1990).
5. R. Weissleder, U. Mahmood: Molecular imaging. *Radiology*, **219**, 316–333 (2001).
6. T.F. Massoud, S.S. Gambhir: Molecular imaging in living subjects: seeing fundamental biological processes in a new light. *Genes Dev.*, **17**, 545–580 (2003).
7. R. Weissleder, V. Ntziachristos: Shedding light onto live molecular targets. *Nat. Med.*, **9**, 123–128 (2003).
8. S.R. Cherry: In vivo molecular and genomic imaging: new challenges for imaging physics. *Phys. Med. Biol.*, **49**, R13–R48 (2004).
9. M. Rudin: *Molecular Imaging* (Imperial College Press, London, 2005).
10. G.S. Sandhu, L. Solorio, A.-M. Broome, N. Salem, J. Kolthammer, T. Shah, C. Flask, J.L. Duerk: Whole animal imaging. *Wiley Interdiscipl. Rev. Syst. Biol. Med.*, **2**, 398–421 (2010).
11. G. Choy, P. Choyke, S.K. Libutti: Current advances in molecular imaging: noninvasive in vivo bioluminescent and fluorescent optical imaging of cancer. *Mol. Imag.*, **2**, 303–312 (2003).
12. V. Ntziachristos, J. Ripoll, L.V. Wang, R. Weissleder: Looking and listening to light: the evolution of whole-body photonic imaging. *Nature Biotech.*, **23**, 313–320 (2005).
13. W. Du, Y. Wang, Q. Luo, B.-F. Liu: Optical molecular imaging for systems biology: from molecule to organism. *Anal. Bioanal. Chem.*, **386(3)**, 444–457 (2006).
14. T. Krucker, B.S. Sandanaraj: Optical imaging for the new grammar of drug discovery. *Phil. Trans. R. Soc. A*, **369**, 4651–4665 (2011).
15. R.Y. Tsien: The green fluorescent protein. *Annu. Rev. Biochem.*, **67**, 509–544 (1998).
16. N.C. Shaner, R.E. Campbell, P.A Steinbach, B.N.G. Giepmans, A.E. Palmer, R.Y. Tsien: Improved monomeric red, orange and yeallow fluorescent proteins derived from Discosoma sp. red fluorescent protein. *Nature Biotech.*, **22**, 1567–1572 (2004).
17. R.M. Hoffman: The multiple uses of fluorescent proteins to visualize cancer in vivo. *Nat. Rev. Cancer*, **5**, 796–806 (2005).
18. G.-J. Kremers, J. Goedhart, D.J. van den Heuvel, H.C. Gerritsen, T.W.J. Gadella: Improved green and blue fluorescent proteins for expression in bacteria and mammalian cells. *Biochemistry*, **46(12)**, 3775–3783 (2007).
19. C.J. Daly, J.C. McGrath: Fluorescent ligands, antibodies, and proteins for the study of receptors. *Pharmacol. Therap.*, **100**, 101–118 (2003).
20. R. Weissleder, C.H. Tung, U. Mahmood, A. Bogdanov: In vivo imaging of tumors with protease-activated near-infrared fluorescent probes. *Nature Biotech.*, **17**, 375–378 (1999).
21. C.H. Tung: Fluorescent peptide probes for in vivo diagnostic imaging. *Biopolymers*, **76**, 391–403 (2004).
22. X. Michalet, F.F Pinaud, L.A. Bentolila, J.M. Tsay, S. Doose, J.J. Li, G. Sundaresan, A.M. Wu, S.S. Gambhir, S. Weiss: Quantum dots for live cells, in vivo imaging, and diagnostics. *Science*, **307**, 538–544 (2005).

23. J.V. Frangioni: In vivo near-infrared fluorescence imaging. *Curr. Opin. Chem. Biol.*, **7**, 626–634 (2003).
24. J.R. Lakowicz: *Principles of Fluorescence Spectroscopy* (Springer, New York, 2004).
25. V. Ntziachristos: Fluorescence molecular imaging. *Annu. Rev. Biomed. Eng.*, **8**, 1–33 (2006).
26. J. Rao, A. Dragulescu-Andrasi, H. Yao: Fluorescence imaging in vivo: recent advances. *Curr. Opin. Biotech.*, **18**, 17–25 (2007).
27. C.H. Contag, M.H. Bachmann: Advances in in vivo bioluminescence imaging of gene expression. *Ann. Rev. Biomed. Engi.*, **4**, 235–260 (2002).
28. B.W. Rice, M.D. Cable, M.B. Nelson: In vivo imaging of light-emitting probes. *J. Biomed. Opt.*, **6(4)**, 432–440 (2001).
29. D.K. Welsh, S.A. Kay: Bioluminescence imaging in living organisms. *Curr. Opin. Biotechnol.*, **16**, 73–78 (2005).
30. T. Troy, D. Jekic-McMullen, L. Sambucetti, B. Rice: Quantitative comparison of the sensitivity of detection of fluorescent and bioluminescent reporters in animal models. *Molecular Imaging*, **3(1)**, 9–23 (2004).
31. H. Zhao, T.C. Doyle, O. Coquoz, F. Kalish, B. Rice, C.H. Contag: Emission spectra of bioluminescent reporters and interaction with mammalian tissue determine the sensitivity of detection in vivo. *J. Biomed. Opt.*, **10(4)**, 041210-1–9 (2005).
32. W.M. Burch: Cerenkov light from 32-P as an aid to diagnosis of eye tumours. *Nature*, **234**, 358 (1971).
33. R. Robertson, M. S. Germanos, C. Li, G. S. Mitchell, S. R. Cherry, M. D. Silva: Optical imaging of Cerenkov light generation from positron-emitting radiotracers. *Phys. Med. Biol.*, **54(16)**, N355–N365 (2009).
34. J.S. Cho, R. Taschereau, S. Olma, K. Liu, Y.-Ch. Chen, C.K.-F. Shen, R. M. van Dam, A.F. Chatziioannou: Cerenkov radiation imaging as a method for quantitative measurements of beta particles in a microfluidic chip. *Phys. Med. Biol.*, **54**, 6757–6771 (2009).
35. P.A. Cerenkov: Visible radiation produced by electrons moving in a medium with velocities exceeding that of light. *Phys. Rev.*, **52**, 378–379 (1937).
36. W.-F. Cheong, S.A. Prahl, A.J. Welch: A review of the optical properties of biological tissues. *IEEE Journal of Quantum Electronics*, **26(12)**, 2166–2185 (1990).
37. B. Beauvoit, S.M. Evans, T.W. Jenkins, E.E. Miller, B. Chance: Correlation between the light scattering and the mitochondrial content of normal tissues and transplantable rodent tumors. *Analyt. Biochem.*, **226**, 167–174 (1995).
38. J.R. Mourant, J.P. Freyer, A.H. Hielscher, A.A. Eick, D. Shen, T.M. Johnson: Mechanisms of light scattering from biological cells relevant to noninvasive optical-tissue diagnosis. *Appl. Opt.*, **37(16)**, 3586–3593 (1998).
39. J.R. Mourant, M. Canpolat, C. Brocker, O. Esponda-Ramos, T.M. Johnson, A. Matanock, K. Stetter, J.P. Freyer: Light scattering from cells: The contribution of the nucleus and the effects of proliferative status. *J. Biomed. Opt.*, **5**, 131–137 (2000).
40. A.D. Klose, B.J. Beattie, H. Dehghani, L. Vider, C. Le, V. Ponomarev, R. Blasberg: In vivo bioluminescence tomography with a blocking-off finite-difference SP3 method and MRI/CT coregistration. *Med. Phys.*, **37** 329–338 (2010).
41. K.M. Yoo, F. Liu, R.R. Alfano: When does the diffusion approximation fail to describe photon transport in random media? *Phys. Rev. Lett.*, **64(22)**, 2647–2650 (1990).
42. A.D. Kim, A. Ishimaru: Optical diffusion of continuous-wave, pulsed, and density waves in scattering media and comparisons with radiative transfer. *Appl. Opt.*, **37**, 5313–5319 (1998).
43. B. Chen, K. Stamnes, J.J. Stamnes: Validity of the diffusion approximation in bio-optical imaging. *Applied Optics*, **40**, 6356–6366 (2001).

44. S. Chandrasekhar: *Radiative Transfer* (Oxford University Press, London, 1960).
45. J.J. Duderstadt, W.R. Martin: *Transport Theory* (John Wiley, New York, 1979).
46. K.M. Case, P.F. Zweifer: *Linear transport Theory* (Addison-Wesley, Reading, 1967).
47. A.D. Klose: Radiative transfer of luminescence light in biological tissue. In: *Light Scattering Reviews 4, Single Light Scattering and Radiative Transfer*. Ed.: A.A. Kokhanovsky (Springer-Praxis, Chichester, 2009), 293–345.
48. F. Natterer: Mathematical models for medical imaging. In: *Computational Radiology and Imaging: Therapy and Diagnostics. The IMA Volumes in Mathematics and its Applications* Ed.: Ch. Boergers, F. Natterer (Springer, New York, 1999), 17–32.
49. H.C. Yi, R. Sanchez, N J. McCormick: Bioluminescence estimation from ocean in situ irradiances. *Appl. Opt.* **31(6)**, 822–830 (1992).
50. Z. Tao, N.J. McCormick, R. Sanchez: Ocean source and optical property estimation from explicit and implicit algorithms. *Appl. Opt.*, **33(15)**, 3265–3275 (1994).
51. L.K. Sundman, R. Sanchez, N.J. McCormick: Ocean optical source estimation with widely spaced irradiance measurements. *Appl. Opt.*, **37(18)**, 3793–3803 (1998).
52. S. Stephany, H.F. de Campos Velho, F.M. Ramos, C.D. Mobley: Identification of inherent optical properties and bioluminescence source term in a hydrologic optics problem. *J. Quant. Spetrosc. Radiat. Transfer*, **67**, 113–123 (2000).
53. H.Y. Li, M.N. Ozisik: Estimation of the radiation source term with conjugate-gradient method of inverse analysis. *J. Quant. Spectrosc. Radiat. Transfer*, **48(3)**, 237–244 (1992).
54. A.D. Klose, E.W. Larsen: Light transport in biological tissue based on the simplified spherical harmonics equations. *J. Comp. Phys.*, **220**, 441–470 (2006).
55. M. Lemanska: On the simplified P_n method in the 2-D diffusion code EXTERMINATOR. *Atomkernenergie*, **37**, 173–175 (1981).
56. D.I. Tomasevic, E.W. Larsen: The simplified P_2 approximation. *Nucl. Sci. Eng.*, **122**, 309–325 (1996).
57. E.W. Larsen, J.E. Morel, J.M. McGhee: Asymptotic derivation of the multigroup P1 and simplified P_N equations with anisotropic scattering. *Nucl. Sci. Eng.*, **123**, 328–342 (1996).
58. J.E. Morel, J.M. McGhee, E.W. Larsen: A three-dimensional time-dependent unstructured tetrahedral-mesh SP_N method. *Nucl. Sci. Eng.*, **123**, 319–327 (1996).
59. E.E. Lewis, G. Palmiotti: Simplified spherical harmonics in the variational nodal method. *Nucl. Sci. Eng.*, **126**, 48–58 (1997).
60. P.S. Brantley, E.W. Larsen: The simplified P3 approximation. *Nucl. Sci. Eng.*, **134**, 1–21 (2000).
61. P. Kotiluoto: Fast tree multigrid transport application for the simplified P_3 approximation. *Nucl. Sci. Eng.* **138**, 269–278 (2001).
62. P. Kotiluoto, P. Hiismáki: Application of the new MultiTrans SP_3 radiation transport code in BNCT dose planning. *Med. Phys.*, **28(9)**, 1905–1910 (2001).
63. R. Ciolini, G.G.M. Coppa B. Montagnini, P. Ravetto: Simplified P_N and A_N methods in neutron transport. *Prog. Nucl. Energy*, **40(2)**, 237–264 (2002).
64. E.W. Larsen: Asymptotic diffusion and simplified PN approximations for diffusive and deep penetration problems. Part 1: *Theory. Transport Theory and Statistical Physics*, **39(2)**, 110–163 (2010).
65. A. Liemert, A. Kienle: Analytical solutions of the simplified spherical harmonics equations. *Opt. Lett.*, **35(20)**, 3507-3509 (2010).
66. A. Liemert, A. Kienle: Analytical Green's function of the radiative transfer radiance for the infinite medium. *Physical Review E*, **83**, 036605 (2011).

67. A.D. Klose: Hyperspectral excitation-resolved fluorescence tomography of quantum dots. *Opt. Lett.*, **34** 2477–9 (2009).
68. A.D. Klose, T. Pöschinger: Excitation-resolved fluorescence tomography with simplified spherical harmonics equations. *Phys. Med. Biol.*, **56**, 1443–1469 (2011).
69. Y. Lu, A.F. Chatziioannou: A parallel adaptive finite element method for the simulation of photon migration with the radiative-transfer-based model. *Commun. Numer. Methods Eng.*, **25(6)**, 751–770 (2009).
70. M. Chu, K. Vishwanath, A.D. Klose, H. Dehghani: Light transport in biological tissue using three-dimensional frequency-domain simplified spherical harmonics equations. *Phys. Med. Biol.*, **54** 2493–2509 (2009).
71. M. Chu, H. Dehghani: Image reconstruction in diffuse optical tomography based on simplified spherical harmonics approximation. *Optics Express*, **17(26)**, 24208-24223 (2009).
72. J.B. Dominguez, Yves Bérubé-Lauzire: Diffuse light propagation in biological media by a time-domain parabolic simplified spherical harmonics approximation with ray-divergence effects. *Appl. Opt.*, **49(8)**, 1414–1429 (2010).
73. Y. Lu, B. Zhu, H. Shen, J.C. Rasmussen, G. Wang, E.M. Sevick-Muraca: A parallel adaptive finite element simplified spherical harmonics approximation solver for frequency domain fluorescence molecular imaging. *Phys. Med. Biol.*, **55**, 4625–4645 (2010).
74. D. Han, J. Tian, K. Liu, J. Feng, B. Zhang, X. Ma, C. Qin: Sparsity-promoting tomographic fluorescence imaging with simplified spherical harmonics approximation. *IEEE Trans. Biomed. Eng.*, **57(10)**, 2564–2567 (2010).
75. K. Liu, Y. Lu, C. Qin, X. Yang, S. Zhu, X. Yang, Q. Gao, D. Han: Evaluation of the simplified spherical harmonics approximation in bioluminescence tomography through heterogeneous mouse models. *Optics Express*, **18(20)**, 20988–21002 (2010).
76. J.B. Dominguez, Yves Bérubé-Lauzire: Light propagation from fluorescent probes in biological tissues by coupled time-dependent parabolic simplified spherical harmonics. *Biomed. Optics Express*, **2(4)**, 817–837 (2011).
77. H. Zheng, W. Han: On simplified spherical harmonics equations for the radiative transfer equation. *Journal of Mathematical Chemistry*, **49**, 1–13 (2011).
78. M.N. Wernick, J.N. Aarsvold: *Emission Tomography. The Fundamentals of PET and SPECT* (Elsevier Academic Press, San Diego, 2004).
79. G.L. Zeng: *Medical Image Reconstruction* (Springer, Heidelberg, 2010).
80. R. Gordon, R. Bender, G. T. Herman: Algebraic reconstruction techniques (ART) for three-dimensional electron microscopy and x-ray photography. *J. Theor. Biol.*, **29**, 471–481 (1970).
81. G.T. Herman: *Fundamentals of computerized tomography: Image reconstruction from projection*, 2nd edition (Springer, New York, 2009).
82. L.A. Shepp, Y. Vardi: Maximum likelihood reconstruction for emission tomography. *IEEE Transactions on Medical Imaging*, **1**, 113 (1982).
83. K. Lange, R. Carson: EM reconstruction algorithms for emission and transmission tomography. *J. Comput. Assist. Tomogr.*. **8(2)**, 306–316 (1984).
84. J. Nocedal, S.J. Wright: *Numerical Optimization* (Springer, New York,1999).
85. S. Kaczmarz: Angenäherte Auflösung von Systemen linearer Gleichungen. *Bulletin International de l'Académie Polonaise des Sciences et des Lettres. Classe des Sciences Mathématiques et Naturelles. Série A, Sciences Mathématiques*, **35** 355–357 (1937).
86. W.H. Richardson: Bayesian-based iterative method of image restoration. *JOSA*, **62(1)**, 55–59 (1972).
87. L.B. Lucy: An iterative technique for the rectification of observed distributions. *Astronomical Journal*, **79(6)**, 745–754 (1974).

88. D.G. Luenberger: *Linear and Nonlinear Programming* (Addison-Wesley, Reading, MA, 1984).
89. S.G. Nash: *Linear and Nonlinear Programming* (McGraw-Hill, New York, 1996).
90. A.D. Klose, A.H. Hielscher: Quasi-Newton methods in optical tomographic imaging. *Inverse Problems*, **19**, 387–409 (2003).
91. A.D. Klose, V. Ntziachristos, A.H. Hielscher: The inverse source problem based on the radiative transfer equation in optical molecular imaging. *Journal of Computational Physics*, **202**, 323–345 (2005).
92. G. Wang, Y. Li, M. Jiang: Uniqueness theorems in bioluminescence tomography. *Med. Phys.*, **31**, 2289–2299 (2004).
93. A.D. Klose: Transport-theory based stochastic image reconstruction of bioluminescent sources. *Journal of the Optical Society of America A*, **24(6)**, 1601–1608 (2007).
94. S. Kirkpatrick, C. D. Gelatt, M. P. Vecchi, Optimization by simulated annealing, *Science*, **220**, 671–680 (1983).
95. Z. Michalewicz: *Genetic Algorithms + Data Structures = Evolution Programs* (Springer, New York, 1998).
96. H.-G. Beyer, H.-P. Schwefel: Evolution strategies. *Nat. Comput.*, **1**, 3–52 (2002).
97. H.-P. Schwefel: Evolution and Optimum Seeking (Wiley, New York, 1995).
98. G. Alexandrakis, F.R. Rannou, A.F. Chatziioannou: Tomographic bioluminescence imaging by use of a combined optical-PET (OPET) system: a computer simulation feasibility study. *Phys. Med. Biol.*, **50**, 4225–4241 (2005).
99. A.J. Chaudhari, F. Darvas, J.R. Bading, R.A. Moats, P.S. Conti, D.J. Smith, S.R. Cherry, R.M. Leahy: Hyperspectral and multispectral bioluminescence optical tomography for small animal imaging. *Phys. Med. Biol.*, **50**, 5421–5441 (2005).
100. H. Dehghani, S.C. Davis, S. Jiang, B.W. Pogue, K.D. Paulsen: Spectrally resolved bioluminescence optical tomography. *Opt. Lett.*, **31(3)**, 365–367 (2006).
101. A.X. Cong, G. Wang: Multi-spectral bioluminescence tomography: Methodology and simulation. *Int. J. Biomed. Imaging*, 1–7 (2006).
102. C. Kuo, O. Coquoz, T.L. Troy, H. Xu, B.W. Rice: Three-dimensional reconstruction of in vivo bioluminescent sources based on multispectral imaging. *J. Biomed. Opt.*, **12(2)**, 024007-1–12 (2007).
103. J. Virostko, A.C. Powers, E.D. Jansen: Validation of luminescent source reconstruction using single-view spectrally resolved bioluminescence images. *Appl. Opt.*, **46(13)**, 2540–2547 (2007).
104. Y. Lv, J. Tian, W. Cong, G. Wang, W. Yang, C. Qin, M. Xu: Spectrally resolved bioluminescence tomography with adaptive finite element analysis: methodology and simulation. *Phys. Med. Biol.*, **52**, 4497–4512 (2007).
105. W. Han, G. Wang: Theoretical and numerical analysis on multispectral bioluminescence tomography. *IMA Journal of Applied Mathematics*, **72(1)**, 67–85 (2007).
106. J. Feng, K. Jia, G. Yan, S. Zhu, C. Qin, Y. Lv, and J. Tian: An optimal permissible source region strategy for multispectral bioluminescence tomography. *Optics Express*, **16**, 15640–15654 (2008).
107. S. Ahn, A.J. Chaudhari, F. Darvas, C.A. Bouman, R.M. Leahy: Fast iterative image reconstruction methods for fully 3D multispectral bioluminescence tomography. *Phys. Med. Biol.*, **53** 3921 (2008).
108. Y. Lu, H.B. Machado, A. Douraghy, D. Stout, H. Herschman, A.F. Chatziioannou: Experimental bioluminescence tomography with fully parallel radiative-transfer-based reconstruction framework. *Optics Express*, **17 (19)**, 16681–16695 (2009).
109. C. Qin, J. Tian, X. Yang, J. Feng, K. Liu, J. Liu, G. Yan, S. Zhu, M. Xu: Adaptive improved element free Galerkin method for quasi or multi spectral bioluminescence tomography. *Optics Express*, **17**, 21925–21934 (2009).

110. Y. Lu, H.B. Machado, Q. Bao, D. Stout, H. Herschman, A.F. Chatziioannou: In vivo mouse bioluminescence tomography with radionuclide-based imaging validation. *Molecular Imaging and Biology*, **13(1)**, 53–58 (2010).
111. K. Liu, X. Yang, D. Liu, C. Qin, J. Liu, Z. Chang, M. Xu, J. Tian: Spectrally resolved three-dimensional bioluminescence tomography with a level-set strategy. *JOSA A*, **27(6)**, 1413–1423 (2010).
112. Y. Lu, A. Douraghy, H. Machado, D. Stout, H. Herschman, A. Chatziioannou: Spectrally resolved bioluminescence tomography with the third-order simplified spherical harmonics approximation. *Phys. Med. Biol.*, **54** 6477–6493 (2009).
113. Y. Lu, H.B. Machado, Q. Bao et al.: In vivo mouse bioluminescence tomography with radionuclide-based imaging validation. *Mol Imaging Biol*, **13**, 53–58 (2011).
114. A. Ruggiero, J. P. Holland, J.S. Lewis, J. Grimm. Cerenkov luminescence imaging of medical isotopes. *J. Nucl. Med.*, **51**, 1123–1130 (2010).
115. J.P. Holland, G. Normand, A. Ruggiero, J. S. Lewis, J. Grimm: Intraoperative imaging of positron emission tomographic radiotracers using Cerenkov luminescence emissions. *Molecular Imaging*, **10(3)**, 177–186 (2010).
116. H. Liu, G. Ren, Z. Miao, X. Zhang, X. Tang, P. Han, S.S. Gambhir, Z. Cheng: Molecular optical imaging with radioactive probes. *PLoS ONE*, **5(3)**, e9470. doi:10.1371/journal.pone.0009470 (2010).
117. A.E. Spinelli, D. D'Ambrosio, L. Calderan, M. Marengo, A. Sbarbati, F. Boschi: Cerenkov radiation allows in vivo optical imaging of positron emitting radiotracers. *Phys. Med. Biol.*, **55**, 483–495 (2010).
118. R. S. Dothager, R. J. Goiffon, E. Jackson, S. Harpstrite, D. Piwnica-Worms. Cerenkov radiation energy transfer (CRET) imaging: A novel method for optical imaging of PET isotopes. In Biological Systems, *PLoS ONE*, **5(10)**, e13300, (2010).
119. M.A. Lewis, V.D. Kodibagkar, O.K., R.P. Mason: On the potential for molecular imaging with Cerenkov luminescence. *Opt. Lett.*, **35(23)**, 3889–3891 (2010).
120. H. Liu, G. Ren, S. Liu et al: Optical imaging of reporter gene expression using a positron-emission-tomography probe. *J Biomed Opt*, 15:060505 (2010).
121. H. Liu, X. Zhang, B. Xing, P. Han, S.S. Gambhir, Z. Cheng: Radiation-luminescence-excited quantum dots for in vivo multiplexed optical imaging. *Small*, 6:1087–1091 (2010).
122. J. Ch. Park, G. Il An, Se-Il Park, J. Oh, H. J. Kim, Y. Su Ha, E. K. Wang, K. M. Kim, J. Y. Kim, J. Lee, M.J. Welch, J. Yoo: Luminescence imaging using radionuclides: a potential application in molecular imaging. *Nuclear Medicine and Biology*, **38(3)**, 321–329 (2011).
123. S.Y. Jeong, M.H. Hwang, J.E. Kim, S. Kang, J.C. Park, J. Yoo, J.H. Ha, S.W. Lee, B.C. Ahn, J. Lee: Combined Cerenkov luminescence and nuclear imaging of radioiodine in the thyroid gland and thyroid cancer cells expressing sodium iodide symporter: Initial feasibility study. *Endocr J.*, **58**(7), 575–583 (2011).
124. G. Lucignani: Cerenkov radioactive optical imaging: a promising new strategy. *Eur. J. Nucl. Med. Mol. Imaging*, **38(3)**, 592–595 (2011).
125. C. Ran, Z. Zhang, J. Hooker, A. Moore: In vivo photoactivation without light: use of Cherenkov radiation to overcome the penetration limit of light. *Mol. Imaging Biol.*, **14**(2), 156–162 (2011).
126. F. Boschi, L. Calderan, D. D'Ambrosio et al.: In vivo 18F-FDG tumour uptake measurements in small animals using Cerenkov radiation. *Eur. J. Nucl. Med. Mol. Imaging*, **38**, 120–127 (2011).
127. C. Li, G. S. Mitchell, S.R. Cherry: Cerenkov luminescence tomography for small-animal imaging. *Opt. Lett.*, **35 (7)**, 1109–1111 (2010).

128. Z. Hu, J. Liang, W. Yang, W. Fan, C. Li, X. Ma, X. Chen, X. Ma, X. Li, X. Qu, J. Wang, F. Cao, J. Tian: Experimental Cerenkov luminescence tomography of the mouse model with SPECT imaging validation. *Optics Express*, **18 (24)**, 24441 (2010).
129. J. Zhong, J. Tian, X. Yang, C. Qin: Whole-body Cerenkov luminescence tomography with the finite element SP3 method. *Ann. Biomed. Eng.*, **39 (6)**, 1728–1735 (2011).
130. J. Zhong, C. Qin, X. Yang, S. Zhu, X. Zhang, J. Tian: Cerenkov luminescence tomography for in vivo radiopharmaceutical imaging. *Int. J. Biomed. Imaging*, **2011**, Article ID 641618 (2011).
131. J. Zhong, C. Qin, X. Yang, Z. Chen, X. Yang, J. Tian: Fast-specific tomography imaging via cerenkov emission. *Mol. Imaging Biol.*, (2011).
132. A.E. Spinelli, C. Kuo, B.W. Rice, R. Calandrino, P. Marzola, A. Sbarbati, F. Boschi: Multispectral Cerenkov luminescence tomography for small animal optical imaging. *Optics Express*, **19(13)**, 12605–12618 (2011).
133. J. Chang, H.L. Graber, R.L. Barbour: Luminescence optical tomography of dense scattering media. *J. Opt. Soc. Am. A*, **14(1)**, 288–299 (1997).
134. D.Y. Paithankar, A.U. Chen, B.W. Pogue, M.S. Patterson, E.M. Sevick-Muraca: Imaging of fluorescent yield and lifetime from multiply scattered light reemitted from random media. *Appl. Opt.*, **36**, 2260–2272 (1997).
135. V. Ntziachristos, R. Weissleder: Experimental three-dimensional fluorescence reconstruction of diffuse media by use of a normalized Born approximation. *Opt. Lett.*, **26**, 893–895 (2001).
136. J. Lee, E.M. Sevick-Muraca: Three-dimensional fluorescence enhanced optical tomography using referenced frequency-domain photon migration measurements at emission and excitation wavelengths. *J. Opt. Soc. Am. A*, **19**, 759–771 (2002).
137. A.B. Milstein, S. Oh, K.J. Webb, C.A. Bouman, Q. Zhang, D.A. Boas, R.P. Millane: Fluorescence optical diffusion tomography. *Appl. Optics*, **42**, 3081–3094 (2003).
138. A.J. Chaudhari, S. Ahn, R. Levenson, R.D. Badawi, S.R. Cherry, R.M. Leahy: Excitation spectroscopy in multispectral optical fluorescence tomography: methodology, feasibility and computer simulation studies. *Phys. Med. Biol.*, **54**, 4687–704 (2009).

3 Markovian approach and its applications in a cloudy atmosphere

Evgueni Kassianov, Dana E. Lane-Veron, Larry K. Berg, Mikhail Ovchinnikov, and Pavlos Kollias

3.1 Introduction

New ideas in mathematics, even when they are seemingly insignificant or impractical, can have unforeseen and important consequences that may not become apparent at the time they were suggested and occasionally not even for several decades. One such idea was proposed by the gifted Russian mathematician, A. A. Markov (1856–1922) at the beginning of the 20th century (e.g., Markov, 1907). Radically innovative for that time, it advanced substantially 19th-century works on probability theory by generalizing various limit laws established for independent random variables (e.g., the law of large numbers) to dependent ones. It stated that most probable prediction of a future event depends on what happens 'today,' and any information from 'past' events is irrelevant in such predictions. Markov's generalization was inspired by the internal demands of probability theory (Basharin et al., 2004). What knowledge of nature can possibly be obtained in the framework of this generalized approach? Although Markov never discussed potential applications of his approach to the physical science, the suggested concept of dependent random variables or Markov chains (processes), as it is called in modern probability theory, has a strong connection to physical phenomena. 'Conceptually, a Markov process is the probabilistic analogue of the processes of classical mechanics, where the future development is completely determined by the present state and is independent of the way in which the present state has developed' (Feller, 1950, p. 369). Thus, it favors the well-known thesis of the renowned 19th-century mathematician and physicist Laplace that 'We ought then to regard the present state of the universe as the effect of its anterior state and as the cause of the one which is to follow' (Laplace, 1902, p. 4). The Markovian approach does not reveal what causes transitions between events, but it permits probabilistic predictions to be made for the outcome of a great variety of events.

Eventually, the proposed concept of statistically dependent variables fell into the hands of 20th-century physicists who struggled to deepen the understanding and description of a wide diversity of natural phenomena and to solve important long-standing physical problems. It seems appropriate to start this chapter with a few historical examples. Perhaps the best-known early application of Markov chains (Appendix A) goes back to the work by Metropolis et al. (Metropolis et al.,

1953), which was aimed at effectively computing multi-dimensional integrals with Boltzmann weights by using a random walk (Markov chain). We should note that Metropolis 'researched nuclear reactors with Enrico Fermi and Edward Teller, and, through his work with such noteworthy scientists, he came to the attention of J. Robert Oppenheimer, head of the Manhattan Project-the United States government's plan to build the first atomic bomb.' (Hitchcock, 2003, p. 254). A more general version of the Metropolis algorithm was introduced later by Hastings (1970). The extreme versatility of the Metropolis–Hastings algorithm (Chib and Greenberg, 1995) makes possible simulations of complex one- and multi-dimensional distributions; its impressive numerous applications can be found in physics, computational science, and image analysis (Kendall et al., 2005). The Markovian nature of nearest-neighbor interactions was tackled by one of the greatest mathematical physicists of the 20th century, Chandrasekhar. In his paper 'Stochastic Problems in Physics and Astronomy,' Chandrasekhar applied the Markovian approach to derive the statistics of the gravitational field arising from a random distribution of stars (Chandrasekhar, 1943). Chandrasekhar's work proposed that the force acting on a star depends on its neighboring stars and grasped the great potential of the Markovian approach: 'In the general form the problem can be solved by using a method originally devised by A. A. Markoff. Now, Markoff's method is of such extreme generality that is actually enables us to solve [...] fundamental problems.' (Chandrasekhar, 1943, p. 8). Chandrasekhar's paper sheds light on essential problems in astronomy, and it was the precursor to important present-day accomplishments, such as the application of the Markovian approach to the development of statistical methods for gravitational lensing (Deguchi and Watson, 1988).

A few words about Markov's work give some indication of how both mathematics and art served to brighten the path to significant progress in probability theory at that time (e.g., Markov, 1913). A passage of text can be treated as a sequence of random variables with two states (vowel and consonant). Assuming that a sequence of letters is a chain, the corresponding probabilities can be obtained. To illustrate this, Markov considered two large samples of text. The first sample includes the sequences of 20,000 letters from Pushkin's novel in verse *Eugene Onegin* (a classic of Russian literature). The second sample includes the sequences of 100,000 letters from Aksakov's novel *Childhood of Bagrov the Grandson* (a major text in the Russian pastoral tradition). Markov performed calculations (by hand!) and illustrated that while the probabilities of vowels occurring were comparable (0.432 and 0.449), the probabilities of a vowel following a vowel diverged considerably (0.128 and 0.552). These striking results demonstrated that a passage of text can be viewed as a collection of dependent entities and that the language/author can be specified quite accurately when the relationship between these entities is identified. The Markovian approach provides a powerful tool for describing the communication phenomena; this tool has been applied successfully in different language-related studies, including cryptography (Diaconis, 2009). This approach was suggested in the early 1900s and had a chance to percolate for several decades before stringent demands were made on it.

We started with a few examples where the Markovian approach illuminates the description of statistically dependent events in diverse research areas such as statistical physics and astronomy. Our tour of such applications will continue and include stochastic radiative transfer in a cloudy atmosphere. No attempt will be made to provide a comprehensive survey of all available methods and approaches, and perhaps several relevant and influential works on these topics are not included here. Instead, we will take a short detour and discuss a small number of landmark ideas and their role in solving important problems. There will be a couple of intermediate stops on the way, where we acknowledge major efforts of passionate scientists after whom several popular mathematical models are named. A substantial part of this chapter concentrates on a theoretical basis of these models, illustrates their 'aesthetic' mathematical value and physical significance, and discusses how they can be applied to current challenges in atmospheric and climate sciences.

3.2 Stochastic radiative transfer

Radiative transfer (RT) theory has been extensively developed over a century with numerous important applications (e.g., Liou, 2002; Davis and Marshak, 2010; Mishchenko et al., 2011, and references therein). Somewhat ironically, its development was initiated by phenomenological concepts that made both the kinetic theory and atmospheric science communities realize the large gap that separated the RT equation (RTE) from the macroscopic Maxwell equations. Establishing a fundamental physical link between RT theory and electromagnetic theory has received enormous attention in recent years, and an innovative microphysical approach has been suggested to bridge this long-standing gap and provide explicit physical meaning of RTE parameters, such as the extinction and phase matrices (e.g., Mishchenko, 2011, and references therein).

For numerous practical applications, these RTE parameters are assumed to be deterministic functions of space and time. Occasionally, the lack of a complete description of these parameters leads to a statistical view: although we do not know their exact values at each point in space and time, we do know their feasible values and the corresponding probabilities. Broken clouds are a perfect example of this situation (Fig. 3.1). Anyone can enjoy watching clouds ... but perhaps no one can describe them in a deterministic way accurately and precisely. The location, geometry, and optical properties of each individual cloud are not known in a deterministic fashion. Thus, they can be considered as random variables. The radiation transmitted and reflected by such a cloud field also becomes random, and the cloud–radiation interaction needs a statistical description. In general, the cloud statistics describing three-dimensional (3D) cloud structure are sufficient for accurate estimations of averaged radiative properties (e.g., Barker et al., 1999; Han and Ellingson, 2000; Petty, 2002; Zuev and Titov, 1995). The statistical information of clouds can be obtained from observations (e.g., Stokes and Schwartz, 1994; Ackerman and Stokes, 2003; Xie et al., 2010) or outputs from numerical cloud models (e.g., Oreopoulos and Khairoutdinov, 2003; Alexandrov et al., 2010a,b). The model outputs mimic real clouds, thus, the simulated cloud fields are a good source for obtaining cloud statistics. In addition to the model simulations, cloud observations that integrate capabilities of the passive and active remote sensing can also offer statistics of 3D clouds (sections 3.4 and 3.5).

For broken clouds, these statistics define bulk geometry parameters, such as the directional cloud fraction $N(\theta) = 1 - P(\theta)$ (e.g., Zhao and Di Girolamo, 2004;

Fig. 3.1. Cumulus clouds with irregular cauliflower shape redistribute incoming sunlight non-uniformly. They block the surface from direct-beam sunlight (dark cloud shadows) and bounce some sunlight down to the surface. Courtesy of the ARM Program.

Taylor and Ellingson, 2008), where $P(\theta)$ is the probability that a line of sight at a given zenith angle θ can pass through a cloud field without intersecting a cloud, the so-called probability of clear line of sight. This probability is a function of the total nadir-view cloud fraction (CF) and cloud aspect ratio. The nadir-view CF and cloud aspect ratio are typically defined as the horizontal area fraction covered by clouds as viewed from nadir (e.g., Del Genio et al., 1996) and ratio of the mean vertical and horizontal cloud sizes (e.g., Titov, 1990), respectively. A simple example of $N(\theta)$ is the fractional area of cloud shadows determined by general cloud contours if θ is specified as a direction of sunlight direct beam. To describe the shortwave RT effects of $N(\theta)$ changes associated with diurnal variations of solar zenith angle (SZA), the tilted independent pixel approximation (TIPA) was suggested (Varnai and Davies, 1999). For broken clouds with small-to-moderate CF, the mean radiative properties depend mostly on $N(\theta)$, while small-scale fluctuations of cloud properties including irregular geometry of an individual cloud are of secondary importance (e.g., Han and Ellingson, 2000; Zuev and Titov, 1995).

This section outlines several approaches for establishing the relationship between the statistical parameters of clouds and radiation. Before we embark on that important endeavor, let us take a look at the stochastic RTE with random parameters (e.g., Pomraning, 1991)

$$\omega \nabla I(\mathbf{r},\omega) + \sigma(\mathbf{r})\, I(\mathbf{r},\omega) = \sigma_s(\mathbf{r}) \int_{4\pi} g(\mathbf{r},\omega,\omega')\, I(\mathbf{r},\omega')\, d\omega' + S(\mathbf{r},\omega)\,. \quad (3.1)$$

Here $I(\mathbf{r}, \boldsymbol{\omega})$ is the radiance[1] (e.g., $\mathrm{W\,m^{-2}\,sr^{-1}\,nm^{-1}}$) at the point $\mathbf{r} = (x, y, z)$ in direction $\boldsymbol{\omega} = (a, b, c)$, $S(\mathbf{r}, \boldsymbol{\omega})$ is the source, $\sigma(\mathbf{r})$, $\sigma_s(\mathbf{r})$ and $g(\mathbf{r}, \boldsymbol{\omega}, \boldsymbol{\omega}')$ are the extinction coefficient (total cross section), the scattering coefficienti (scattering cross-section), and the scattering phase function, respectively. Optical properties σ and σ_s are assumed to be random variables with known statistical properties. Following Pomraning (1991), we want to distinguish among three approaches for estimating the mean radiative properties. Although, these overlapping approaches are considered separately, they can be combined for solving a specific problem.

The first technique is a *numerical averaging*, in which the goal is to estimate the mean radiative properties using the following main steps. The numerical averaging starts with obtaining a large number of cloud realizations (a thousand is not atypical) using assumed/specified probability distribution of $\sigma(\mathbf{r})$ and $\sigma_s(\mathbf{r})$. These realizations can be generated by several stochastic algorithms (e.g., Evans and Wiscombe, 2004; Prigarin and Marshak, 2009; Venema et al., 2010), or obtained from observations (e.g., Wen et al., 2006, 2007) and cloud model results (e.g., Barker et al., 2004; Kassianov et al., 2009). Typically, these algorithms incorporate one- and two-point statistics to generate simple (Fig. 3.2) or more complicated realizations of cloud fields. The next step is to solve the deterministic RTE for each of the realizations using well-known techniques (e.g., Liou, 2002; Marshak and Davis, 2005) and provide the radiative properties of interest (e.g., surface irradiance and albedo). Finally, the ensemble of solutions is processed to obtain the desired statistics of radiative properties. The numerical averaging is very simple to implement (as long as the radiative properties are computed), and it is of great importance for producing benchmark results (e.g., Adams et al., 1989; Zuchuat et al., 1994; Zuev and Titov, 1995; Petty, 2002). However, the computational burden of cloud simulation and RTE solution is large, which makes direct implementation of the numerical averaging for practical applications problematic.

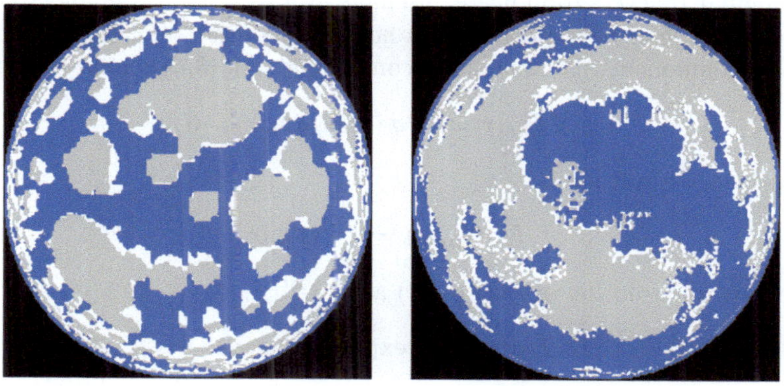

Fig. 3.2. Computer realizations of hemispherical images (after Kassianov et al., 2005c) generated by the stochastic models: cumulus (left) and stratocumulus (right) clouds; cloud bases (gray), sides (white).

[1] Sometimes *radiance* is called *intensity* (e.g., Liou, 2002, p. 122). Note that the term *intensity* has five different meanings in optics (Palmer, 1993).

The second approach is based on the statistical approximations and offers an alternative way around the computational cost issue mentioned above. Instead of calculating the mean radiative properties directly, it links closed-form one-dimensional (1D) RT solutions with statistics of cloud optical properties. Typically, these solutions are represented as a function of cloud optical depth τ, and the averaged radiative properties are defined using the corresponding distribution $f(\tau)$ (e.g., Cahalan et al., 1994; Barker, 1996). For example, analytical expressions of mean direct-beam transmittance can obtained quite easily for many commonly used probability density functions $f(\tau)$ (e.g., Xiu and Karniadakis, 2002). Note that such an approach also makes it possible to derive the distribution functions of radiative fluxes as done by Stephens et al. (1991). These methods are quite close to those where the mean cloud optical depth is rescaled (e.g., Davis et al., 1990; Gabriel et al., 1990) or the 'effective thickness approximation' (ETA) is applied (e.g., Cahalan, 1994; Cahalan et al., 1994). The rescaling/ETA allows one to approximate the average radiative properties of the 3D clouds (e.g., mean albedo) by their plane-parallel counterparts calculated at the rescaled/effective values of cloud optical depth. The application of the rescaling/ETA becomes more apparent when we consider clouds with large horizontal extent such as marine stratocumulus (e.g., Bauml et al., 2004; de Roode and Los, 2008).

Finally, there are several studies with focus on obtaining new equations for the mean radiance from the stochastic RTE. These studies apply different assumptions related to basic cloud statistics, such as mean, variance, and correlation scale (e.g., Stephens, 1988; Cairns et al., 2000; Gabriel and Evans, 1996; Davis and Marshak, 2004, 2010). These assumptions are required to address an important issue associated with the cross-correlation of radiance and extinction. To illustrate this, we consider the simplest source-free case ($S = 0$) for a purely absorbing ($\sigma_s(\mathbf{r}) = 0$) random medium. We wish now to obtain the solution for the ensemble-averaged direct-beam transmittance $\langle I_{dir}(\mathbf{r}, \boldsymbol{\omega})\rangle$ from Eq. (3.1). Here and below, we use angular brackets to indicate ensemble averages. The ensemble-averaged radiance $\langle I(\mathbf{r}, \boldsymbol{\omega})\rangle$ can be expressed as a sum of the direct $\langle I_{dir}(\mathbf{r}, \boldsymbol{\omega})\rangle$ and diffuse $\langle I_{dif}(\mathbf{r}, \boldsymbol{\omega})\rangle$components. For the direct component, Eq. (3.1) should be replaced by

$$\boldsymbol{\omega}\, \nabla I_{dir}(\mathbf{r}, \boldsymbol{\omega}) + \sigma\,(\mathbf{r})\, I_{dir}(\mathbf{r}, \boldsymbol{\omega}) = 0\,. \tag{3.2a}$$

If the boundary condition is

$$I_{dir}(\mathbf{r}_0, \boldsymbol{\omega}) = 1; \quad c < 0\,, \tag{3.2b}$$

then we have the solution of Eqs. (3.2a) and (3.2b)

$$I_{dir}(\mathbf{r}, \boldsymbol{\omega}) = \exp\left[-\tau(\mathbf{r}, \boldsymbol{\omega})\right]\,, \tag{3.3}$$

where $\tau(\mathbf{r}, \mathbf{r}_0)$ is the random optical depth

$$\tau(\mathbf{r}, \boldsymbol{\omega}) = \int_0^{|\mathbf{r}-\mathbf{r}_0|} \sigma(\mathbf{r} - s\boldsymbol{\omega})\, ds\,, \tag{3.4}$$

where s is a spatial coordinate in the direction $\boldsymbol{\omega}$. Eq. (3.4) represents the well-known Beer's law of exponential direct-beam transmittance. Since optical depth τ is random, the transmittance I_{dir} is random as well.

To perform averaging of Eq. (3.2a), let us represent the extinction coefficient σ and direct-beam transmittance I_{dir} as $\sigma = \langle\sigma\rangle + \tilde{\sigma}$ and $I_{dir} = \langle I_{dir}\rangle + \tilde{I}_{dir}$, where the deviations from ensemble means, $\tilde{\sigma}$ and $\tilde{I}_{dir}$, are random quantities with zero expected values ($\langle\tilde{\sigma}\rangle = \langle\tilde{I}_{dir}\rangle = 0$). With such representation, the ensemble averaging of Eq. (3.2a) gives

$$\omega\nabla\langle I_{dir}\rangle + \langle\sigma\rangle\langle I_{dir}\rangle + \langle\tilde{\sigma}\,\tilde{I}_{dir}\rangle = 0 \tag{3.5}$$

This equation is open, since it contains the unknown term $\langle\tilde{\sigma}\,\tilde{I}_{dir}\rangle$, which defines the so-called statistical correction to the transport description. The random transmittance $\tilde{I}_{dir}$ is dependent on the values of the field $\tilde{\sigma}(\mathbf{r})$ on the segment of the straight line between points $\mathbf{r}_0$ and $\mathbf{r}$. Thus, the term $\langle\tilde{\sigma}\,\tilde{I}_{dir}\rangle$ cannot be separated into $\langle\tilde{\sigma}\rangle\,\langle\tilde{I}_{dir}\rangle$. In general, it is described by a hierarchy of equations with multipoint spatial correlations $\langle\tilde{\sigma}(\mathbf{r}_1)\,\tilde{\sigma}(\mathbf{r}_2)\ldots\tilde{\sigma}(\mathbf{r}_{n-1})\rangle$ (e.g., Pomraning, 1991). However, such description is a formidable task for practical applications. Thus, we need to terminate the hierarchy at a specified low-order term. This termination is equivalent to the closure procedures (e.g., Stephens, 1988; Cairns et al., 2000; Gabriel and Evans, 1996). Here is an example of one of the simplest forms for the statistical correlation (Stephens, 1988):

$$\langle\tilde{\sigma}\,\tilde{I}_{dir}\rangle = C_{\sigma I}\,\langle\sigma\rangle\,\langle I_{dir}\rangle\,, \tag{3.6}$$

where the dimensionless measure $C_{\sigma I}$ describes the correlation between the radiance and extinction coefficient. If we further assume that $C_{\sigma I}$ is ω-independent, then the solution of Eqs. (3.5) and (3.6) for $\langle I_{dir}\rangle$ will be defined by Eqs. (3.3) and (3.4) with replacement of the random σ by $\langle\sigma\rangle(1 + C_{\sigma I})$. In certain respects, this example is similar to the concept for the rescaling of mean cloud optical depth described above. After looking at this example, we should note that the complicated description of the statistical correction becomes simpler when fluctuations of the cloud extinction coefficient are approximated by some popular statistical models (e.g., Pomraning, 1991; Anisimov and Fukshansky, 1992; Evans, 1993; Borovoi, 2002). Below, we consider two relevant models and their application to stochastic RT in a cloudy atmosphere.

3.3 Markovian cloud models

To select an appropriate statistical model, we should consider possibilities for: (i) determining the input parameters of the model from observations and (ii) for deriving relatively simple analytical equations, which link the ensemble-averaged radiance with cloud statistics. Two Markovian models introduced independently by both the particle transport and atmospheric science communities satisfy these criteria. The following sections outline these models and show how they can be applied to solve the stochastic RTE.

3.3.1 Levermore–Pomraning model

Originally, Levermore, Pomraning, and their colleagues in the particle transport community introduced the Markovian model, known as the Levermore–Pomraning

(LP) model, for describing particle transport[2] through a two-component (liquid water and vapor) stochastic mixture (Levermore et al., 1986). 'They were motivated to find an accurate transport description of a two-fluid turbulent mixture to support the laser fusion design codes of the inertially confined fusion program, but realized such a technique would have much wider application' (Byrne, 2005, p. 389). This approach has been successfully applied to the study of neutron transport through randomly distributed lumps and absorbers (Williams, 1974; Cassell and Williams, 2008), and of radiative transfer through random media with Rayleigh scattering (El-Wakil et al., 2004) and clouds (e.g., Su and Pomraning, 1994; Foster and Veron, 2008, and references therein).

To describe the geometry of broken clouds, Pomraning and his colleagues adopted a two-state homogeneous Markov process (e.g., Malvagi et al., 1993). Unless otherwise stated (section 3.3.3), all considered cases have homogeneous Markovian statistics, which are invariant by translation. Thus, the component properties (e.g., extinction coefficient) and the statistical properties (e.g., mean chord lengths and probability for a component to be at a given location) are independent on position. In particular, the LP model approximates the clouds and clear sky as a two-component random mixture with exponentially distributed chord lengths. To cite an example, the extinction coefficient σ is considered a discrete random variable with two states, $S = \{\sigma_0, \sigma_1\}$, where σ_0 and σ_1 define its values for clear sky and clouds, respectively. Each realization of σ is composed of alternating segments of the two components with corresponding mean chord lengths λ_0 and λ_1. The probability of finding component i along the extinction path is $p_i = \lambda_i/(\lambda_0 + \lambda_1)$, where $i = 1, 2$. The conditional probabilities have a common exponential term. For example,

$$P_{11}(s', s) = \exp(-A|s' - s|)(1 - p_1) + p_1 \,, \tag{3.7}$$

where parameter $A = \lambda_0\lambda_1(\lambda_0 + \lambda_1)^{-1}$ and A^{-1} is the correlation length. In other words, the conditional probability of the cloud presence $P_{11}(s', s)$ is the conditional probability of the event (point s is in cloud), given that another event (point s' is in cloud) occurs.

The LP model offers an exact solution for the pure absorbing (no scattering) limit with Markovian statistics

$$\langle I_{dir}(s)\rangle = \sum_{i=1}^{2} C_i \exp\{-\chi_i s\} \,, \tag{3.8}$$

where coefficients C_i and χ_i, $i = 1, 2$ are functions of Markovian model parameters. The obtained solution applies the well-known assumption that alternating segments of clouds and clear sky along a given direction $\boldsymbol{\omega}$, so-called chord lengths of clouds and clear-sky, are distributed exponentially.

In the presence of scattering, an exact set of two coupled equations can be obtained (e.g., Adams et al., 1989; Sanchez, 1989)

$$\boldsymbol{\omega}\,\nabla(p_i I_i) + \sigma_i p_i I_i = \sigma_{si} \int_{4\pi} g_i(\boldsymbol{\omega}, \boldsymbol{\omega}') p_i I_i(\boldsymbol{\omega}')\, d\boldsymbol{\omega}' + p_i S_i + \frac{p_j I_{ji}}{\lambda_j} - \frac{p_i I_{ij}}{\lambda_i} \,, \tag{3.9}$$

[2]The transport equation describes interaction (absorption and scattering) of neutral particles with a background material but not with themselves (e.g., Pomraning, 1991).

where $i, j = 0, 1$, $j \neq i$, p_i is the probability of finding component i at a given spatial location, and the I_{ij} is the conditional ensemble-averaged intensity conditioned upon position $\mathbf{r}$ lying at an interface between component (or material) i and component j, with component i to the left of the interface (vector $\boldsymbol{\omega}$ points from left to right). In the particle transport theory, these averages I_i and I_{ij} are often referred to as the 'material' and 'interface' ensemble-averaged fluxes, respectively (e.g., Zuchuat et al., 1994).

The unconditional ensemble-averaged intensity is defined by

$$\langle I \rangle = \sum_{i=0}^{1} p_i I_i \tag{3.10}$$

The two coupled equations (3.9) are open, since they contain four unknowns I_0, I_1, I_{01} and I_{10}. Thus, a closure is needed for solving these equations. Several closures have been suggested (e.g., Adams et al., 1989; Sanchez, 1989; Pomraning, 1991), and the simplest low-order closure is given by the following relation

$$I_{ij} = I_i \tag{3.11}$$

The application of Eq. (3.11) to the Eqs. (3.9) results in the closed set of LP equations (e.g., Sanchez, 1989). The low-order closure, described by Eq. (3.11), is known to be exact only for purely absorbing (no scattering) medium with Markovian statistics, which apply the exponentially distributed chord lengths. Note that the LP equations can treat both the homogeneous and inhomogeneous Markovian statistic and several derivations of the LP model exist (e.g., Pomraning, 1998; Sanchez, 2008, and references therein). For medium with arbitrary non-Markovian statistics (i.e., the possibility of infinite chord lengths), where every component is characterized by a given chord length distribution (not necessarily exponential), the theory of alternating renewal processes can be applied for obtaining coupled integral (rather than the differential) equations (e.g., Levermore et al., 1988; Sanchez and Pomraning, 1991; Zuchuat et al., 1994). The renewal equations were discussed in more details by Sanchez (1989) and approximations related to different closures were compared with the benchmark results (e.g., Zuchuat et al., 1994).

3.3.2 Titov model

Titov and his colleagues developed a Markovian model for the atmospheric science community (e.g., Titov, 1990; Zuev and Titov, 1995; Titov et al., 1997). Such development was motivated by early and interesting studies of Avaste and Vainikko (1974). This model is the special case ($\sigma_1 \gg \sigma_0$) of the more general LP model (Pomraning, 1991). An alternate derivation of the two coupled integral equations of the ensemble-averaged intensity was provided by Titov (1990) and Pomraning (1999):

$$\langle I(s, \boldsymbol{\omega}) \rangle = \frac{\sigma_{s1}}{\sigma_1} \int_0^s \varphi(s, s')\, ds' \int_{4\pi} g_1(\boldsymbol{\omega}, \boldsymbol{\omega}')\, f_1(s', \boldsymbol{\omega}')\, d\boldsymbol{\omega}' + \langle I_{dir}(s, \boldsymbol{\omega}) \rangle\,, \tag{3.12}$$

where $\langle I_{dir}(s, \boldsymbol{\omega}) \rangle$ is the direct component defined by Eq. (3.8), and $f_1(s, \boldsymbol{\omega}) = \sigma_1 p_1 I_1(s, \boldsymbol{\omega})$ is the mean collision density. By converting the double (line and angle) integrals over $ds'\,d\boldsymbol{\omega}'$ to the volume one, the corresponding integral equation can be written as

$$f_1(\mathbf{x}) = \int_X k(\mathbf{x},\mathbf{x}')\, f_1(\mathbf{x}')\, d\mathbf{x}' + \Psi(\mathbf{x}) \tag{3.13}$$

and

$$k(\mathbf{x},\mathbf{x}') = \frac{\sigma_{s1}\, g_1(\boldsymbol{\omega},\boldsymbol{\omega}')\, \eta(|\mathbf{r}-\mathbf{r}'|)}{\sigma_1 2\pi |\mathbf{r}-\mathbf{r}'|^2}\, \delta(\frac{\mathbf{r}-\mathbf{r}'}{|\mathbf{r}-\mathbf{r}'|} - \boldsymbol{\omega})\,, \tag{3.14}$$

where X is the phase space of coordinates and directions, $\mathbf{x} = (\mathbf{r},\, \boldsymbol{\omega})$, and δ is the surface delta function such as $\int \delta(\boldsymbol{\omega}-\boldsymbol{\omega}')\rho(\boldsymbol{\omega}')\, d\boldsymbol{\omega}' = \rho(\boldsymbol{\omega})$. Thus, the delta function in Eq. (3.14) indicates that the points $\mathbf{r}$ and $\mathbf{r}'$ lie on a line along the direction $\boldsymbol{\omega}$: $\mathbf{r} = \mathbf{r}' + s\boldsymbol{\omega}$. Similar to the LP equations, the obtained equations (3.12) to (3.14) represent a 1D model, where the ensemble averages $\langle I \rangle$ and f_1 are independent of two spatial coordinates (x and y). Advantages of both the LP and Titov models is that they allow one to calculate the mean radiative properties (e.g., mean albedo) without requiring an exact description of the 3D cloud geometry.

The ensemble-averaged equations (3.12) to (3.14) have three functions associated with 'direct-beam' exponential components (Table B.1). These functions can be described in terms of the Monte Carlo (MC) method. The first two functions $\Psi(s)$ and $\eta(s)$ represent the probability density of the free path length of a photon[3] between two successive collisions (along a given direction). The third function $\varphi(s)$ is a 'direct-beam' component of the contribution function from the popular 'local estimation' scheme (e.g., Marchuck et al., 1980).

Moreover, the ensemble-averaged equations (3.12) to (3.14) resemble the well-known equations obtained for radiance and collision density (the product of extinction coefficient and radiance) from the deterministic RTE (e.g., Marchuk et al., 1980; Marshak and Davis, 2005). Thus, they can be solved using traditional approaches, including the Monte Carlo (MC) method. Available MC algorithms developed for other applications (e.g., Buras and Mayer, 2011; Emde et al., 2011) can be implemented with minor changes for solving the obtained equations (Appendix B). In particular, we developed MC algorithms for simulating the ensemble-averaged reflectance for different cloud types (e.g., Kasyanov and Titov, 1994). Figure 3.3 shows an example of simulated reflectances for cumulus and equivalent stratus clouds. The equivalence is understood as follows: the fields of cumulus and stratus clouds have identical optical and geometrical characteristics, and differ in a single parameter, namely the cloud aspect ratio. The assumed values of this parameter are 1 and 0 for cumulus and stratus clouds, respectively. The qualitatively different dependence of the mean reflectance (cumulus versus stratus) on the viewing zenith angle (Fig. 3.3) illustrates the importance of the cloud aspect ratio to the ensemble-averaged reflectance.

Under certain atmospheric conditions, such as cirrus clouds overlying a stratus field, the LP and Titov models may be applied at multiple levels in the atmosphere (Foster and Veron, 2008). These models may also have applications in situations with non-Markovian statistics, such as cloud fields that are not described by exponential chord length distributions (Veron et al., 2009). However, in general, the Markovian models considered do not represent strong vertical variability of cloud

[3] Mishchenko (2009) debunked the common 'photonic' misinterpretations.

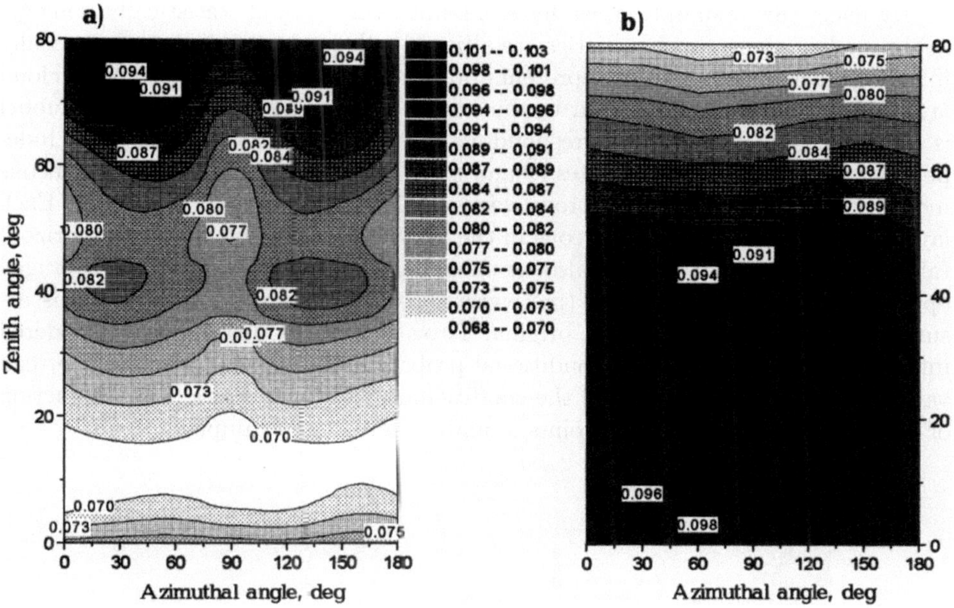

Fig. 3.3. Two-dimensional diagrams of the ensemble-averaged reflectance (visible spectral range) simulated for cumulus clouds (a) and equivalent stratus clouds (b) over a dark surface (after Kasyanov et al., 1994). The equivalent stratus clouds represent an asymptotic case (Appendix B.2) where the averaged cloud chord length (CCL) is very large compared to the photon mean-free-path (PMFP).

properties in addition to the substantial horizontal changes of them. Thus, the extension of the Markovian approach to random fields where the cloud statistics are altitude-dependent, the so-called inhomogeneous statistics, is highly desirable. For the inhomogeneous model, the component properties (e.g., extinction coefficient) and the statistical properties (e.g., mean chord lengths and probability of a component being at a given location) are not invariant to translation and depend on the vertical position. Moreover, such extension is a necessary step for improving RT calculations in climate models (section 3.6).

3.3.3 Generalized Titov model

To generalize the Markovian approach to statistically inhomogeneous broken clouds, we need to prescribe a statistical relationship between cloud layers. An example of this relationship is the observed vertical distribution of liquid/ice in cloud fields. To get this information, we take advantage of results from observational and model-based studies (e.g., Hogan and Illingworth, 2000; Oreopoulos and Khairoutdinov, 2003), which demonstrated that (i) the overlap of clouds at two levels tends to decrease rapidly as their vertical separation is increased and (ii) the degree of overlap as a function of level separation can be described by a simple decaying exponential expression. Thus, in the generalized model we assume that the statistical relationship between two adjusted layers is described by an inverse-exponential relation.

By using this assumption, we have extended the original statistically homogeneous model, where cloud statistics are altitude-independent, to a new statistically inhomogeneous model, which represents broken clouds as a set of correlated cloud layers (Kassianov, 2003). Similar to the original model, the inhomogeneous model is one-dimensional, and the corresponding ensemble averages $\langle I \rangle$ and f are independent of two spatial coordinates (x and y). The advantage of the inhomogeneous model is that it accounts the strong vertical variability of cloud properties. Each layer is assumed to be homogeneous in the vertical but inhomogeneous in horizontal dimensions (Fig. 3.4). It is also assumed that for the each layer, the averaged optical properties are constant (piecewise constant approximation), and this assumption is consistent with the original Titov and LP model. For the considered inhomogeneous model, the unconditional probability of the cloud presence p_1 can vary strongly with altitude, and the conditional probability $P_{11}(s', s)$ is a function of the relative position of two points s' and s. Here is an example:

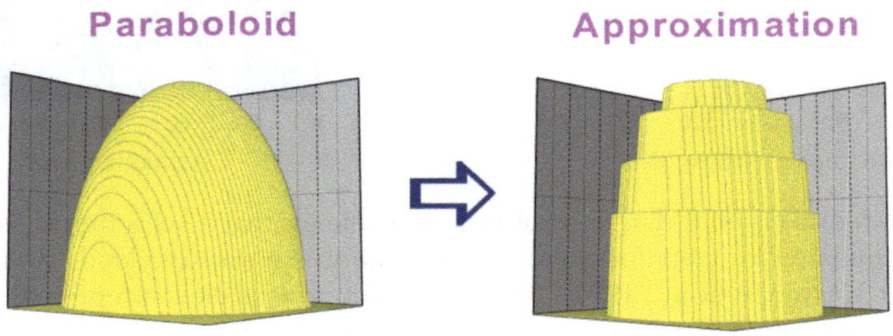

Fig. 3.4. Schematic diagram (after Kassianov, 2003) illustrates the approximation of a truncated paraboloid of revolution by a set of cylinders with different diameters. Similar approximation can be applied for cloud layers with different statistical properties (e.g., different and altitude-dependent values of mean chord length).

(1) if the points s' and s belong to the same kth layer, then

$$P_{11}(s', s) = \exp(-A_k|s' - s|)(1 - p_{1k}) + p_{1k}\,; \tag{3.15}$$

(2) if the points s' and s belong to different adjacent layers, namely kth and mth layers, then

$$P_{11}(s', s) = \exp(-A_{km}|s^* - s|)(P_{11}(s', s^*) - p_{1m}) + p_{1m}\,, \tag{3.16}$$

where the parameter A_{km} determines the statistical relationship between the kth and mth layers, the points s', $s*$ and s lie on a line along a given direction $\boldsymbol{\omega}$. The conditional probability $P_{11}(s', s)$ in Eq. (3.16) is computed in terms of two processes. The first process $P_{11}(s', s^*)$ goes, within kth layer, from the interior point s' of the kth layer to the intersection point $s*$. The latter defines the intersection of the trajectory from s' to s with the interface between the layer kth and its neighbor layer mth. The second process $P_{11}(s^*, s)$ goes from $s*$ to s within mth layer. Note that all these parameters, A_k and A_{km}, depend on both the 3D cloud structure and the positions of points s' and s. By changing the values of these parameters, one can describe different combinations of maximum and random overlaps for adjusted cloud layers (Kassianov, 2003), and thus the total CF (e.g., Tompkins and Di Giuseppe, 2007).

The statistically inhomogeneous model is a logical extension of the statistically homogeneous models described in the previous sections. Thus, the outlined generalization took advantage of an elegant theory and effective numerical methods, which were developed for the statistically homogeneous models outlined above. The obtained generalized equations for the radiance and mean collision density have the same form as those from the previous section Eqs. (3.12) to (3.14). The only difference is that the functions η, φ, and Ψ in the obtained equations (inhomogeneous model) are piecewise constant functions and depend on s' and s positions. For example, if s' and s belong to the same kth layer, then one should use equations with coefficients defined for this layer (Table B.1). If s' and s belong to the different layers kth and mth layers, then one should use other equations where coefficients depend on parameters of these adjusted layers and the parameter A_{km} as well (Kassianov, 2003). These generalized functions are similar in form to those given in Table B.1. Thus, the numerical solution of these equations (accounting for scattering) can be performed using the same straightforward framework described in Appendix B.

For cases without scattering (e.g., the direct-beam transmittance), a simple equation (Kassianov, 2003) can be applied for calculating. This equation is a generalized version of Eq. (3.8) and accounts for the vertical changes of cloud statistics (e.g., CF). This generalized equation has been shown to be quite accurate (Kassianov et al., 2003). In particular, this equation illustrated the importance of the cloud overlap assumption using the ensemble-averaged direct-beam transmittance calculated for two cases (Fig. 3.5). These two cases have identical cloud optical and geometrical properties, such as the vertical profiles of CF and mean horizontal size (Fig. 3.5(a)), and differ in the cloud overlap assumption (maximum versus random) only. The cloud overlap assumption has strong impact on the directional cloud fraction, and consequently on the mean direct-beam transmittance (Fig. 3.5). Similar results have been obtained in several studies (e.g., Barker et al., 1999).

These generalized equations agree with the corresponding equations obtained for (i) the statistically homogeneous Titov model and (ii) the vertically inhomogeneous overcast cloud field (Liou, 1980/2002). To illustrate that, one should set the parameters properly. For example, to get the equations for the vertically inhomogeneous overcast cloud field, one should set $A_{km} = 0$ (maximum overlap) and $p_k = 1$ (overcast case) in the obtained equations (inhomogeneous model). The accuracy and robustness of these equations for diffuse radiance was estimated (Kassianov et al., 2003; 2005a) by comparing the mean radiative properties obtained by numerical and analytical averaging (section 3.3.1). The numerical averaging provided a *reference* case. For a given 3D cloud field we calculated the ensemble- and domain-averaged radiative properties. Since the full 3D cloud geometry was used in the radiative calculations, the calculated mean radiative properties were considered the reference values. The analytical averaging provided an *approximate* case. Approximate equations for the mean radiance, which have been derived by analytically averaging the stochastic radiative transfer equation, were also used for estimating ensemble-averaged radiative properties. Since only the bulk cloud statistics were applied in the RT calculations, the mean radiative properties obtained by this method are considered as approximations of the true radiative proper-

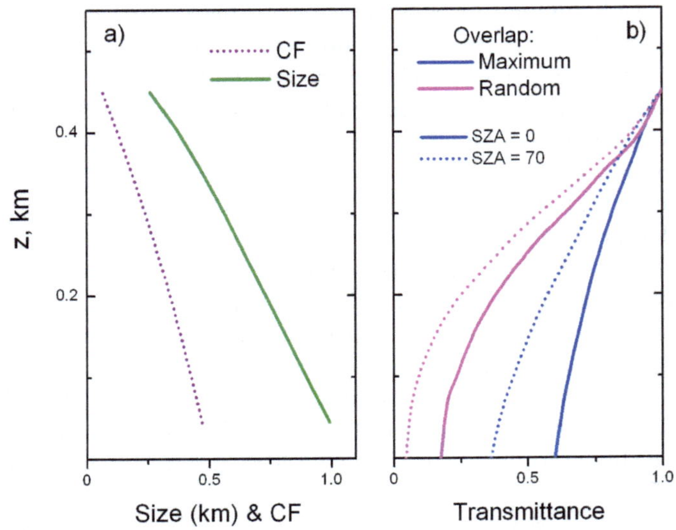

Fig. 3.5. Vertical distributions of (a) the mean horizontal cloud size and cloud fraction (CF), and (b) the ensemble-averaged direct-beam transmittance (after Kassianov et al., 2003). The mean transmittance is calculated using two cloud overlap assumptions (maximum and random overlap) and two values of solar zenith angle (SZA). These geometrical and radiative properties are obtained for an ensemble of cloud realization formed by a group of truncated paraboloids of revolution (Fig. 3.4). The model parameters (e.g., CF) are set to values typical for small marine cumulus clouds.

ties. The approximate equations provided reasonable accuracy (~15%) for both the ensemble-averaged and domain-averaged radiative properties. This is similar to the accuracy found in the traditional LP model when an approximate closure, which accounts for non-Markovian chord characteristics, is applied (Levermore et al., 1988; Malvagi et al., 1993). Also, the angular distribution histograms and the photon path length distributions of the mean albedo and transmittance, which have been obtained by exact and approximated methods, agreed qualitatively and quantitatively. It should be emphasized that application of the approximated methods reduces substantially the computational burden of RT calculations. Such computational burden reduction is case-dependent, but the approximated methods speed up the RT calculations typically by factors of 10–100.

As already mentioned in section 3.2, the averaged radiative properties are mostly governed by the bulk geometry of broken clouds, and this geometry can be described by a few cloud statistics, such as total CF and mean horizontal cloud size. These statistics have a strong link with parameters of the Markovian cloud models. For the statistically homogeneous cloud models (sections 3.3.1 and 3.3.2), the unconditional probability of cloud presence defines the total CF, while the conditional probability can be used to specify the mean horizontal cloud size. Likewise, for the statistically inhomogeneous model (section 3.3.3), the vertical distributions of the unconditional and conditional (both in horizontal and vertical directions) probabilities are related to the vertical profiles of CF and mean horizontal cloud size and a parameter describing overlap between cloud layers.

3.4 Estimation of model parameters

The ability to obtain necessary input parameters from observations/simulations is an attractive feature of Markovian models. In this section, we will illustrate how the parameters needed for the model of broken clouds can be estimated from ground-based passive and active remote sensing. In particular, we consider estimation of horizontal cloud sizes and CF. Synthesis of passive and active satellite remote sensing can offer a great potential for estimating these parameters from space (e.g., Kato et al., 2010; Barker et al., 2011). For example, the CF can be obtained from satellite observations, such as the Moderate Resolution Imaging Spectroradiometer (MODIS) Cloud Product (MOD06), while the CloudSat (Stephens et al., 2002) and the Cloud-Aerosol Lidar with Orthogonal Polarization (CALIOP) (e.g., Winker et al., 2007) can provide information about the vertical structure of the cloud field.

3.4.1 Vertically-integrated statistics

The majority of climate-related studies describe the cloud amount in terms of the fractional sky cover (FSC), which is a hemispherical measure of CF. The estimation of FSC has been performed typically by a well-established empirical method (Long et al., 2006) based on measured broadband shortwave all-sky fluxes or sky images (e.g., Kassianov et al., 2005b). Spectrally-resolved measurements by multi-filter rotating shadowband radiometers (MFRSRs) can also be applied successfully for FSC estimation (Min et al., 2008; Kassianov et al., 2011a). Since these broadband and spectrally-resolved fluxes represent hemispherical observations that may include contributions from overlapping clouds and cloud sides (Fig. 3.2), the FSC can overestimate the CF. Such overestimation is more pronounced for clouds with comparable mean cloud vertical and horizontal sizes (Kassianov et al., 2005b). Using sky imagers with a 'limited' 100-degree field-of-view (FOV) or quite simple corrections can improve the CF estimations (Kassianov et al., 2005b).

In addition to the FSC, directional CF $N(\theta)$ can be defined from passive ground-based observations. The directional CF is defined as a fraction of time in which an instrument would detect clouds for a given direction (e.g., to the sun). For example, Kaufman and Koren (2006) applied data from the Aerosol Robotic Network (AERONET) for obtaining the directional CF around the globe for each month during a 12-month period. Similarly, Lane et al. (2002) estimated $N(\theta)$ of small scattered cumuli from MFRSR observations at the US Department of Energy Atmospheric Radiation Measurement (ARM) Climate Research Facility Southern Great Plains (SGP) site and demonstrated that changes of the directional CF are in a quite narrow range (from 0 to 0.3). This range is consistent with those obtained from other studies of fair-weather cumuli (e.g., Plank, 1969; Wielicki and Welch, 1986; Hozumi et al., 1982). Observational data from the whole sky imager at the ARM Tropical Western Pacific (TWP) site were used to define $P(\theta) = 1 - N(\theta)$ of tropical marine cumuli (Taylor and Ellingson, 2008). The estimated probabilities $P(\theta)$ were contrasted to those obtained from several computationally convenient parameterizations with different cloud shape parameters.

Lane et al. (2002) gained the benefit of combined measurements at the SGP site for estimating the horizontal scale clouds and spacing between them. They combined the MFRSR measurements with observations of wind speed and cloud-base height. The MFRSR data were applied to calculate the duration of the direct normal radiation blocked by a cloud. To estimate cloud size, they multiply this duration time by the wind speed at the observed cloud-base height. Since the instrument traces a path from the leading edge to the trailing edge of the cloud overhead, the so-called cloud chord length (CCL), the estimated cloud size represents a 1D cross-section of a cloud along wind direction. A similar approach was applied for estimating cloud size from ground-based active remote sensing (section 3.5). For example, Berg and Kassianov (2008) utilized high-temporal resolution (10 s) observations provided by zenith-pointing lidar and radar instead of the MFRSR observations.

Typical wind speed at the cloud base height during the time of interest was about 10 m s^{-1}, and the temporal resolution of MFRSR observations is 20 s. Therefore, these combined observations do not allow one to capture the small-scale cloud variability (e.g., less than 0.2 km). Lane et al. (2002) used the same approach for estimating the spacing between clouds. The obtained distributions of cloud-size and cloud spacing (Fig. 3.6) have peaks at the smallest observable clouds (less than 0.2 km), which are in agreement with previous studies (e.g., Plank, 1969; Hozumi et al., 1982). In addition, these distributions are approximated reasonably

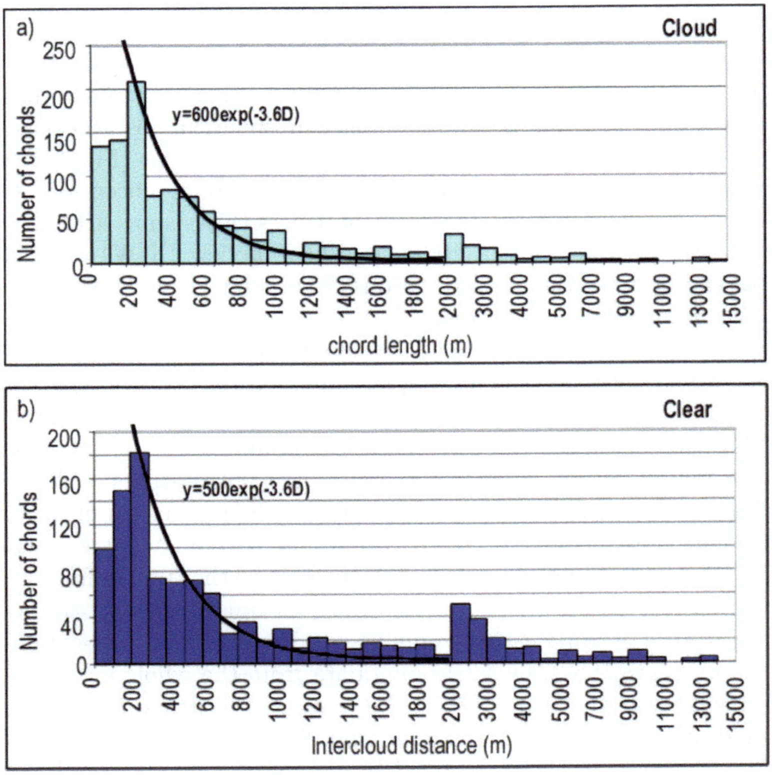

Fig. 3.6. Histograms of (a) cloud chord length and (b) distance between clouds obtained from MFRSR observations (after Lane et al., 2002) and the corresponding exponential fits (Plank, 1969).

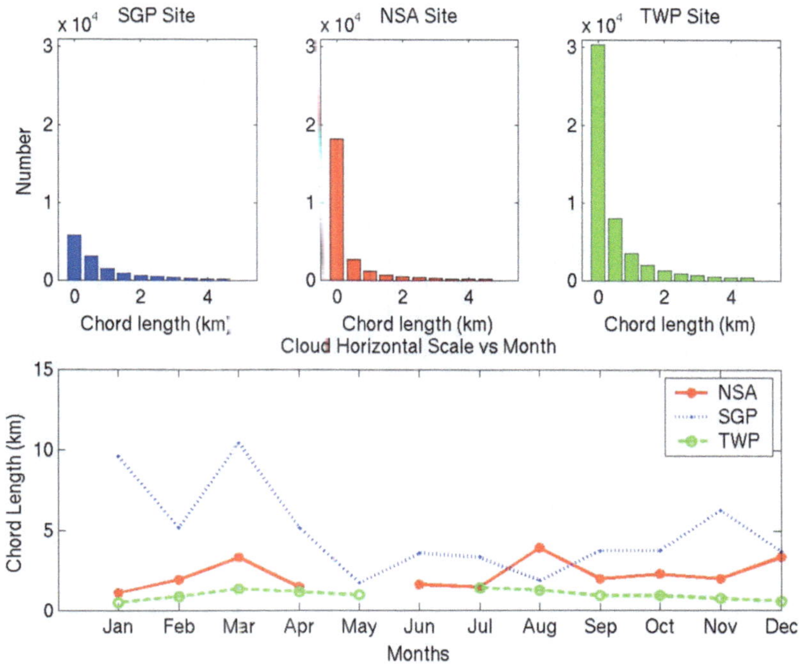

Fig. 3.7. Histograms of CCL (top) for all cloudy hours from the year 2000 at three different ARM sites and (bottom) the monthly mean CCL as function of location (after Veron et al., 2011). These statistics represent combinations of different cloud types (up to three cloudy layers) including FWC. Note that the CCL histograms (top) can be approximated by exponential fits even for such complicated situations.

well by exponential fits provided by Plank (1969), who used aerial photography to determine cumuli statistics over Florida. Further studies by Lane-Veron applying the Infrared Thermometer in place of the MFRSR, with a higher temporal, and therefore spatial, resolution indicated that the shape of the distribution remains the same, with better resolution of the smallest CCL bin. Moreover, these studies demonstrated that CCL histogram depends on cloud type, season, and location (Fig. 3.7).

3.4.2 Vertically-resolved statistics

Ground-based active remote sensing instruments, such as lidar and radar, have been used successfully to sample the vertical profiles of cloud properties (e.g., Clothiaux et al., 2000; Kollias et al., 2007). Here, we illustrate how radar observations can provide cloud statistics required for the Markovian models (Appendix C). These statistics include (i) the cloud fraction (the unconditional probability of cloud presence) for each layer, (ii) the conditional probabilities of cloud presence for each layer, and (iii) the conditional probabilities of cloud presence for two adjacent layers in both the zenith and nadir directions.

The following example (Kassianov et al., 2005a) demonstrates an assessment of these probabilities from radar-derived data collected during observations of fair-

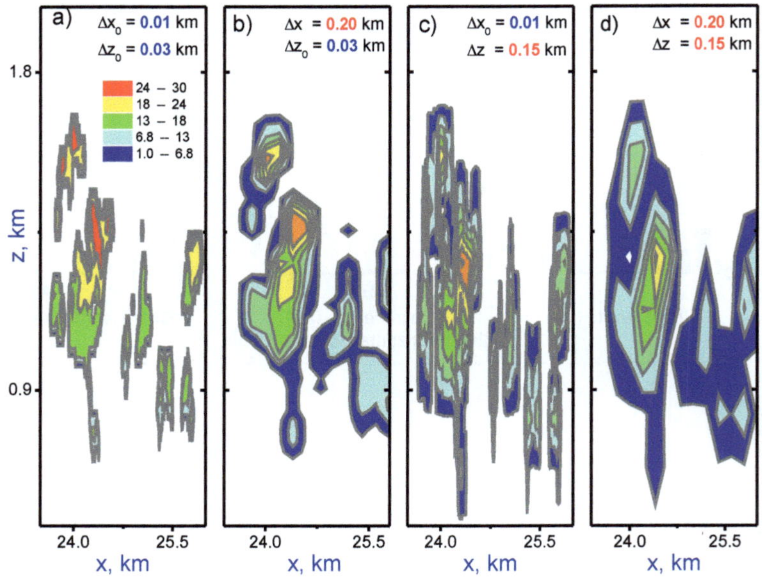

Fig. 3.8. Cross-section (the horizontal and vertical dimensions) of the extinction coefficient obtained from radar observations (after Kassianov et al., 2005a) at (a) high resolution and (b, c, d) degraded resolution in (b) the x-direction, (c) the z-direction, and (d) both x- and z-directions.

weather cumulus by the University of Miami 94-GHz cloud radar. The original dataset includes calculated values of cloud extinction coefficient converted from radar measurements at very high resolution (0.03-km vertical and 0.01-km horizontal resolution). Three additional datasets of extinction coefficient are obtained from the original one by lowering the resolution in the horizontal, vertical, and horizontal and vertical directions. Fig. 3.8 shows examples of these cloud fields. Note that a 'perfect cloud detector' approach (Di Girolamo and Davies, 1997) is applied for separation of cloudy pixels from clear-sky ones for all datasets. According to this approach, a pixel is considered to be a cloudy pixel if the extinction coefficient value for this pixel is distinct from zero (cloud threshold is scale-independent). We set the extinction coefficient threshold equal to 0.1 km^{-1}. The original field and three additional datasets have the same domain-averaged optical depth. However, other statistics are different (Figs. 3.9 and 3.10). For example, coarser resolution in the z-direction does not change the nadir-view CF (Fig. 3.8) but modifies its vertical profile (Figs. 3.9 and 3.10).

Here we illustrate how the input parameters of Markovian models can be estimated from ground-based passive (section 3.4.1) and active (section 3.4.2) observations. The included examples represent case studies with a limited number of cloudy days. To evaluate the Markovian approach more thoroughly, we need statistics of clouds and radiative properties from extended and integrated datasets, as we will now discuss.

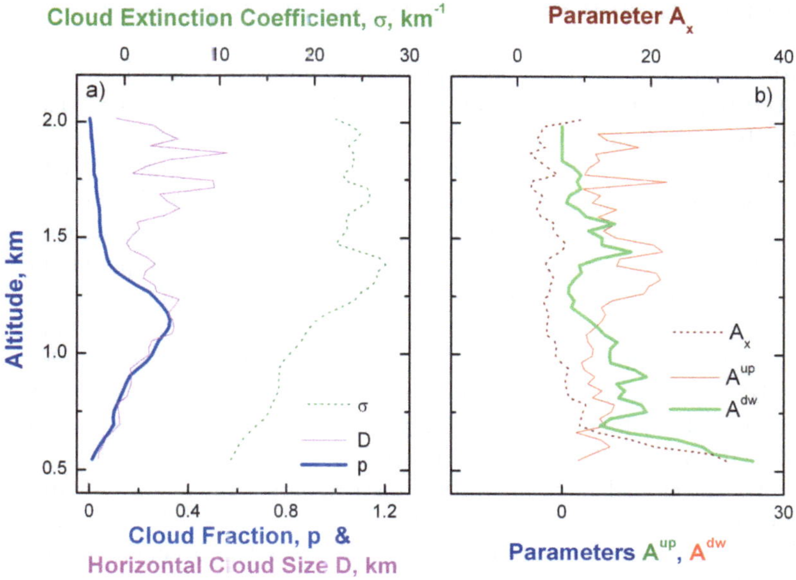

Fig. 3.9. The bulk statistics that corresponded to the high-resolution data (after Kassianov et al., 2005a): the vertical profiles of (a) the cloud fraction N, mean horizontal cloud size D, and cloud extinction coefficient σ, and (b) parameters A_x (x-direction), A^{up} (zenith direction), A^{dw} (nadir direction).

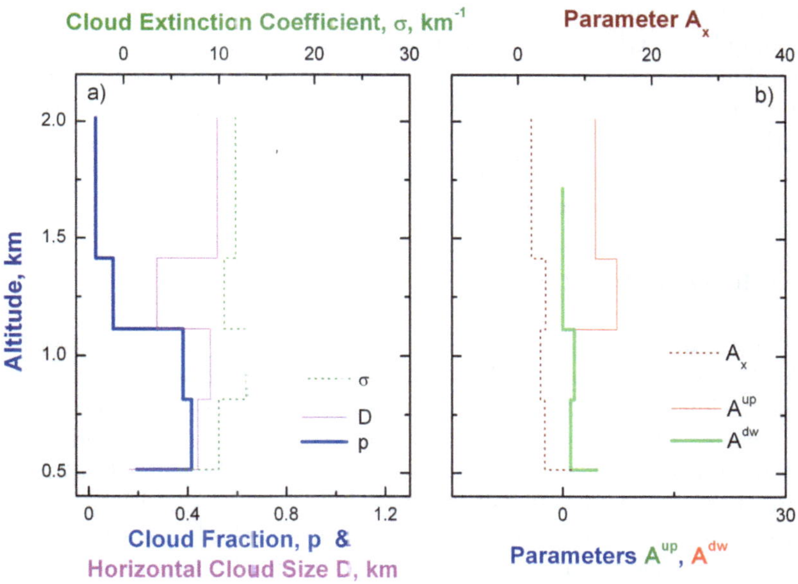

Fig. 3.10. The same as Fig. 3.9, except that these bulk statistics correspond to the field at degraded resolution in the z-direction (after Kassianov et al., 2005a).

3.5 Long-term and enhanced observational datasets

We started the development of the integrated datasets from observations at the ARM SGP (e.g., Berg and Kassianov, 2008; Berg et al., 2009, 2011; Kassianov et al., 2011a,b) and at the TWP and North Slope of Alaska (NSA) sites (e.g, Foster and Veron, 2010; Veron et al., 2011). The TWP site is composed of the Manus, Darwin, and Nauru Island observational facilities. So far, these datasets have leaned heavily on data offered by the zenith-pointing active remote sensing instruments (radar, lidar) with a narrow FOV. These instruments can sample clouds directly above them. In general, the cloud statistics obtained from such transects along wind direction may not be representative of a larger area surrounding these instruments. Lane et al. (2002) considered this issue for the fair-weather cumuli (FWC) observed at the SGP site from the multiple locations with MFRSR installations throughout a large region. They found that the zenith-view cloud statistics derived for a given location held up for distances up to 95 km.

Recently developed next-generation instruments with scanning capability can provide cloud statistics beyond the 1D 'soda straw' view of clouds. In 2006, the ARM Program initiated feasibility studies to assess the potential to map the 3D structure of clouds and precipitation using scanning radars. The scanning ARM cloud radars (SACRs) are dual-frequency radar systems with Doppler and polarization capabilities. These capabilities allow more accurate probing of a variety of cloud systems (e.g., low cumuli), better correction for attenuation, use of attenuation for liquid water content retrievals, polarimetric characterization of non-spherical particles, and habit identification. As a supplement to the estimations of directional cloud fraction $N(\theta)$, the scanning radar observations are expected to provide angular distribution of important parameters describing overlap between adjacent cloud layers, the so-called decorrelation length scale (e.g., Tompkins and Di Giuseppe, 2007). Up to now, simple parameterizations specified this angular distribution for addressing the 3D radiative transfer challenges in inhomogeneous cloudy atmospheres. Below, we describe multi-year datasets and demonstrate capabilities of scanning radars for sampling cloud properties.

3.5.1 Multi-year statistics

The development of cloud statistics over the ARM SGP site during a ten-year period (2000–2009) involves data from the 915-MHz radar wind profiler and total sky imager, along with data from an ARM-developed algorithm, called the Active Remote Sensing of Clouds (ARSCL), which combines data from radar, lidar, microwave radiometer. The cloud statistics include the CF and CCL measured by zenith-looking instruments. Recall that the CCL is a horizontal length scale representative of the FWC. This length is equal to the length of time that an individual cloud is over the cloud radar, multiplied by the wind speed at cloud base (section 3.4.1). Since the temporal resolution of ASRCL data is 10 s, the corresponding spatial resolution is about 0.07 km for the averaged summertime wind speed of 7 $\mathrm{m\,s^{-1}}$ at the ARM SGP site. Similar to the case studies (section 3.4.1), multi-year histograms of CCL are approximated quite well by exponential fits (Fig. 3.11). These fits are defined as $F = A\ \exp(-Al)$, where F is the normalized frequency

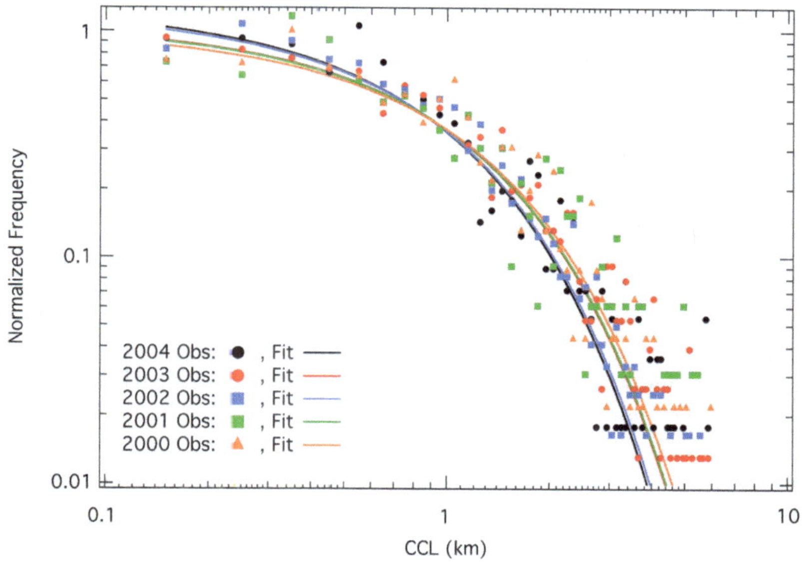

Fig. 3.11. Distributions of cloud chord lengths (CCLs) obtained in Oklahoma, USA, for five summers (after Berg and Kassianov, 2008). Lines correspond to the best-fit exponential distribution for each year. The CCL is a horizontal-length scale representative of the clouds. The CCL is defined as the length of time that an individual cloud is over a ground-based zenith-pointing instrument, multiplied by the wind speed at cloud base.

and A is the best-bit parameter. The value of A determines how quickly the distribution decreases with CCL, and the mean of the distribution is A^{-1}. In our study, the mean CCL ranges from approximately 1.0 km to 1.2 km. We also fit a double power law to the CCL distributions (e.g., Cahalan and Joseph, 1989; Benner and Curry, 1998; Sengupta et al., 1990), but the value of the χ^2 statistic is smaller for the exponential fits, so only results for the exponential fit are shown.

The contribution of clouds with various sizes to the CF has been examined in several studies. For example, Rodts et al. (2003) used aircraft observations over Florida and found based on the 1D sampling that the smallest shallow clouds contributed the most to the CF. In contrast to the results presented by Rodts et al. (2003), but similar to those from other studies (e.g. Plank, 1969; Wielicki and Welch, 1986; Sengupta et al., 1990; Neggers et al., 2003). Berg and Kassianov (2008) found that the contribution to CF is largest for an intermediate cloud size (CCL $\sim$ 1 km). This size is often referred to as the dominant cloud size. Since the spatial resolution of the ARSCL data is about 0.01 km, small gaps between clouds might be missed. As a result, the number of long CCL could be overestimated and vice versa. Our results (Berg and Kassianov, 2008) suggest that the differences in the estimates of dominant cloud size are likely driven by the complicated cloud geometry (Fig. 3.1) rather than by the dimensionality of the measurement (1D versus 2D cases).

The improved mapping of the 3D cloud geometry requires multi-angle observations. In particular, Kassianov et al. (2003) examined the potential of multi-angular satellite-based observations for retrieving the bulk geometry parameters and showed a substantial impact of cloud top/base variations on the directional CF. Note that these variations are related to the vertical profiles of CF and a parameter describing overlap between cloud layers in zenith-view direction (Kassianov, 2003). The next section discusses the potential of multi-angular ground-based active observations for retrieving bulk geometry parameters.

3.5.2 Scanning cloud radar observations

Scanning clouds in 3D has very recently been done in a continuous operating environment. Cloud properties are vastly different for rain and snow shafts that are the primary target of common scanning precipitation radars. While ARM scientists expect to re-use many of the ideas implemented in scanning weather radar systems, the need to detect both low- and high-level stratiform clouds, broken clouds, and multi-layer cloud conditions requires new sampling approaches.

Consequently, the SACR's sampling strategies are composites of Range-Height Indicator (RHI) scan sequences. When scanning in RHI mode, the radar holds its azimuth angle constant but varies its elevation angle. One of the most popular scan strategies is to repeat crosswind direction horizon-to-horizon RHIs. This scan strategy provides 2D slices of clouds at all heights. Using the horizontal wind speed, time can be converted to a third length dimension and thus allow 3D mapping of cloud systems (Fig. 3.12). Another proposed scan strategy performs horizon-to-horizon RHIs at six azimuth angles spaced by 30°. This scan strategy can provide information about the cloud field anisotropy with respect to horizontal wind direc-

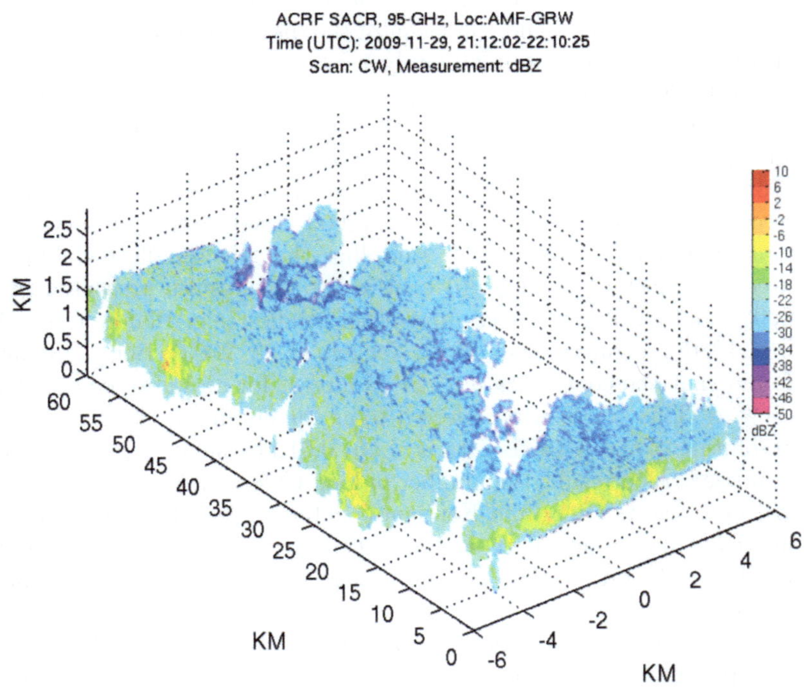

Fig. 3.12. Example of 3D reflectivity obtained from scanning radar observations.

tion or other atmospheric or topography parameters. Finally, a third scan strategy is based on a sequence of horizon-to-zenith RHI scans spaced in azimuth by 2°. The sequence of RHI scans covers a 90° sector usually centered to the wind direction at the cloud level. This scan strategy provides detailed imaging of cloud elements (layer or broken) and can be used to develop a reconstruction of the 3D cloud elements. The aforementioned scan strategy is repeated several times and thus provides 4D documentation (time sequences of 3D cloud elements).

The described sampling strategies are expected to create great opportunities for obtaining the cloud statistics as a function of viewing direction. One of these statistics is the decorrelation length scale for describing the rate at which two cloud layers decorrelate in a given direction as distance between layers increases. For example, studies based on the zenith-viewing ground-based radar observations demonstrated that this rate can be approximated by exponential fit (e.g., Hogan and Illingworth, 2000). To account for the SZA-associated changes of this scale, simple parameterizations have been suggested recently (e.g., Tompkins and Di Giuseppe, 2007). Similar parameterizations can be developed using the outlined multiangular cloud samplings. The latter can be very beneficial for estimation of the directional CF as well. The expected directional cloud statistics with long-term perspective are needed to evaluate and improve properly the Markovian models, and thus to provide basis for their potential widespread application.

3.6 Application of Markovian models

Applications of the Markovian approach to modeling atmospheric and climate processes fall into two broad and overlapping areas. The first area includes studies with focus on the *remote sensing* applications, such as the impact of the 3D radiative effects on the remote sensing of aerosol optical depth under partly cloudy conditions (e.g., Marshak et al., 2008), snow reflection function and albedo (Zhuravleva and Kokhanovsky, 2011), and vegetation canopies (Huang at al., 2008). All these studies assumed that spatial variability of the scattering/absorbing medium (cloud, snow, vegetation) can be parameterized by a few Markovian statistics (section 3.3), and these statistics can be obtained from observations (sections 3.4 and 3.5). The second area includes studies with focus on the climate-related applications where representation of small-scale variability of clouds in large-scale models requires a substantial improvement. It should be mentioned that typical model output includes the domain-averaged cloud water content (CWC), cloud base, and cloud thickness. Recently, Veron et al. (2009) suggested an approach for estimating the distribution of CCLs and the corresponding CF for cloud fields produced by a large-scale model. This approach is threshold-based and involves object identification. The estimation of CCLs and CF includes steps similar to those described in section 3.4. This approach was applied successfully to the CWC fields simulated by the Regional Atmospheric Modeling System (RAMS) within large domains ($\sim$1,000 $\times$ 1,000 km^2) with 3-km horizontal grid spacing. The simulations represent two storm periods (March 2–3 and March 7–8, 2000) observed over the DOE ARM SGP site. Veron et al. (2009) applied the derived statistics as input for the stochastic RT calculations and demonstrated high sensitivity of the domain-averaged radiative properties (e.g., downwelling shortwave irradiance) to the CWC threshold and spatial resolution. These results can be considered as valuable guidance for effective incorporation of subgrid cloud variability in large-scale RT computations.

Development of more accurate representation of small-scale variability of clouds within climate models can combine three basic concepts (e.g., Veron et al., 2009), such as (i) maturity and application of models with higher spatial resolution, (ii) development of a multi-scale modeling framework or a 'superparameterization' approach in which higher-resolution models are run inside each coarse scale grid box, and (iii) development of the so-called 'statistical cloud scheme' methodology when the subgrid statistics (e.g., variance, skewness) of quasi-conserved variables like total water content are explicitly predicted. Below we discuss advantages and limitations of these three approaches as well as linkages between them.

Refining the General Circulation Model (GCM) grid so that essential cloud features are better resolved is the most obvious path to improving the representation of cloud processes in climate simulations. However, only limited benefits can be achieved in the near future with this relatively straightforward approach due to limited computational resources. Since GCMs currently operate at horizontal resolutions on the order of tens to hundreds of kilometers, they can truly resolve only synoptic or meso-scale cloud field patterns, while numerous clouds of smaller horizontal extent remain unresolved. Consider trade wind cumulus clouds, for example, that cover significant areas of the World Ocean, thereby having a direct effect on both the radiative budget at the top of the atmosphere and the surface energy budget and for which half of the cloud fraction is contributed from clouds smaller than 2 km in diameter (Zhao and Di Girolamo, 2007). Horizontal resolution of $\sim$0.1 km is needed to explicitly model the dynamics of deep and shallow convective clouds, respectively. In many cases, vertical and temporal resolutions must be increased proportionally as well. Thus, to support a tenfold refinement in global model resolution, computational power must be amplified by a factor of 1,000. Although short global simulations with a resolution of under 10 km have been conducted, clearly, we are decades away from truly cloud-resolving century-long climate simulations.

A feasible alternative to the global cloud-resolving model has been recently introduced in the form of a Multiscale Modeling Framework (MMF), where a cloud-resolving model (CRM) is embedded into every grid column of a GCM to represent all of the essential cloud processes. Since the CRM does not occupy the whole horizontal extent of the GCM grid, it provides only statistical representation of the cloud fields for the local environment characteristic of the GCM column, making the MMF different from and computationally more efficient than the nested grid approach. Since the CRM does not occupy the whole horizontal extent of the GCM grid, it provides only statistical representation of the cloud fields for the local environment characteristic of the GCM column. This makes the MMF computationally more efficient than the nested grid approach, in which the higher-resolution domain has to fill completely the inside of the coarser grid. Over the last several years, the MMF, originally developed at the Colorado State University (Khairoutdinov et al., 2005; Randall et al., 2003) has been evaluated and expanded (Marchand et al., 2009; Ovtchinnikov et al., 2006; Pritchard and Somerville, 2009; Tao et al., 2009). Most recent modifications included extensive upgrades in aerosol and cloud microphysics treatments on both the GCM and CRM sides (Wang et al., 2011a). Although refined estimates of the aerosol effects on climate have already been obtained (Wang et al., 2011b), there are still scale gaps in the model that need to be bridged. The current CRM horizontal grid size of 4 km implies that significant variability in cloud properties still occurs at sub-grid scales, and it is essential to account for these processes that are not explicitly resolved. One cost-efficient way to do that is by introducing sub-grid scale parameterizations based on probability density functions for prognostic variables. This approach has been in development for CRM and GCM for a number of years (Golaz et al., 2002; Larson and Golaz, 2005; Sommeria and Deardorff, 1977; Tompkins, 2002), and now is being extended to the MMF.

Obtaining information on cloud properties at fine scales is challenging but represents only part of the problem of accounting for the small-scale variability in the large-scale models. Another equally important task is to determine the radiative effects of the inhomogeneous cloud fields. Although studies have shown that realistic distributions of cloud reflectances can be modeled when a 3D radiative transfer model is combined with high-resolution cloud simulations (e.g., Barker et al., 2003; Ovtchinnikov and Marchand, 2007), such RT calculations are strenuous and generally not suited for interactive radiative computations even in a small domain model. Stochastic RT models, such as those discussed earlier in the chapter, can be applied to obtain the needed ensemble- or domain-averaged radiative properties for the coarse grid without processing each cloud field realization individually.

3.7 Summary

In the Nobel lecture by the 1979 Laureate, Professor S. Weinberg wrote: 'Our job in physics is to see things simply, to understand a great many complicated phenomena in a unified way in terms of a few simple principles' (Weinberg, 1980, p. 1). The idea of a quite simple description of statistically dependent variables, which commonly occur in nature, was embedded in several papers of the Russian mathematician A. A. Markov (1856–1922) in the early 1900s. Markov's papers had purely impractical mathematical status and thus did not receive attention from the physical sciences at that time. Several decades later, the Markovian approach finally gained its deserved recognition, and scientific luminaries such as Chandrasekhar began to champion the Markovian approach and to demonstrate its great potential for understanding and describing physical phenomena. In this chapter, we have provided a few examples of use of the Markovian approach in diverse research areas, including astronomy and statistical physics, and emphasized its application to the stochastic RT in a cloudy atmosphere.

Cloud models with Markovian statistics, originally suggested by Pomraning and Titov and their colleagues, have captured attention from their beginning and created the foundation upon which a popular statistical view of RT under cloudy conditions is built. Pomraning and Titov arrived at these models independently by adapting the Markovian concept for the particle transport and the atmospheric science, respectively. The simplicity of these models is an appealing feature that makes them mathematically tractable and the obtaining of analytical solutions of the stochastic RT equation possible. Although these models are rather simple, they represent the complex cloud–radiation interactions quite realistically, capture vital features of the phenomena, and offer essential physical insights. Moreover, the generality of these models can potentially lead to their widespread applications in remote sensing (e.g., snow, vegetation, aerosol) and climate-related studies based on the probabilistic representation of small-scale variability of clouds in large-scale models.

The theoretical progress is ongoing, and the hindrances originate mostly from the fact that the development of cloud models and their detailed evaluations require appropriate observational constrains. The advanced observations of clouds and radiation for days with broken clouds have undergone significant changes. A good example is impressive data provided by next generation ARM cloud radars with Doppler and polarization capabilities. Moreover, these observations are moving rapidly to upgrade substantially their statistical offerings. In the face of sweeping changes of clouds with complex geometry and the corresponding strong variations of radiative properties, the long-term integrated observations will play an important role in helping us to shape reality in terms of a few statistics. 'Mathematical modeling, like painting [...], is an art, requiring proper balance between composition and the ability to convey a message.' (Syski, 1989, p. 377). The statistics obtained from these observations are expected to add bright and delightful colors to the painting.

Acknowledgments

This work has been supported by the National Aeronautics and Space Administration (NASA) through the Radiation Sciences Program (grant NNX08AI72G) and the Office of Biological and Environmental Research (OBER) of the US Department of Energy (DOE) as part of the Atmospheric System Research (ASR) Program. The data were provided by DOE as part of the Atmospheric Radiation Measurement (ARM) Climate Research Facility. The Pacific Northwest National Laboratory (PNNL) is operated by Battelle for the DOE under contract DE-AC06-76RLO 1830. We are grateful to Prof. R. Sanchez and Dr. A. Davis for their insightful comments and suggestions.

Dedication

This paper is dedicated to the memory of two distinguished scientists, Gerald Pomraning (1936–1999) and Georgii Titov (1948–1998), who sparked our interest in the exciting field of stochastic radiative transfer and shared with us their personal and scientific generosity.

Georgii Titov (left) and Gerald Pomraning (right). Courtesy of Lucia Levermore.

Appendix A: Markov processes and fields

This section reviews definitions of the Markov process and Markov random fields. We define $X_t = X(t)$ a sequence of random variables $\{ X_t \}_{t\geq 0} = \{ X_0, X_1, \dots \}$ taking values in a discrete set of states $S = \{s_0, s_1, \dots\}$. A sequence of states $S = \{s_0, s_1, \dots\}$ at successive times can be considered as a realization. In contrast to the simplest model, which assumes that random variables $\{ X_0, X_1, \dots \}$ are independent and identically distributed, a family of random processes describes statistical dependences between variables at nearby times, the so-called *Markov chains*. Specifically, a Markov chain is a random vector $\{ X_0, X_1, \dots \}$ for which the conditional probability satisfies the Markov property

$$P(X_{n+1} = s_j \mid X_m = s_i\,,\ \forall m \leq n) = P(X_{n+1} = s_j \mid X_n = s_i) \qquad \text{(A.1)}$$

In other words, the Markov property says that the conditional probability, given any number of events in the history of the process, depends only on the more recent one. The right-hand side of Eq. (A.1) is referred to as the *transition* probability $p_{ij}(t_n, t_{n+1})$ from s_i to s_j. If the transition probability depends on a time difference $p_{ij}(t_n, t_{n+1}) = p_{ij}(t_{n+1} - t_n)$, the Markov chain is said to be *time-homogeneous.* Below we consider homogeneous Markov chains unless specifically stated otherwise. If times t_n are defined as an arithmetic progression $t_n = n\Delta t$, then for each $\Delta t > 0$ the sequence of random variables $X(n\Delta t)$ represents a discrete-time Markov chain with a one-step transition probability $p_{ij} = p_{ij}(\Delta t)$.

The Poisson process is an example of a continuous-time Markov chain where $X(t)$ defines the total number of events that have occurred in the interval $[0, t)$. This counting process has stationary and independent increments with exponentially distributed intervals between consecutive events

$$p_{jk}(t - \tau) = P(X(t) = k \mid X(\tau) = j) = \frac{(\lambda(t - \tau))^{k-j}}{(k - j)!} \exp(-\lambda(t - \tau)) \qquad \text{(A.2)}$$

where $0 \leq \tau < t$, $j \leq k$ are non-negative integers and $k - j = 0,\ 1,\ 2, \dots$ The Poisson process provides an important framework for modeling multi-dimensional Markov random fields (section 3.3). Note that several natural processes, including radioactive disintegration, can be described by Poisson statistics (e.g., Gurney and Condon, 1929).

Markov random fields (MRF) are a natural extension of Markov chains on spatially distributed random variables. Markov chain formulation can be generalized to a random field setting in a straightforward manner, depending only on the definition of neighborhood structure. Specifically, a discrete MRF is a random vector $\{ X_0, X_1, \dots \}$ for which the conditional probability satisfies the local Markov property

$$P(X_n = s_j \mid X_m = s_i,\ n \neq m) = P(X_n = s_j \mid X_m = s_i,\ m \in \partial(n))\,, \qquad \text{(A.3)}$$

where $\partial(n)$ is a set of neighbors of n. For example, the first-order neighborhood includes two (left and right) or four (left, right, up, and down) nearest sites with the same edge for one- and two-dimensional models, respectively. The celebrated Hammersley–Clifford theorem states (Moussouris, 1974) that the MRF is a Gibbs

Random field (GRF) in relation to a neighborhood $\partial(n)$ if the probability distribution function is given by

$$P(X) = \frac{1}{Z} \exp(-U_{\partial(n)}(X)), \tag{A.4}$$

where Z is a normalization constant and $U_{\partial(n)}(X)$ is an energy function defining the interaction between neighboring sites ($s_i \sim s_j$). It should be mentioned that the corresponding conditional probability does not depend on Z and represents exponential family distributions, introduced as 'auto-models' by Besag (1974):

$$P(X_n = s_j \mid X_m = s_i,\ m \in \partial(n)) \quad \propto \quad \exp(-U_{\partial(n)}(X)), \tag{A.5}$$

Such conditional distributions are called the *local specification* of the MRF. The equivalence between MRFs and GRFs provided by the Hammersley–Clifford theorem suggests that probabilistic tools developed for the GRFs can be applied to MRFs as the need arises (and vice versa).

The Ising model (Brush, 1967), originally suggested in 1920, belongs to the family of GRFs and is identical to the binary MRFs. This model considers an ensemble of interacting magnetic particles (dipoles) arranged onto a regular lattice. These particles have fixed positions, and the spin of each particle is either up (+1) or down (−1). Thus, the random binary variable X has only two possible states $S = \{-1, +1\}$. Each dipole interacts with its nearest neighbor (denoted '$s_i \sim s_j$' as a set). For the simplest one-dimensional model, the particles are arranged in a chain, and the corresponding energy function is defined as (Cipra, 1987)

$$U_{\partial(n)}(X) = \beta \sum_{s_i \sim s_j} s_i s_j, \tag{A.6}$$

where the sum is over all pairs of nearest neighbors and the parameter β determines the strength of the spin interaction. Eq. (A.6) defines the simplest case when the external magnetic field is turned off. More realistic versions of the Ising model and corresponding computational methods are described thoroughly in several articles and textbooks (e.g., Newman and Barkema, 1999). It is interesting that nearly thirty years passed between the creation of the Ising model and its world-wide recognition.

Surveys of a wide variety of algorithms for simulating MRFs are given in several textbooks (e.g., Ross, 2002; Stoyan and Stoyan, 1994), where sound theoretical basis and valuable practical guidance are provided. Here we mention only two of them, which have been the subject of considerable attention and are well-known for performing simulation tasks. The first algorithm is the so-called Gibbs sampler, which is a special case of the Metropolis–Hastings algorithm (Ross, 2002). The Gibbs sampler constructs a Markov chain (having $P(X)$ as its limiting unconditional distribution) by using the *local specification.* The second algorithm is based on simulating Poisson fields (Ross, 2002; Stoyan and Stoyan, 1994). The simulation includes generalization of (i) the number of points within a given area and (ii) the positions of these points. For example, for a rectangular-shape area, these points can be considered as x- and y-coordinates where the distances between subsequent points are independent and exponentially distributed (for each axis). Both homogeneous and non-homogeneous Poisson fields have been used to provide an initial spatial structure for simulations of more complicated models such as the Gibbs field (the so-called Poisson hard core field) and the Boolean models with a binary set of states $S = \{0, 1\}$. We applied a Boolean model (Kassianov et al., 2005c) to simulate cumulus clouds (Fig. 3.2, left panel).

Appendix B: Functions associated with 'direct-beam' exponential components and asymptotic cases

Functions $\Psi(s)$, $\eta(s)$, and $\varphi(s)$ have two exponential terms weighted with coefficients D_i and C_i, $i = 1, 2$ (Table B.1). These coefficients and parameters χ_i depend on cloud statistics. To get a better idea of the flavor of these functions, we included two asymptotic cases (Table B.2) where these functions have the simplest forms. The first asymptotic case describes a situation when the averaged cloud chord length (CCL) is very large compared to the PMFP. This case reduces the 3D RT problem to its 1D counterpart where the averaged radiative properties are computed as a sum of clear-sky and cloudy-sky components weighted with CF. Contrary to the first case, the second one describes a situation when the averaged CCL is very small compared to the photon mean-free-path (PMFP). This case reduces the 3D RT problem to its 1D counterpart with averaged extinction coefficient $\sigma_1 p_1$. Note, this case is well-known as the 'atomic mix limit' in the particle transport theory (e.g., Pomraning, 1991).

The process by which the life history of a photon is simulated by MC method includes three basic steps that mimic the physical phenomena involved in its travel (e.g., Marchuck et al., 1980). These steps determine (i) a collision point, (ii) type of interaction (e.g., clouds or surface), and (iii) result of interaction (absorption or scattering), respectively. The only difference between traditional MC method and its 'stochastic' cousin is the selection of a collision point. For the traditional MC, a collision point is selected from exponential probability density, which is $\sigma \exp(-\sigma s)$ for a homogeneous medium. For the MC method applied to Eqs. (12–14), a collision point is selected from the probability densities $\Psi(s)$ and $\eta(s)$, respectively (e.g., Kasyanov and Titov, 1994). Thus, we can use available MC algorithms developed for important applications (with only minor changes) for numerical solution of Eqs. (3.12) to (3.14).

Before we illustrate a simulation of the free path length of a photon, it should be emphasized that there is a simple relationship between the coefficients (Table B.1)

$$\begin{aligned} D_1 + D_2 &= 1 \\ C_1 + C_2 &= 1\,. \end{aligned} \tag{B.1}$$

They can be considered as probabilities that determine the corresponding probability density. Therefore, the free path length of a photon can be simulated by the composition method (e.g., Ross, 2002), where the distribution function of interest is defined as a mixture of density functions

$$\eta(s) = \begin{cases} \chi_1 \exp(-\chi_1 s), & \alpha < D_1 \\ \chi_2 \exp(-\chi_2 s), & \alpha \geq D_1 \end{cases} \tag{B.2}$$

and

$$\Psi(s) = \begin{cases} \chi_1 \exp(-\chi_1 s), & \alpha < C_1 \\ \chi_2 \exp(-\chi_2 s), & \alpha \geq C_1 \, , \end{cases} \tag{B.3}$$

where α is random number uniformly distributed over the interval (0, 1).Thus, simulation of the free path length of a photon from these functions is straightforward and includes two basic steps. First, we select the probability (e.g., D_1 or C_1). Then the free path length of a photon is sampled from the corresponding probability density.

Table B.1. Functions from Eqs. (3.12) to (3.14) and their deterministic counterparts

Function	Stochastic RTE	Deterministic RTE
$\eta(s)$	$\sum_{i=1}^{2} D_i \, \chi_i \, \exp\{-\chi_i s\}$	$\sigma \exp(-\sigma s)$
$\Psi(s)$	$\sum_{i=1}^{2} C_i \, \chi_i \, \exp\{-\chi_i s\}$	$\sigma \exp(-\sigma s)$
$\varphi(s)$	$\sum_{i=1}^{2} D_i \exp\{-\chi_i s\}$	$\exp(-\sigma s)$

Table B.2. Functions from Eqs. (3.12) to (3.14): Asymptotic cases

Function	CCL $\gg$ PMFP	CCL $\ll$ PMFP
$\eta(s)$	$\sigma_1 \exp(-\sigma_1 s)$	$\sigma_1 p_1 \exp(-\sigma_1 p_1 s)$
$\Psi(s)$	$p_1 \sigma_1 \exp(-\sigma_1 s)$	$\sigma_1 p_1 \exp(-\sigma_1 p_1 s)$
$\varphi(s)$	$\exp(-\sigma_1 s)$	$\exp(-\sigma_1 p_1 s)$

Appendix C: Estimation of cloud statistics

For a given 1D cross-section (e.g. kth layer, along x-direction), the unconditional p_1 and conditional P_{11} probabilities are estimated as

$$\widehat{p}_1 = \frac{n_1}{n_1 + n_0} \, , \tag{C.1}$$

and

$$\widehat{P}_{11} = \begin{cases} \dfrac{n_{11}}{n_1}, & n_1 > 0 \\ 0, & n_1 = 0 \end{cases} \tag{C.2}$$

where n_{11} is the number of observed transitions from a cloudy pixel to the next cloudy pixel, and n_1 and n_0 are the total number cloudy and clear-sky pixels, respectively. Viewed in this way, conditional probabilities P_{11} can be estimated for two adjusted layers (e.g., nadir direction).

Let us consider a 1D realization with four pixels (Fig. 3.13). Two cloudy pixels can be distributed differently along the x-axis (Cases A and B). It follows from Eq. (C.1) that the cloud fraction p is 0.5 for both cases. The number of transitions from cloudy pixel to next cloudy pixel is 0 and 1 for case A and case B, receptively. The corresponding conditional probabilities follow immediately from Eq. (C.2): they are 0 and 0.5. For two adjusted layers with different vertical distribution of cloudy pixels (Fig. 3.14), they become 0 and 1, respectively.

We derive parameters A_k for each kth layer using obtained probabilities and Eq. (3.15)

$$A_k = -\ln\left(\frac{P_{11}(\mathbf{r}', \mathbf{r}) - p_{1k}}{1 - p_{1k}}\right) \Big/ |\mathbf{r} - \mathbf{r}'| , \tag{C.3}$$

Similarly, other parameters A_{km}, A_{kl} (upward and downward directions) are obtained for each kth layer by using obtained probabilities and Eq. (3.16). For zenith and nadir directions, the conditional probability $P_{11}(\mathbf{r}', \mathbf{r}^*) = 1$, and one can derive A_{km} (nadir direction) as

$$A_{km} = -\ln\left(\frac{P_{11}(\mathbf{r}', \mathbf{r}) - p_{1m}}{1 - p_{1m}}\right) \Big/ |\mathbf{r} - \mathbf{r}^*| , \tag{C.4}$$

In a similar way, A_{kl} (zenith direction) can be calculated. If $|\mathbf{r} - \mathbf{r}'| = \Delta x$ or $|\mathbf{r} - \mathbf{r}^*| = \Delta z$ where Δx and Δz are horizontal and vertical resolution of data/model output, then P_{11} can be considered as a one-step transition probability and is interpreted as the probability that a cloudy pixel is followed by cloudy pixel in one spatial step.

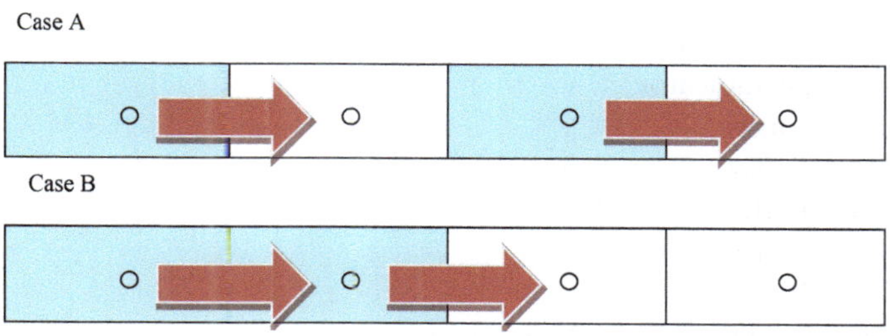

Fig. 3.13. Graphical illustration of cloudy-sky realizations. These realizations include cloudy (blue) and clear-sky (white) pixels. One-step transitions from a cloudy pixel to the next pixel are indicated by arrows.

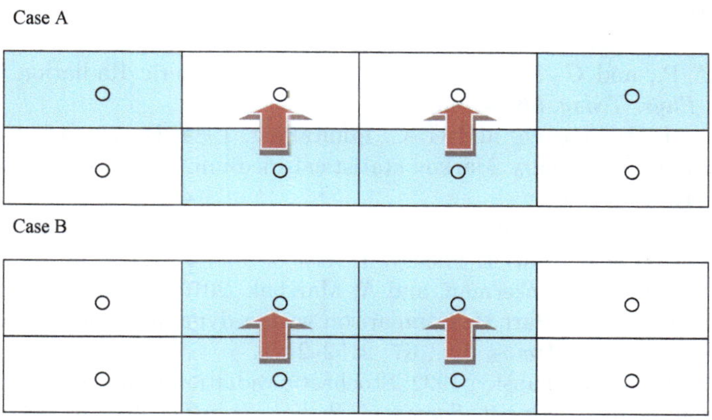

Fig. 3.14. The same as in Fig. 3.13, except that this diagram illustrates different relationships between two adjusted layers (nadir direction).

List of abbreviations

1D/3D/4D – one-/ three-/ four-dimensional
ARM – Atmospheric Radiation Measurement
ARSCL – Active Remote Sensing of Clouds
CALIOP – Cloud-Aerosol Lidar with Orthogonal Polarization
CCL – Cloud Chord Length
CF – Cloud Fraction
CRM – Cloud-Resolving Model
CWC – Cloud Water Content
ETA – Effective Thickness Approximation
FSC – Fractional Sky Cover
FWC – Fair-Weather Cumuli
GCM – General Circulation Model
GRF – Gibbs Random Field
LP – Levermore–Pomraning
MC – Monte Carlo
MFRSR – Multi-Filter Rotating Shadowband Radiometer
MMF – Multiscale Modeling Framework
MODIS – Moderate Resolution Imaging Spectroradiometer
MRF – Markov Random Field
NSA – North Slope of Alaska
PMFP – Photon Mean-Free-Path
RHI – Range-Height Indicator
RT – Radiative Transfer
RTE – Radiative Transfer Equation
SACRs – Scanning ARM Cloud Radars
SGP – Southern Great Plains
SZA – Solar Zenith Angle
TIPA – Tilted Independent Pixel Approximation
TWP – Tropical Western Pacific

References

Ackerman, T. P., and G. M. Stokes, 2003: The Atmospheric Radiation Measurement program. *Phys. Today*, **56**, 38–44.

Adams, M. L., E. W. Larsen, and G. C. Pomraning, 1989: Benchmark results for particle transport in a binary Markov statistical medium. *J. Quant. Spectrosc. Radiat. Transfer*, **42**, 253–266.

Alexandrov, M. D., A. Marshak, and A. S. Ackerman, 2010a: Cellular statistical models of broken cloud fields. Part I: Theory. *J. Atmos. Sci.*, **67**, 2125–2151.

Alexandrov, M. D., A. S. Ackerman, and A. Marshak, 2010b: Cellular statistical models of broken cloud fields. Part II: Comparison with a dynamical model and statistics of diverse ensembles. *J. Atmos. Sci.*, **67**, 2152–2170.

Anisimov, O., and L. Fukshansky, 1992: Stochastic radiation in macroheterogeneous random optical media. *J. Quant. Spectrosc. Radiat. Transfer*, **48**, 169–186.

Avaste, O. A., and G. M. Vainikko, 1974: Solar radiation transfer in broken clouds. *Izv. Acad. Sci. USSR Atmos. Oceanic Phys.*, **10**, 1054–1061.

Barker, H. W., M. P. Jerg, T. Wehr, S. Kato, D. P. Donovan, and R. J. Hogan, 2011: A 3D cloud-construction algorithm for the EarthCARE satellite mission. *Quart. J. Roy. Meteor. Soc.*, **137**: n/a. doi: 10.1002/qj.824.

Barker, H. W., 1996: A parameterization for computing grid-averaged solar fluxes for inhomogeneous marine boundary layer clouds: I. Methodology and homogeneous biases. *J. Atmos. Sci.*, **53**,2289–2303.

Barker, H. W., C. Pavloski, M. Ovtchinnikov, and E. E. Clothiaux, 2004: Assessing a cloud optical depth retrieval algorithm with model-generated data and the frozen turbulence assumption. *J. Atmos. Sci.*, **61**, 2951–2956.

Barker, H. W., G. L. Stephens, and Q. Fu, 1999: The sensitivity of domain averaged solar fluxes to assumptions about cloud geometry. *Quart. J. Roy. Meteor. Soc.*, **125**, 2127–2152.

Barker, H. W., R. K. Goldstein, and D. E. Stevens, 2003: placeMonte Carlo simulation of solar reflectances for cloudy atmospheres. *J. Atmos. Sci.*, **60**, 1881–1894.

Basharin, G., A. Langville, and V. Naumov, 2004: The life and work of A. A. Markov. *Linear Algebra Appl.*, **386**, 3–26.

Bauml, G., Chlond A., and placeE. Roeckner, 2004: Estimating the PPH-bias for simulations of convective and stratiform clouds. *Atmos. Res.*, **72**, 317–328.

Benner, T. C., and J. A. Curry, 1998: Characteristics of small tropical cumulus clouds and their impact on the environment. *J. Geophys. Res.*, **103**, 28753–28767.

Berg, L. K., and E. I. Kassianov, 2008: Temporal variability of fair-weather cumulus statistics at the ACRF SGP site. *J. Climate*, **21**, 3344–3358.

Berg, L. K., C. M. Berkowitz, J. A. Ogren, C. A. Hostetler, and co-authors, 2009: Overview of the cumulus humilis aerosol processing study (CHAPS). *Bull. Amer. Meteor. Soc.*, **90**, 1653–1667.

Berg, L. K., E. I. Kassianov, C. N. Long, and D. L. Mills Jr., 2011: Surface summertime radiative forcing by shallow cumuli at the Atmospheric Radiation Measurement Southern Great Plains site. *J. Geophys. Res.*, **116**, D01202, doi:10.1029/2010JD014593.

Besag, J., 1974: Spatial interaction and the statistical analysis of lattice systems. With discussion. *J. Royal Statist. Soc.*, **36**, 192–326.

Borovoi, A., 2002: On the extinction of radiation by a homogeneous but spatially correlated *random medium*: comment. *J. Opt. Soc. Am. A*, **19**, 2517–2520.

Brush, S., 1967: History of the Lenz–Ising model. *Rev. Modern Phys.*, **39**, 883–895.

Buras, R., and B. Mayer, 2011: Efficient unbiased variance reduction techniques for placeMonte Carlo simulations of radiative transfer in cloudy atmospheres: the solution. *J. Quant. Spectrosc. Radiat. Transfer*, **112**, 434–447.

Byrne, N., 2005: 3D radiative transfer in stochastic media. *3D Radiative Transfer in Cloudy Atmospheres*, Marshak A., and CityDavis A. B. (Eds.). Springer, placeStateNew York.

Cahalan, R. F., 1994: Bounded cascade clouds: Albedo and effective thickness. *Nonlinear Proc. in Geophys.,* **1**, 156–167.

Cahalan, R. F., and J. H. Joseph, 1989: Fractal statistics of cloud fields. *Mon. Wea. Rev.*, **117**, 261–272.

Cahalan, R. F., W. Ridgway, W. J. Wiscombe, S. Gollmer, and Harshvardhan, 1994: Independent pixel and Monte-Carlo estimates of stratocumulus albedo. *J. Atmos. Sci.*, **51**, 3776–3790.

Cairns, B., A. A. Lacis, and B. E. Carlson, 2000: Absorption within inhomogeneous clouds and its parameterization in general circulation models. *J. Atmos. Sci.*, **57**, 700–714.

Cassell, J. S., and M. M. R. Williams, 2008: An approximate method for solving radiation and neutron transport problems in spatially stochastic media. Annals of Nuclear Energy, **35**, 790–803.

Chandrasekhar, S., 1943: Stochastic problems in physics and astronomy. *Rev. Modern Phys.*, **15**, 2–89.

Chib, S. and E. Greenberg, 1995: Understanding the Metropolis–Hastings Algorithm. *Am. Statist*, **49**, 327–335.

Cipra, B. A., 1987: An introduction to the Ising model. *Am. Math. Monthly*, **94**, 937–959.

Clothiaux, E. E., T. P. Ackerman, G. Mace, K. P. Moran, R. T. Marchand, M. Miller, and B. E. Martner, 2000: Objective determination of cloud heights and radar reflectivities using a combination of active remote sensors at the ARM CART sites. *J. Appl. Meteor.*, **39**, 645–665.

Davis, A. B., and A. Marshak, 2004: Photon propagation in heterogeneous optical media with spatial correlations: enhanced mean-free-paths and wider-than-exponential free-path distributions. *J. Quant. Spectrosc. Radiat. Transfer*, **84**, 3–34.

Davis, A. B., and A. Marshak, 2010: Solar radiation transport in the cloudy atmosphere: a 3D perspective on observations and climate impacts. *Reports on Progress in Physics*, **73**, 026801, doi:10.1088/0034-4885/73/2/026801.

Davis, A., P. Gabriel, S. Lovejoy, D. Schertzer, and G. Austin, 1990: Discrete angle radiative transfer. III. Numerical results and meteorological applications. *J. Geophys. Res.*, **95**, 11729–11742.

de Roode, S. R., and A. Los, 2008: The effect of temperature and humidity fluctuations on the liquid water path of non-precipitating closed-cell stratocumulus. *Q. J. Roy. Meteor. Soc.*, **134**, 403–416, doi:10.1002/qj.222.

Deguchi, S., and W. D. Watson, 1988: An analytic treatment of gravitational microlensing for sources of finite size at large optical depths. *Astrophysical Journal*, **335**, 67–73.

Del Genio, A. D., M. S. Yao, W. Kovari, and K. K. W. Lo, 1996: A prognostic cloud water parameterization for global climate models. *J. Climate*, **9**, 270–304.

Di Girolamo, L., and R. Davies, 1997: Cloud fraction errors caused by finite resolution measurements. *J. Geophys. Res.*, **102**, 1739–1756.

Diaconis, P., 2009: The Markov chain Monte Carlo revolution. *Bull. Am. Math. Soc.*, **46**, 179–205.

El-Wakil, S. A., A. R. Degheidy, and M. Sallah, 2004: Time-dependent radiation transfer in a semi-infinite stochastic medium with Rayleigh scattering. *J. Quant. Spectrosc. Radiat. Transfer*, **85**, 13–24.

Emde, C., R. Buras, and B. Mayer, 2011: ALIS: an efficient method to compute high spectral resolution polarized solar radiances using the Monte Carlo approach. *J. Quant. Spectrosc. Radiat. Transfer*, doi:10. 1016/j.jqsrt.2011.03.018., in press.

Evans, K. F., 1993: A general solution for stochastic radiative transfer. *Geophys. Res. Lett.*, **20**, 2075–2078.

Evans, K. F., and W. J. Wiscombe, 2004: An algorithm for generating stochastic cloud fields from radar profile statistics. *Atmos. Res.*, **72**, 263–289.

Feller, W., 1950: *An Introduction to Probability Theory and its Applications.* John Wiley & Sons, New York.

Foster, M. J., and D. E. Veron, 2008: Evaluating the stochastic approach to shortwave radiative transfer in the tropical western Pacific. *J. Geophys. Res.*, **113**, D22205, doi:10.1029/2007JD009581.

Foster, M. J., and D. E. Veron, 2010: Employing cluster analysis to detect significant cloud 3D RT effect indicators. *J. Atmos. Sci.*, **67**, 2226–2239.

Gabriel, P., and K. F. Evans, 1996: Simple radiative transfer methods for calculating domain-averaged solar fluxes in inhomogeneous clouds. *J. Atmos. Sci.*, **53**, 858–877.

Gabriel, P., S. Lovejoy, A. Davis., D. Schertzer, and G. Austin, 1990: Discrete angle radiative transfer, 2. Renormalization approach for homogeneous and fractal clouds. *J. Geophys. Res.*, **95**, 11717–11728.

Golaz, J. C., V. E. Larson, and W. R. Cotton, 2002: A PDF-based model for boundary layer clouds. Part I: Method and model description. *J. Atmos. Sci.*, **59**, 3540–3551.

Gurney, R.W., and E.U. Condon, 1929: Quantum mechanics and radioactive disintegration. *Physical Review*, **33**, 127-139.

Han, D., and R. G. Ellingson, 2000: An experimental technique for testing the validity of cumulus cloud parameterizations for longwave radiation calculations. *J. Appl. Meteor.*, **39**, 1147–1159.

CityHastings, W., 1970: Monte Carlo sampling methods using Markov chains and their application. *Biometrika*, **57**, 97–109.

Hitchcock, D. B., 2003: A history of the Metropolis–Hastings algorithm. *Am. Statist.*, **57**, 254–257.

Hogan, R. J., and A. J. Illingworth 2000: Deriving cloud overlap statistics from radar. *Q. J. Roy. Meteor. Soc.*, **126**, 2903–2909.

Hozumi, K., T. Harimaya, and C. Magono, 1982: The size distribution of cumulus clouds as a function of cloud amount. *J. Meteor. Soc. Japan*, **60**, 691–699.

Huang, C., Y. Knyazikhin, W. Wang, D. W. Deering, P. Stenberg, N. Shabanov, B. Tan, and R. B. Myneni, 2008: Stochastic transport theory for investigating the three-dimensional canopy structure from space measurements. *Remote Sensing of Environment*, **112**, 35–50.

Kassianov, E. I., 2003: Stochastic radiative transfer in multilayer broken clouds. Part I: Markovian approach. *J. Quant. Spectrosc. Radiat. Transfer*, **77**, 373–393.

Kassianov, E., T. Ackerman, R. Marchand, and M. Ovtchinnikov, 2003: Satellite multiangle cumulus geometry retrieval: Case study. *J. Geophys. Res.*, **108**, 4117, doi:10.1029/2002JD002350.

Kassianov, E.., T. P. Ackerman, R. T. Marchand, and M. Ovtchinnikov, 2003: Stochastic radiative transfer in multilayer broken clouds. Part II: Validation tests. *J. Quant. Spectrosc. Radiat. Transfer*, **77**, 395–416.

Kassianov, E.., T. Ackerman, and P. Kollias, 2005a: The role of cloud-scale resolution on radiative properties of oceanic cumulus clouds. *J. Quant. Spectrosc. Radiat. Transfer*, **91**, 211–226.

Kassianov, E., C. N. Long, and M. Ovtchinnikov, 2005b: Cloud sky cover versus cloud fraction: whole-sky simulations and observations. *J. Appl. Meteorol.*, **44**, 86–98.

Kassianov, E., C. N. Long, and J. E. Christy, 2005c: Cloud-base-height estimation from paired ground-based hemispherical observations. *Applied Meteorology*, **44**, 1221–1233.

Kassianov, E., M. Ovchinnikov, L. K. Berg, and co-authors, 2009: Retrieval of aerosol optical depth in vicinity of broken clouds from reflectance ratios: sensitivity study. *J. Quant. Spectrosc. Radiat. Transfer*, **110**, 1677–1689.

Kassianov, E., J. Barnard, L. K. Berg C. Flynn, and C. N. Long, 2011a: Sky cover from MFRSR observations: cumulus clouds. *Atmos. Meas. Tech. Discuss.*, **4**, 715–735.

Kassianov, E., J. Barnard, L. K. Berg, C. N. Long, and C. Flynn, 2011b: Shortwave spectral radiative forcing of cumulus clouds from surface observations. *Geophys. Res. Lett.*, **38**, L07801, doi:10.1029/2010GL046282.

Kasyanov, E.I., and G.A. Titov, 1994: Brightness field of reflected solar radiation under conditions of broken clouds. Part I. Algorithm of calculation, *Atmos. Oceanic Opt.*, **7**, 629-633.

Kasyanov, E.I., G.A. Titov, and D.A. Zimin, 1994: Brightness field of reflected solar radiation under conditions of broken clouds. Part II. Calculated results, *Atmos. Oceanic Opt.*, **7**, 634-639.

Kato, S., S. Sun-Mack, W. F. Miller, F G. Rose, Y. Chen, P. Minnis, and B. A. Wielicki, 2010: Relationships among cloud occurrence frequency, overlap, and effective thickness derived from CALIPSO and CloudSat merged cloud vertical profiles. *J. Geophys. Res.*, **115**, D00H28, doi:10.1029/2009JD012277.

Kaufman, Y. J., and I. Koren, 2006: Smoke and pollution aerosol effect on cloud cover. *Science*, **313**, 655–658.

Kendall, W., F. Liang, J. S. Wang (Eds.), 2005: *Markov Chain Monte Carlo: Innovations and Applications*, World Scientific Publishing.

Khairoutdinov, M., D. Randall, and C. DeMott, 2005: Simulations of the atmospheric general circulation using a cloud-resolving model as a superparameterization of physical processes. *J. Atmos. Sci.*, **62**, 2136–2154.

Kollias, P., G. Tselioudis, and B. A. Albrecht, 2007: Cloud climatology at the Southern Great Plains and the layer structure, drizzle, and atmospheric modes of continental stratus. *J. Geophys. Res.*, **112**, D09116, doi:10.1029/2006JD007307.

Lane, D. E., K. Goris, and R. C. J. Somerville, 2002: Radiative transfer through broken clouds: Observations and model validation. *J. Climate*, **15**, 2921–2933.

Laplace, P. S., 1902: *Philosophical Essay on Probabilities*, (Trans. by F. W. Truscott and F. L. Emory), New York: John Wiley & Sons; London: Chapman & Hall.

Larson, V. E., and J. C. Golaz, 2005: Using probability density functions to derive consistent closure relationships among higher-order moments, *Mon. Weather Rev.*, **133**, 1023–1042.

Levermore, C. D., G. C. Pomraning, D. L. Sanzo, and J. Wong, 1986: Linear transport theory in a random medium. *J. Math. Phys.*, **27**, 2526–2536.

Levermore, C. D., J. Wong, and G. C. Pomraning, 1988: Renewal theory for transport processes in binary statistical mixtures. *J. Math. Phys.*, **29**, 995–1004.

Liou, K. N., 2002: *An Introduction to Atmospheric Radiation.* Academic Press, New York.

Long, C. N., T. P. Ackerman, K. L. Gaustad, and J. N. S. Cole, 2006: Estimation of fractional sky cover from broadband shortwave radiometer measurements. *J. Geophys. Res.*, **111**, D11204, doi:10.1029/2005JD006475.

Malvagi, F., N. Byrne, G. C. Pomraning, and R. C. J. Somerville, 1993: Stochastic radiative transfer in a partially cloudy atmosphere. *J. Atmos. Sci.*, **50**, 2146–2158.

Marchand, R., N. Beagley, and T. P. Ackerman, 2009: Evaluation of hydrometeor occurrence profiles in the multiscale modeling framework climate model using atmospheric classification. *Journal of Climate*, **22**, 4557–4573.

Marchuk, G. I., G. A. Mikhailov, M. A. Nazaraliev, R. A. Darbinjan, B. A. Kargin, and B. S. Elepov, 1980: *The Monte Carlo Methods in Atmospheric Optics.* Springer-Verlag, Heidelberg.

Markov, A. A., 1907: Issledovanie zamechatel'nogo sluchaya zavisimyh ispytanij, *Izv. Akad. Nauk*, SPb, VI, **1**, 61–80 (Translated into French, Recherches sur un cas remarquable d'epreuves dependantes, Acta Math., Stockholm 33 (1910) 87–104).

Markov, A. A., 1913: Primer statisticheskogo issledovaniya nad tekstom "Evgeniya Onegina", illyustriruyuschij svyaz' ispytanij v cep', *Izv. Akad. Nauk*, SPb, VI, **7**, 153–162.

Marshak, A., and A. B. Davis (Eds.), 2005: *3D Radiative Transfer in Cloudy Atmospheres.* Springer, New York.

Marshak, A., G. Wen, J. Coakley, L. Remer, N. G. Loeb, and R. F. Cahalan, 2008: A simple model for the cloud adjacency effect and the apparent bluing of aerosols near clouds. *J. Geophys. Res.*, **113**, D14S17, doi:10.1029/2007JD009196.

Metropolis, N., A. W. Rosenbluth, M. N. Rosenbluth, A. H. Teller, and E. Teller, 1953: Equations of state calculations by fast computing machines. *J. Chem. Phys.*, **21**, 1087–1091.

Min, Q., T. Wang, C. N. Long, and M. Duan, 2008: Estimating fractional sky cover from spectral measurements. *J. Geophys. Res.*, **113**, D20208, doi:10.1029/2008JD010278.

Mishchenko, M. I., 2009: Gustav Mie and the fundamental concept of electromagnetic scattering by particles: a perspective. *J. Quant. Spectrosc. Radiat. Transfer*, **110**, 1210–1222.

Mishchenko, M.I., 2011: Directional radiometry and radiative transfer: a new paradigm, *J. Quant. Spectrosc. Radiat. Transfer*, (in press).

Mishchenko, M. I., G. Videen, V. K. Rosenbush, and Y. Yatskiv (Eds.), 2011: *Polarimetric Detection, Characterization, and Remote Sensing*, Springer, New York.

Moussouris, J., 1974: Gibbs and Markov systems with constraints. *J. Stat. Phys.*, **10**, 11–33.

Neggers, R. A. J., H. J. J. Jonker, and A. P. Siebesma, 2003: Size statistics of cumulus cloud populations in large-eddy simulation. *J. Atmos. Sci.*, **60**, 1060–1074.

Newman, M., and G. Barkema, 1999: *Monte Carlo Methods in Statistical Physics*. Clarendon Press, Oxford.

Oreopoulos, L., and M. Khairoutdinov, 2003: Overlap properties of clouds generated by a cloud-resolving model. *J. Geophys. Res.*, **108**, 4479, doi:10.1029/2002JD003329.

Ovtchinnikov, M., and R. T. Marchand, 2007: Cloud model evaluation using radiometric measurements from the airborne multiangle imaging spectroradiometer (AirMISR). *Remote Sensing of Environment*, **107**, 185–193.

Ovtchinnikov, M., T. Ackerman, R. Marchand, and M. Khairoutdinov, 2006: Evaluation of the multiscale modeling framework using data from the atmospheric radiation measurement program. *Journal of Climate*, **19**, 1716–1729.

Palmer, J.M., 1993: Getting intense on intensity. *Metrologia*, **30**, 371–372.

Petty, G., 2002: Area-average solar radiative transfer in three-dimensionally inhomogeneous clouds: The independently scattering cloudlet model. *J. Atmos. Sci.*, **59**, 2910–2929.

Plank, V. G., 1969: The size distribution of cumulus clouds in representative Florida populations. *J. Appl. Meteor.*, **8**, 46–67.

Pomraning, G. C., 1998: Radiative transfer and transport phenomena in stochastic media. *Int. J. Engineering Sci.*, **36**, 1595–1621.

Pomraning, G., 1991: *Linear Kinetic Theory and Particle Transport in Stochastic Mixtures*. World Scientific Publishing.

Pomraning, G., 1999: Cloud–radiation interactions: The Titov and other models. *J. Atmos. Ocean Opt.*, **12**, 204–210.

Prigarin, S. M., and A. Marshak, 2009: A simple stochastic model for generating broken cloud optical depth and cloud-top height fields. *J. Atmos. Sci.*, **66**, 92–104.

Pritchard, M. S., and R. C. J. Somerville, 2009: Empirical orthogonal function analysis of the diurnal cycle of precipitation in a multi-scale climate model. *Geophys. Res. Lett.*, **36**, L05812, doi:10.1029/2008GL036964.

Randall, D., M. Khairoutdinov, A. Arakawa, and W. Grabowski, 2003: Breaking the cloud parameterization deadlock. *Bull. Amer. Meteorol. Soc.*, **84**, 1547–1564.

Rodts, S. M. A., P. G. Duynkerke, and H. J. J. Jonker, 2003: Size distributions and dynamical properties of shallow cumulus clouds from aircraft observations and satellite data. *J. Atmos. Sci.*, **60**, 1895–1912.

Ross, S. M., 2002: *Simulation*. Academic Press, New York.

Sanchez, R., 1989: Linear kinetic theory in stochastic media, *J. Math. Physics*, **30**, 2948-2511.

Sanchez, R., 2008: A critique of the modified Levermore–Pomraning equations. *Ann. Nucl. Energy*, **35**, 446–457.

Sanchez, R., and G. C. Pomraning, 1991: A statistical analysis of the double heterogeneity problem. *Ann. Nucl. Energy*, **18**, 371–395.

Sengupta, S. K., M. S. Navar, D. W. Chen, R. M. Welch, and T. A. Berendes, 1990: Cumulus cloud field morphology and spatial patterns derived from high spatial resolution Landsat imagery. *J. Appl. Meteor.*, **29**, 1245–1267.

Sommeria, G., and J. W. Deardorff, 1977: Subgrid-scale condensation in models of non-precipitating clouds. *J. Atmos. Sci.*, **34**, 344–355.

Stephens, G. L., 1988: Radiative transfer through arbitrarily shaped optical media. Part II: Group theory and simple closures. *J. Atmos. Sci.*, **45**, 1837–1849.

Stephens, G. L., P. Gabriel, S. C. Tsay, 1991: Statistical radiative transport in one-dimensional media and its application to the terrestrial atmosphere. *J. Transport Th. Statist. Phys.*, **20**, 139–175.

Stephens, G., and co-authors, 2002: The CloudSat mission and the A-train. *Bull. Am. Meteorol. Soc.*, **83**, 1771–1790.

Stokes, G. M., and S. E. Schwartz, 1994: The atmospheric radiation measurement (ARM) program: Programmatic background and design of the Cloud and Radiation Test Bed. *Bull. Amer. Meteor. Soc.*, **75**, 1201–1221.

Stoyan, D., and H. Stoyan, 1994: *Fractals, Random Shapes and Point Fields.* John Wiley & Sons, New York.

Su, B., and G. C. Pomraning, 1994: A stochastic description of a broken cloud field. *J. Atmos. Sci.*, **51**, 1969–1977.

Syski, R., 1989: *Random Processes.* Marcel Dekker, New York.

Tao, W. K., D. Anderson, J. Chern, J. Entin, and co-authors, 2009: The Goddard multi-scale modeling system with unified physics. *Ann. Geophys.*, **27**, 3055–3064.

Taylor, P. C., and R. G. Ellingson, 2008: A study of the probability of clear line of sight through single-layer cumulus cloud fields in the Tropical Western Pacific. *J. Atmos. Sci.*, **65**, 3497–3512, doi: 10.1175/2008JAS2620.1

Titov, G., 1990: Statistical description of radiation transfer in clouds. *J. Atmos. Sci.*, **47**, 24–38.

Titov, G. A., T. B. Zhuravleva, and V. E. Zuev, 1997: Mean radiation fluxes in the near-IR spectral range: Algorithms for calculation. *J. Geophys. Res.*, **102**, 1819–1832.

Tompkins, A. M., 2002: A prognostic parameterization for the subgrid-scale variability of water vapor and clouds in large-scale models and its use to diagnose cloud cover. *J. Atmos. Sci.*, **59**, 1917–1942.

Tompkins, A. M., and F. Di Giuseppe, 2007: Generalising cloud overlap treatment to include solar zenith angle effects on cloud geometry. *J. Atmos. Sci.*, **64**, 2116–2125.

Varnai, T., and R. Davies, 1999: Effects of cloud heterogeneities on shortwave radiation: Comparison of cloud-top variability and internal heterogeneity. *J. Atmos. Sci.*, **56**, 4206–4223.

Venema, V., S. G. Garcia, and C. Simmer, 2010: A new algorithm for the downscaling of cloud fields. *Quart. J. R. Meteorol. Soc.*, **136**, 91–106.

Veron, D. E., C. P. Weaver, F. Veron, and M. J. Foster, 2009: Stochastic radiative transfer on modeled cloud fields. *IEEE Geoscience Remote Sensing Lett.*, **6**, 184–188.

Veron, D. E., J. Secora, and M. Foster, 2012: Deriving cloud statistics for stochastic radiative transfer from long-term observations. *J. Climate* (under revision).

Wang, M., S. Ghan, R. Easter, M. Ovchinnikov, and co-authors, 2011a: The multi-scale aerosol-climate model PNNL-MMF: model description and evaluation. *Geosci. Model Dev.*, **4**, 137–168.

Wang, M., S. Ghan, M. Ovchinnikov, X. Liu, and co-authors, 2011b: Aerosol indirect effects in a multi-scale aerosol-climate model PNNL-MMF. *Atmos. Chem. Phys. Discuss.*, **11**, 3399-3459, doi:10.5194/acpd-11-3399-2011.

Weinberg, S., 1980: Conceptual foundations of the unified theory of weak and electromagnetic interactions. *Review of Modern Physics*, **52**, 515–523.

Wen, G., A. Marshak, and R. Cahalan, 2006: Impact of 3-D clouds on clear sky reflectance and aerosol retrieval in a biomass burning region of Brazil. *IEEE Geosci. Remote Sens. Lett.*, **3**, 169–172.

Wen, G., A. Marshak, R. F. Cahalan, et al., 2007: 3D aerosol-cloud radiative interaction observed in collocated MODIS and ASTER images of cumulus cloud fields. *J. Geophys. Res.*, **112**, D13204.

Wielicki, B. A., and R. M. Welch, 1986: Cumulus cloud properties derived using Landsat satellite data. *J. Climate Appl. Meteor.*, **25**, 261–276.

Williams, M. M. R., 1974: *Random Processes in Nuclear Reactors.* Pergamon Press, State-New York.

Winker, D. M., W. H. Hunt, and M. J. McGill, 2007: Initial performance assessment of CALIOP. *Geophys. Res. Lett.*, **34**, L19803, doi:10.1029/2007GL030135.

Xie, S., R. McCoy, S. A. Klein, R. T. Cederwall, and co-authors, 2010: ARM climate modeling best estimate data: A new data product for climate studies. *Bull. Am. Meteorol. Soc.*, **91**, 13–20.

Xiu, D., and G. E. Karniadakis, 2002: The Wiener–Askey polynomial chaos for stochastic differential equations. *SIAM J. Sci. Comput.*, **24**, 619–644.

Zhao, G., and L. Di Girolamo, 2004: A cloud fraction versus view angle technique for automatic in-scene evaluation of the MISR cloud mask. *J. Appl. Meteor.*, **43**, 860–869.

Zhao, G., and L. Di Girolamo, 2007 Statistics on the macrophysical properties of trade wind cumuli over the tropical western Atlantic. *J. Geophys. Res.*, **112**, D10204, doi:10.1029/2006JD007371.

Zhuravleva, T. B., and A. A. Kokhanovsky, 2011: Influence of surface roughness on the reflective properties of snow. *J. Quant. Spectrosc. Radiat. Transfer*, **112**, 1353–1368.

Zuchuat, O., R. Sanchez, I. Zmijarevic, and F. Malvagi, 1994: Transport in renewal statistical media: Benchmarking and comparison with models. *J. Quant. Spectrosc. Radiat. Transfer*, **51**, 689–722.

Zuev, V. E., and G. Titov, 1995: Radiative transfer in cloud fields with random geometry. *J. Atmos. Sci.*, **52**, 176–190.

4 Database of optical and structural data for the validation of forest radiative transfer models

Andres Kuusk, Mait Lang, and Joel Kuusk

4.1 Introduction

Recent advances in airborne and spaceborne scanner technologies have been providing vast amounts of multispectral and multi-angular remote sensing data of different spatial and radiometric resolution over the Earth's vegetation. Remote sensing of forests has been one of the major interest in the remote sensing of environment, because forests impact climate, provide different kind of resources for the economy, are related to biodiversity and, on the other hand, are threatened by several anthropogenic factors and disturbances. There is no realistic alternative to the remote sensing for collecting data frequently and over the large forest areas. However, the extraction and analysis of the information about the forest structure from the forest reflectance recorded on a satellite image is not a trivial task. Mapping the abrupt changes like clear cuttings from the satellite images can be considered as a practical tool for controlling state-wide databases (Lang et al., 2006), but small disturbances like light thinning cuttings or damages or change in foliage mass and its properties are often difficult to detect.

The formulation of the spectral reflectance of a forest stand can be investigated by using physically based radiative transfer (RT) models for forest canopies. Since several approaches and approximations are used, there have been initiatives to compare radiative transfer models in controlled environments. One of these efforts is the RAdiation Model Intercomparison (RAMI) (Widlowski et al., 2007). In the first three phases of RAMI the models were tested with simulated data where forest stands and their spectral signatures were generated by combining reflectance models of Monte Carlo type and allometric regression models (Widlowski et al., 2003, 2007). However, to evaluate radiative transfer models in more realistic media, the use of real forest stands together with the actual reflectance data should be considered. In addition, the availability of empirical forest structure data accompanied by airborne and spaceborne reflectance measurements will enable the tests of model inversion. Therefore, we present detailed forest structure and reflectance data collected from three real hemi-boreal stands in Järvselja, Estonia. The database of the forest structure and optical properties is intended for forest radiative transfer modeling experiments. To access the database the authors of this chapter should be contacted.

4.2 Study site

There is a forest study area at Järvselja, Estonia which serves as a field base for forestry students of the Estonian University of Life Sciences. The Järvselja Training and Experimental Forestry District is located at 58.324°N and 27.268°E, Fig. 4.1. A 10 × 10-km test site for the POLDER mission (Deschamps et al., 1994) was there. Several satellite images (Landsat TM, SPOT) of the area have been collected since 1985. The test site serves as a ground truth site for the CHRIS/PROBA mission (Barnsley et al., 2004) as well. Thorough ground measurements on the test site (forest inventory data, LAI, fish-eye images, ground vegetation reflectance spectra) have been performed several times in the framework of the VALERI project (VALERI, 2005).

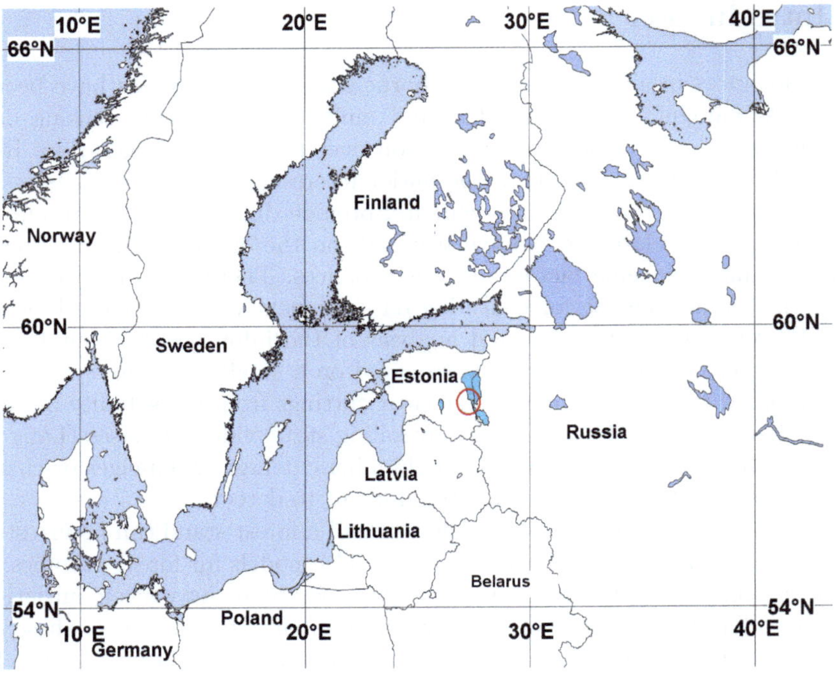

Fig. 4.1. Järvselja Training and Experimental Forestry District, Estonia, 58.324°N and 27.268°E. State borders from (ESRI, 2007).

Järvselja forests are typical for the hemi-boreal zone – dominant tree species are Scots pine (*Pinus sylvestris*), Norway spruce (*Picea abies*), Silver and White birch (*Betula pendula, Betula pubescens*), aspen (*Populus tremula*), Common and White alder (*Alnus glutinosa, Alnus incana*). All these species can grow in pure or mixed stands. Growth conditions range from poor where H_{100} (stand height at the stand age of 100 years) is less than 10 meters to very good where H_{100} can be over 35 meters. All forests are covered by the regular forest inventory from the year 2011. More information on the Järvselja Training and Experimental Forest District can be found on the Internet (Järvselja, 2007).

Three 100 × 100-m sample plots were selected for the detailed study in summer 2007:

- a 124-year-old *Pinus sylvestris* stand,
- a 59-year-old *Picea abies* stand, and
- a 49-year-old *Betula pendula* stand.

Optical measurements at the test site were carried out in July 2006, August 2007, and July 2008. Successful CHRIS/PROBA acquisitions over the test site were obtained on 10 July 2005 and 5 July 2010.

4.3 Instrumentation

4.3.1 PROBA/CHRIS imaging spectrometer

CHRIS is a hyperspectral imager onboard of the experimental satellite PROBA (Barnsley et al., 2004). CHRIS is used to measure directional spectral reflectance of land areas, thus providing new biophysical and biochemical data, and information on land surfaces.

CHRIS properties:

- multiple imaging of same target area under different viewing and illumination geometries
- spectral range: 415–1050 nm
- spectral resolution: 5–12 nm
- spatial resolution: 17 m at nadir
- swath width: 13 km

The number and bandwidths of spectral bands depend on the selected mode. The Mode 3 (18 spectral bands for land and aerosol studies) was used for the Järvselja acquisitions. Spectral bands are listed in Table 4.8.

4.3.2 Airborne spectrometer UAVSpec

The airborne spectrometer system UAVSpec (Kuusk, 2011) is based on the 256-band NIR enhanced miniature spectrometer module MMS-1 by Carl Zeiss Jena GmbH with the front-end-electronics (FEE) by Tec-5 AG Sensorik und Systemtechnik. It has a wavelength range of 306–1140 nm, 15-bit digital output, and noise level of 2–3 bits. The spectrometer is controlled by an Atmel microcontroller ATmega88. The fore-optics restricts the field-of-view to 2° in airborne measurements. The spectrometer system comprises web camera and a GPS receiver for position tracking. Data from the spectrometer are collected by a PC/104-Plus single board computer Puma by VersaLogic Corp. The spectrometer was also used for the measurements of ground vegetation reflectance with the field-of-view of 8°, for the measurements of incident spectral fluxes during ground vegetation measurements using the ASD's remote cosine receptor (RCR), and for the measurements of leaf and needle reflectance and leaf transmittance with a AvaSphere-50-REFL integrating sphere by Avantes BV.

4.3.3 Spectrometer FieldSpec-Pro VNIR

The FieldSpec-Pro VNIR spectrometer is a portable 512-channel photodiode array spectroradiometer by Analytical Spectral Devices, Inc., with fiber-optic input covering the 350–1050 nm region. The sampling interval is 1.4 nm and the resolution is 3 nm at full width half maximum (FWHM) at 700 nm. The spectrometer was used during CHRIS acquisition and airborne measurements for the irradiance measurements using a remote cosine receptor, and for the measurements of leaf reflectance and transmittance using the AvaSphere-50-REFL integrating sphere.

4.3.4 Spectrometer GER-2600

The GER Corporation 2600 Spectroradiometer for 350–2500 nm spectral range uses a fixed-grating array-based design with two linear arrays – 512 Si detectors and 128 PbS detectors. Spectral bandwidth is 1.5 nm in the spectral range 350–1050 nm, and 11.5 nm in the spectral range 1050–2500 nm, field-of-view 3°. The spectrometer was used for the stem bark reflectance measurements in July 2001.

4.3.5 LAI-2000 plant canopy analyzer (Li-Cor)

The LAI-2000 plant canopy analyzer uses a fish-eye light sensor that measures diffuse radiation in five distinct angular bands about the zenith point. The sensor of the LAI-2000 plant canopy analyzer consists of five concentric photodiodes. The hemispheric image is projected onto these rings, allowing each of them to measure the radiation in a band at a known zenith angle. The transmitted radiation is restricted below 490 nm, minimizing the contribution of light that has been scattered by foliage. Gap fractions at five zenith angles can be measured by making a reference reading above the canopy (in a forest clearing), and other readings beneath the canopy.

4.3.6 Coolpix-4500 digital camera

The Coolpix-4500 digital camera was equipped with the Nikon fish-eye converter FC-E8, image size 2272×1704 TIFF-RGB. The fish-eye converter has 183° FOV and equidistant projection.

4.3.7 Nikon total station DTM-332

Tree coordinates were measured with a Nikon DTM-332 Total station (Nikon-Trimble Co., Ltd, Tokyo, Japan). This device is designed for land survey and is equipped with a laser distance meter, and records horizontal and vertical angles. A reflector prism was used for measuring distances to the trees with the laser distance meter. The total station has one axis tilt sensor and leveling compensation.

4.3.8 Leica ALS50-II airborne laser scanner

Airborne laser scanning of stands was carried out with a Leica ALS50-II airborne laser scanner onboard the special mission aircraft Cessna 208B Grand Caravan by the Estonian Land Board on 30 July 2009.

The laser scanner ALS50-II (Leica Geosystems AG, St. Gallen, Switzerland) is a compact laser-based system designed for the acquisition of topographic and return signal intensity data from a variety of airborne platforms. The data is computed using range and return signal intensity measurements recorded in flight along with position and attitude data derived from airborne GPS and inertial subsystems. The lidar collects heights of four return signals. System FOV is adjustable over the range of 0–75 degrees, in 1-degree increments. The scan rate is user-selectable from 0 to 90 Hz in 0.1 Hz increments. The system provides a sinusoidal scan pattern in a plane nominally orthogonal to the longitudinal axis of the scanner, nominally centered about nadir. Output beam divergence is 0.22 mr nominal, measured at the $1/e^2$ point.

4.4 Measurements

4.4.1 Stand structure

Ground measurements

Detailed structure measurements of the three selected stands were carried out in summer 2007. A square plot of 100 × 100 m was marked in every stand, and exact positions of all trees with the trunk diameter at breast height $d_{1.3}$ larger than 4 cm were tallied and their locations were mapped using a Nikon DTM-332 Total Station. The obtained tree location coordinates are in the radius of 5–20 cm of their true position. Trees were callipered using the electronic caliper Masser Racal (Savcor Group Ltd, OY, FIN-50100 Mikkeli, Finland), and the average of two perpendicular trunk diameter measurements was used as the $d_{1.3}$ estimate. For each tree its social status within the stand was assessed according to its properties and relative size to its neighbors and forest stand average characteristics. Four status classes were distinguished: (1) upper layer, (2) second layer, (3) regeneration, and (4) understorey. In order to create allometric models of the tree height and crown dimensions a series of sample trees were measured. The distribution of $d_{1.3}$ and final expert guess were used to distribute sample trees between 1 cm wide $d_{1.3}$ classes according to the total tree count in a particular diameter class. Expert guess was used to add sample trees into the $d_{1.3}$ classes of large or small trees to assure representativeness of the sample for all species and social status classes. The total number of sample trees measured was 73 in the birch stand, 77 in the spruce stand, and 45 in the pine stand. Random sampling from the database was used to identify sample trees. Tree height h, height to live crown base h_{lcb} and two perpendicular crown diameters d_{cr} were measured on each of the sample trees.

Effective leaf area index $\mathrm{LAI}_{\mathrm{eff}}$ was measured with a LAI-2000 instrument on the regular grid of 9 sample plots L1–L9, the grid step being 30 m (Table 4.1). Measurements were done with two LAI-2000 instruments during overcast weather conditions. Three readings of LAI-2000 were taken at every LAI point L1–L9, and mean gap fractions and LAI were calculated for every Li. In the birch stand effective LAI was measured and hemispherical images were taken twice – with full foliage in July and with no leaves in winter in order to estimate the share that stems and branches have in forming the canopy cover.

Table 4.1. The grid of LAI measurements on sample plots

Point ID	Position (x, y), m
L1	(20, 20)
L2	(20, 50)
L3	(20, 80)
L4	(50, 20)
L5	(50, 50)
L6	(50, 80)
L7	(80, 20)
L8	(80, 50)
L9	(80, 80)

Hemispherical color (RGB) images were taken at every LAI point early in the morning with low sun and no clouds. The blue images were thresholded for estimating gap fractions.

The Cajanus sighting tube was used to obtain canopy cover and crown cover estimate (Jennings et al., 1999) according to the methodology by Korhonen et al. (2006). Canopy cover is the proportion of canopy overlying the forest floor and crown cover is the ratio of total area of crown projections to the plot area. Crown cover and canopy cover of upper and lower tree layers was measured taking 345–366 readings in a stand. Tree crown envelopes in each stand were visually judged.

Airborne measurements

Airborne laser scanning of stands was carried out with the Leica ALS50-II Airborne Laser Scanner onboard the special mission aircraft, Cessna 208B Grand Caravan, by the Estonian Land Board on 30 July 2009. Flying height was 500 m above ground level, point spacing across track 0.24 m, point spacing along track 0.73 m, average point area 0.05 m^2, nadir point density 11 pts/m^2. Four returns per pulse were recorded. Every stand was measured twice flying parallel to stand borders over the stand center at perpendicular azimuths, thus the total point density is about 20 pts/m^2.

4.4.2 Spectroscopic measurements

Reflectance spectra of phytoelements

Directional-hemispherical reflectance and transmittance of birch, alder and aspen leaves in the birch stand, and reflectance of spruce and pine needles was measured using the UAVSpec spectrometer equipped with the integrating sphere AvaSphere-50-REFL by Avantes BV. The AvaLight-HAL tungsten halogen light source by Avantes BV was used. The measurements setup UAVSpec and AvaSphere-50-REFL together with AvaLight-HAL was calibrated using measurements of Spectralon references SRT-99-100, SRT-20-120, and Zenith Ultrawhite Reflectance Standard 10 cm × 20%, light trap, and white reference tile WS-2 by Avantes BV.

Birch, alder and aspen branches from the upper part of tree crowns were shot down and brought to the spectrometer in less than half an hour. Reflectances of both adaxial and abaxial surface of a leaf attached to the branch and having black background were measured. Several leaves were measured, and mean reflectance spectra and variance were calculated. Pine and spruce branches were cut by climbing trees. Reflectance of a bundle of needles in a live shoot was measured. Reflectance spectra of stem and branch bark were measured similar to leaf reflectance. Stem bark directional reflectance of some species was measured previously with a GER-2600 spectrometer using natural illumination (Lang et al., 2002).

Reflectance of ground vegetation

Reflectance spectra of ground vegetation were measured with the UAVSpec spectrometer at LAI points L1–L9 in cloudless conditions. At every LAI point radiance spectra of ground vegetation were measured vertically downward from a height of about 1 m using a 8° field restrictor, walking along a nearly circular transect of 5 m radius around the LAI point. Then the field restrictor was replaced by the cosine receptor looking upward, and the measurement was repeated. Fore-optics were leveled using a self-leveling mount. Data from the spectrometer were collected 8 times per second. The reflectance spectrum for a LAI point is the ratio of these two mean values, calibrated to reflectance factor using the Spectralon reference panel. The mean values and standard deviations reported are those over nine LAI points. Sun angles during measurements are listed in Table 4.2.

Table 4.2. Sun angles during understorey measurements

Date	Time, GMT	SZA	SAA
Birch stand			
22.07.2007	12:01–12:42	42.4–46.1	217.5–230.1
25.07.2008	10:06–10:44	38.8–39.1	175.8–189.9
Pine stand			
22.07.2007	08:11–08:52	44.3–41.0	135.4–148.8
05.08.2007	08:22–08:53	46.4–44.1	140.5–150.6
25.07.2008	07:25–08:02	49.7–45.9	122.5–132.9
Spruce stand			
21.07.2007	13:50–14:28	53.6–58.4	248.5–257.7
05.08.2007	10:21–12:07	41.3–46.0	181.6–217.9
25.07.2008	08:57–09:31	41.4–39.7	150.8–162.6

SZA – Sun zenith angle, deg.
SAA – Sun azimuth angle, deg.

Airborne measurements

Hyperspectral reflectance of stands was measured with the UAVSpec spectrometer system (Kuusk, 2011). The spectrometer was mounted rigidly at the chassis of a Robinson R-22 helicopter, looking in the nadir direction during straight flight at a constant speed. Data from the spectrometer were collected approximately 3–5 times per second in 2006, and 8 times per second in 2007 and 2008. Web camera images, position data from the GPS receiver, and FEE temperature were recorded once per second. During helicopter measurements the spectrometer was equipped with a fore-optics which restricted the field-of-view to 2°. Measurements were made in cloudless conditions from the height of 100 m above ground level in July 2006 and from the height of 80 m in 2007 and 2008 measurements. The flight speed was 60 km/h. The measurement conditions are reported in Table 4.3.

Table 4.3. Airborne measurements of the test site

Date	26 July 2006	8 August 2007	24 July 2008
Time, GMT	9:19–9:31	9:57–10:46	8:05–8:30
Sun zenith angle	40.3°–39.8°	42.3°–42.4°	45.4°–43.2°
Sun azimuth angle	158.4°–162.7°	172.6°–189.4°	133.8°–141.6°
D_{550}/Q_{550}	0.097	0.262	0.097
Observation nadir angle	0°	0°	0°
Platform altitude	100 m	80 m	80 m

D_{550}/Q_{550} – ratio of downward diffuse and total spectral fluxes at 550 nm.

CHRIS images

Images were acquired for one near nadir and two backscattering (hot-spot side) viewing geometries on 10 July 2005. Acquisition details are listed in Table 4.4.

Table 4.4. CHRIS acquisition details, 10 July 2005

Scene number:	5703	5705	5707
Sun zenith angle		36.6°	
Sun azimuth angle		167.2°	
Platform altitude		562 km	
Image size, pixels		748×744	
Number of spectral bands[1]		18	
Time, GMT	9:43:39	9:44:28	9:45:18
Observation zenith angle	7.62°	37.23°	56.71°
Observation azimuth angle[2]	−22.46°	19.79°	23.43°
Ground resolution, m	17 × 17	21 × 19	28 × 19

[1] Wavelengths and bandwidths see in Table 4.8.
[2] Relative to the sun azimuth.

Illumination conditions

Downward total Q_λ and diffuse D_λ spectral fluxes were measured at the test site with the FieldSpec Pro spectrometer equipped with a cosine receptor. For the measurement of diffuse sky flux a screening disc was used which screened about 9° of sky in sun direction. At Tartu Observatory, 45 km from the test site an AERONET sun photometer (Holben et al., 1998) is measuring optical properties of the atmosphere: aerosol optical thickness, water vapor content from direct sun measurements, and sky radiance distribution along the solar principal plane and along the solar almucantar for retrieving size distribution and phase function of aerosol.

4.5 Data processing

4.5.1 Stand structure

The nonlinear least squares regression method (NLS) of the *R* statistical software (R-project, 2007) was used to estimate parameter values for the allometric models of tree height, crown length, and crown radius (Table 4.7). The arguments for the models were breast height diameter $d_{1.3}$, tree height h in some models, and the relative diameter d_r of a given tree with respect to its neighbors located within radius of 4 m. For the *Fraxinus excelsior* and *Acer platanoides* trees in the birch stand the social status code (1–4) was informative for the crown radius model. Inclusion of arguments into models was judged by the probability value p of their parameter using the criterion $p \leq 0.05$. The aim of the models was mainly to get best fit on the sample tree data and not to search for general-purpose model formulas. Goodness of fit r^2 for regression models was calculated as

$$r^2 = 1 - \frac{SSE(n-1)}{SSY(n-k)}, \tag{4.1}$$

where $SSE = \sum_{i=1}^{n}(x_i - \hat{x})^2$ is the sum of squared differences between model $\hat{x}$ and observations x_i, $SSY = \sum_{i=1}^{n}(x_i - x)^2$ is the sum of squared deviation of observations from the mean x of the observations, n is the number of observations in the sample, and k is the number of estimated parameters. Standard errors of regression models for tree height, crown length and crown radius are 1.6 m, 1.7 m and 0.28 m, respectively. Foliage mass for all individual trees was estimated using the equations listed in Table 4.5, and foliage area estimates were obtained using specific leaf weights as given in Table 4.6. Finally, the allometric leaf area index $\mathrm{LAI_{all}}$ of the stands was calculated.

4.5.2 Leaf and needle optical properties

The AvaSphere-50-REFL integrating sphere has only one sample port, the sample reflectance is measured by substitution method. The directional-hemispherical reflectance of a sample r_s is calculated from the spectrometer readings in digital

Table 4.5. Equations for foliage mass m

Eq. no.	Foliage mass (kg/tree)	Species[1]	References and notes
1	$m = 3.13 + 0.05947d_{1.3}^2 - 0.801d_{1.3}$	HB	(Tamm, 2000), fresh foliage
2	$m = \exp(-2.6024 + 9.8471(d_{1.3}/(d_{1.3}+7)) + 0.026h - 1.6717\ln(d_{1.3}) + 1.0419\ln(l - 0.0123Loc))$	MA	(Marklund, 1988), $Loc = 65$
3	$m = 1.0394\exp(-3.8719 + 2.1141\ln(D_{\mathrm{lcb}}))$	KS	(Hoffmann and Usoltsev, 2002), D_{lcb} is trunk diameter at live crown base
4	$m = 0.001 \cdot 10^{(0.229+2.507\log_{10}(d_{1.3}))}$	PN, JA	(Wang, 2006), equation for *Amur linden*
5	$m = 3 \cdot 10^{-6}(10d_{1.3})^{2.5470}$	LM	(Johansson, 1999)
6	$m = 1.064\exp(-1.5732 + 8.4127d_{1.3}/(d_{1.3}+12) - 1.5628\ln(h) + 1.4032\ln(l))$	KU	(Marklund, 1988)
7	$m = 1.052 \cdot 10^{(-1.62+1.778\log_{10}(d_{1.3}))}$	others	(Martin et al., 1998), equation for *Acer rubrum*

[1] Species codes are defined in Table 4.6

Table 4.6. Specific leaf weight (SLW)

Species	Code	SLW (g/m^2)	Reference
Acer platanoides	VA	69.0	(Martin et al., 1998)
Alnus glutinosa	LM	77.4	(Niinemets and Kull, 1994)
Betula pendula	KS	76.0	(Niinemets and Kull, 1994)
Fraxinus excelsior	SA	69.0	(Martin et al., 1998)
Pinus sylvestris	MA	160.0	(Pensa and Sellin, 2002)
Picea abies	KU	247.0	(Sellin, 2000)
Populus tremula	HB	190.5	(Tamm, 2000), fresh leaves
Tilia cordata	PN	25.5	(Niinemets and Kull, 1994)
Ulmus glabra	JA	38.5	(Niinemets and Kull, 1994)

counts measuring the sample B_{s} and reference B_{st} (Wendlandt and Hecht, 1966),

$$r_{\mathrm{s}} = \frac{B_{\mathrm{s}}}{B_{\mathrm{st}}} \frac{\eta_{\mathrm{st}}}{\eta_{\mathrm{s}}} r_{\mathrm{st}}. \tag{4.2}$$

Here, r_{st} is the reflectance of reflectance standard (reference), and η_{x} is the sphere efficiency,

$$\eta_{\mathrm{x}} = \frac{1}{(1-\hat{r}_{\mathrm{x}})} \frac{b}{S}; \quad x = s, st, \tag{4.3}$$

Table 4.7. Equations for the tree height, crown radius and crown length

No.	Equation	Stand	Species[1]	r^2(%)	RSE (m)	n
1	$h = 1.3 + 37.83(d_{1.3}/(d_{1.3} - 1))^{-8.23}$	Birch	KS	86.2	1.89	30
2	$h = 1.3 + 37.79(d_{1.3}/(d_{1.3} + 1))^{8.70}$	Birch	HB	94.1	1.15	6
3	$h = 1.3 + 28.91(d_{1.3}/(d_{1.3} + 1))^{12.66}$	Birch	KU	99.2	0.36	3
4	$h = 1.3 + 36.26(d_{1.3}/(d_{1.3} + 4))^{3.05}$	Birch	LM (PJ, JA)	87.9	1.44	14
5	$h = 1.3 + 29.19(d_{1.3}/(d_{1.3} - 4))^{-1.85}$	Birch	PN	74.9	2.39	14
6	$h = 1.3 + 33.79(d_{1.3}/(d_{1.3} + 1))^{9.99}$	Birch	SA, VA	70.4	2.31	6
7	$h = 1.3 + 33.28(d_{1.3}/(d_{1.3} - 1))^{-6.32}$	Spruce	KS	89.5	1.59	14
8	$h = 1.3 + 37.15(d_{1.3}/(d_{1.3} + 1))^{12.66}$	Spruce	KU	93.5	1.95	63
9	$h = 1.3 + 22.31(d_{1.3}/(d_{1.3} + 8))^{1.15}$	Pine	MA	58.4	1.36	45
10	$h_{lcb} = 26.49(d_{1.3}/(d_{1.3} + 4))^{4.054}$	Birch	LM	50.2	2.91	14
11	$l = 0.3782d_{1.3}$	Birch	HB	94.6	1.10	6
12	$l = 17.93(d_{1.3}/(d_{1.3} + 8))^{2.049}$	Birch	KS	77.7	2.40	29
13	$l = 31.075(d_{1.3}/(d_{1.3} + 4))^{4.909}$	Birch	PN	80.3	2.12	14
14	$l = 0.5221h$	Birch	SA, VA	63.0	2.58	6
15	$l = 16.16(d_{1.3}/(d_{1.3} - 1))^{11.154}$	Spruce	KS	92.6	1.34	14
16	$l = (0.1241hd_{1.3}^{1.333})/d_{1.3} + 2.523d_{r}$	Spruce	KU	94.1	1.87	63
17	$l = 12.71(d_{1.3}/(d_{1.3} + 8))^{2.982}$	Pine	MA	86.1	0.79	44
18	$d_{cr} = 1.461 + 0.0899d_{1.3}d_r$	Birch	HB	98.6	0.57	6
19	$d_{cr} = 1.469 + 0.07193d_{1.3}d_r$	Birch	KS	98.7	0.58	30
20	$d_{cr} = 1.856 + 0.08579d_{1.3}d_r$	Birch	LM	98.5	0.51	14
21	$d_{cr} = 0.1836d_{1.3} + d_r^{-1.088}$	Birch	PN	98.4	0.60	14
22	$d_{cr} = 0.1549d_{1.3} \times Status$	Birch	SA, VA	97.4	0.68	6
23	$d_{cr} = 1.418 + 0.004808d_{1.3}^2$	Spruce	KS	98.9	0.51	14
24	$d_{cr} = 0.2145d_{1.3} - 0.8531$	Pine	MA	95.6	0.44	45

[1] Species codes are defined in Table 4.6

h – tree height, m; $d_{1.3}$ – breast height diameter, cm; h_{lcb} – crown base height, m; l – crown length, m; d_{cr} – crown diameter, m; d_r – relative diameter at breast height; RSE – residual standard error, m; n – number of observations

S is the total area of the sphere, b is the area of the exit port, r_x is the reflecting power of the sphere,

$$\hat{r}_x = \frac{(r\,d + r_x c)}{S}, \tag{4.4}$$

r is the reflectivity of the sphere surface, $d = S - a - b - c$, a is the area of illumination port, c is the area of the sample port.

In the AvaSphere-50-REFL integrating sphere a fiber and collimating lens at the illumination port are used for illumination. Therefore a small part of light $(1 - q)$ is falling at the borders of the sample port and creates additional illumination, so the signal is

$$B_x = (q\,r_x + (1 - q)\,r)\,P\,\eta_x. \tag{4.5}$$

The proportion q can be estimated measuring reference $r_s = r_{st}$ and black hole $r_s = 0$,

$$q = \frac{(1-\xi_0)\, r}{(r_{\rm st} - r)\, \xi_0 + r}, \quad \xi_0 = \frac{B_0\ \eta_{\rm st}}{B_{\rm st}\ \eta_0}. \tag{4.6}$$

Finally,

$$\begin{aligned} r_{\rm s} &= \frac{A-\kappa}{C}, \\ \kappa &= \frac{(1-q)}{q} r, \\ A &= (1 - r\, d\,/\,S)\nu \frac{B_{\rm s}}{B_{\rm st}}, \\ C &= 1 + \frac{\nu\, c}{S} \frac{B_{\rm s}}{B_{\rm st}}, \\ \nu &= \frac{r_{\rm st} + \kappa}{1 - r\, d\,/\,S - r_{\rm st} c\,/\,S}. \end{aligned} \tag{4.7}$$

Leaf transmittance $t_{\rm s}$ is measured using an external light source and comparing the spectrometer signals of sample $B_{\rm s}$ and open sample port B_0,

$$\begin{aligned} t_{\rm s} &= \frac{(1-\hat{r}_{\rm s})}{(1-\hat{r}_0)} \frac{B_{\rm s}}{B_0}, \\ &= \frac{(1 - r\, d\,/\,S - r_{\rm s}\, c\,/\,S)}{(1 - r\, d\,/\,S)} \frac{B_{\rm s}}{B_0}. \end{aligned} \tag{4.8}$$

Here $r_{\rm s}$ is the sample reflectance – the reflectance of the leaf side facing the integrating sphere.

4.5.3 Correction of UAVSpec data

The spectrometer system UAVSpec was used for airborne measurements, for the measurements of understorey vegetation reflectance spectra, and of leaf and needle optical properties. Several technical aspects related to the spectrometer module MMS-1 should be noted. First of all, spectral aliasing affects the NIR signal of the spectrometer. The second-order blocking filter which is directly coated on the sensor is not perfect. It lets some visible light cause aliasing effect in the NIR spectral domain. Secondly, straylight contributes to the signal at wavelengths where the sensitivity of the sensor is low. To overcome these metrological problems, we corrected the measured signal for the spectral aliasing and straylight in the following way. The spectrometer was illuminated through a double monochromator and, while scanning the monochromator over the spectral region of the spectrometer, instrument function was measured for all the bands of the spectrometer. By deconvolution, the original signal was restored for every measured spectrum (Kostkowski, 1997). The aliasing correction allowed the reliable NIR spectral domain of the Zeiss MMS-1 spectrometer module to be extended by more than 100 nm (see Fig. 4.2).

Finally, the dark current drift due to temperature changes was taken into account. Since it was not possible to take dark current readings during airborne

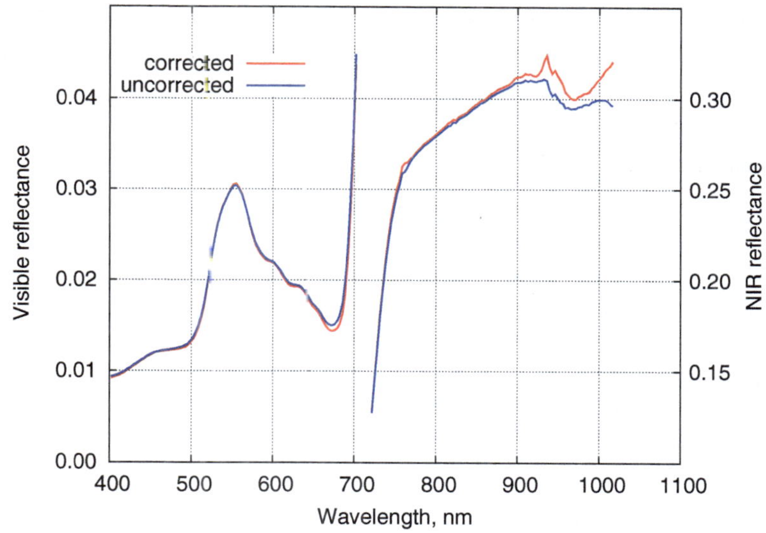

Fig. 4.2. Straylight and aliasing correction of a reflectance spectrum.

measurements, those recorded before the flight were used together with FEE temperature measurements and empirically measured temperature-dependence of dark current,

$$d_\lambda(T) = d_\lambda(T_0) + f(T) - f(T_0), \tag{4.9}$$

where $d_\lambda(T)$ is the dark current in case of FEE temperature T, $d_\lambda(T_0)$ is the dark current measured before the flight, T_0 is the FEE temperature during dark current measurement, and $f(T)$ is the temperature dependence of dark current (Schaepman and Dangel, 2000),

$$f(T) = a + b\,\exp(cT), \tag{4.10}$$

where $a = 193.06$, $b = 0.329218$, and $c = 0.0899416°\mathrm{C}^{-1}$ are empirical constants and T is the FEE temperature in degrees Celsius (°C). There is no integration time in equations (4.9) and (4.10), therefore, the equations are only valid for an integration time of 120 ms. This integration time was used for measurements of the temperature-dependence of dark current, as well as airborne measurements of forest reflectance.

The recorded nadir radiance in digital counts is compared to the radiance of a calibrated Spectralon panel measured in a nearby clearing at the test site just before target measurements. This way the recorded signal is converted to the directional spectral reflectance of targets. To take into account diurnal changes in illumination conditions, downward spectral flux density Q_λ was recorded during measurements with a FieldSpec Pro VNIR spectroradiometer equipped with a cosine receptor,

$$\rho_\lambda(t) = \frac{q_\lambda(t_0)}{n_\lambda(t_0)} \frac{n_\lambda(t)}{q_\lambda(t)} r_\lambda. \tag{4.11}$$

Here, $\rho_\lambda(t)$ is the target spectral directional reflectance measured at time moment t, $q_\lambda(t)$ and $q_\lambda(t_0)$ are the signals of the FieldSpec spectrometer during target measurements and calibration, respectively, $n_\lambda(t)$ and $n_\lambda(t_0)$ are the signals of the UAVSpec spectrometer, and r_λ is the spectral reflectance of the reference panel. All the signals in equation (4.11) are without dark current.

4.5.4 Satellite data

destriping of images

destriping of images was performed assuming that striping is caused mainly by sensor offsets. The destriping function was found for every image as the difference between column mean values and smoothed column mean values using the 9-point Hamming window (Rabiner and Gold, 1975). The mean value of three destriping functions for every band (scenes 5703, 5705, and 5707) was used for all three respective spectral images.

Atmospheric correction

Satellite-based measurements of the radiative signature of terrestrial targets are always affected by the chemical and physical properties of the overlying atmosphere. Highly accurate, reliable and preferably physically based correction schemes are thus required to quantitatively link spaceborne measurements with the structural and spectral characteristics of a given vegetation target.

Several procedures have been developed for the atmospheric correction of satellite remote sensing data. A comprehensive overview of atmospheric correction procedures can be found in (Liang, 2004). Thorough comparisons of various procedures of the atmospheric correction of Landsat Thematic Mapper images are reported by Hadjimitsis et al. (2004), and Song and Woodcock (2003). Here, atmospheric correction is performed with atmospheric RT package 6S (Vermote et al., 1997) using the look-up-table method suggested by Kuusk (1998). Atmospheric correction was performed in two stages. First, with the 6S model a look-up-table (LUT) was generated which links top-of-atmosphere (TOA) radiance to top-of-canopy reflectance. The latter was calculated with the multispectral homogeneous canopy reflectance model MSRM (Kuusk, 1994) by varying ground vegetation parameters. The MSRM model served as the underlying surface in the 6S model, and leaf area index, soil reflectance and leaf optical parameters were varied in a reasonable range characteristic for the test site in order to produce TOA radiance values in the range similar to what we have in CHRIS images. In the calculation of TOA radiance the optical parameters of the atmosphere are needed. An AERONET sun photometer (Holben et al., 1998) is working at Tartu Observatory, 45 km far from the test site, and the 6S model has tools for using AERONET sun-photometer data directly. Aerosol optical thickness, size distribution and refraction index, and amount of water vapor in the atmosphere were determined from the sun-photometer measurements. Level 2 data were used. Ozone data are available from NASA/GSFC Ozone Processing Team at their web-page (McPeters, 2007). For stable weather conditions, and based on an analysis of MODIS images (MODIS, 2005) and air parcel trajectories of HYSPLIT (2007) for 10 July 2005, it was concluded that significant differences in atmosphere properties over a distance of 45 km were not possible – only small changes in the amount of aerosol and water vapor may be present. On this basis the diffuse spectral fluxes in the atmosphere and the diffuse-to-total ratio of downward spectral fluxes D_λ/Q_λ were calculated with the atmospheric RT model. Simultaneous to the CHRIS acquisition the diffuse-to-total ratio of downward spectral fluxes D_λ/Q_λ was measured at the test site with a FieldSpec Pro spectrometer. In Fig. 4.6 we see that the simulated downward diffuse flux using AERONET aerosol optical thickness (at 550 nm $\tau_{550} = 0.08$) exceeds the measured one for some extent. This discrepancy may be caused by differences in the separation of total flux to direct and diffuse components. The field-of-view of the sun-photometer is 1° while the FieldSpec cosine receptor was screened with a disc which screened about 9° in the sun direction during sky flux measurements. In the 6S model the direct flux is collimated.

The created LUTs were approximated by a second-order polynomial separately for every CHRIS band and every view direction,

$$\rho_{\lambda,\mathrm{j}}(b_{\lambda,\mathrm{j}}) = a_{2\,\lambda,\mathrm{j}}\, b^2_{\lambda,\mathrm{j}} + a_{1,\lambda,\mathrm{j}}\, b_{\lambda,\mathrm{j}} + a_{0,\lambda,\mathrm{j}}. \tag{4.12}$$

Here, $b_{\lambda,\mathrm{j}}$ is the TOA radiance, and $\rho_{\lambda,\mathrm{j}}(b_{\lambda,\mathrm{j}})$ is canopy directional reflectance at wavelength λ in the acquisition geometry of CHRIS scenes $j = 5703, 5705, 5707$ (see Table 4.4). The approximation polynomials were used for the calculation of top-of-canopy reflectance using the TOA radiance measured by CHRIS. This procedure was applied separately to every pixel in every spectral image. The advantage of using LUT when compared to the built-in procedure of atmospheric correction in the 6S model is considering the directional dependence of diffuse fluxes scattered from vegetated surface and its variations as a function of wavelength, LAI, and other canopy parameters.

The second step in the atmospheric correction procedure involves the removal of adjacency effects by 2D deconvolution in CHRIS scenes 5703 and 5705. This is because the atmosphere acts as a low-pass filter which degrades satellite images. The recorded image is a convolution of the top-of-canopy radiance pattern and point spread function (PSF) of the atmosphere (or the system atmosphere–fore-optics–sensor) (Banham and Katsaggelos, 1997),

$$g(x, y) = p(\xi, \eta) \otimes f(u, v), \tag{4.13}$$

where $f(u, v)$ is the original (ideal) image, $p(\xi, \eta)$ is the PSF of the atmosphere, and $g(x, y)$ is the recorded (degraded) image, $\otimes$ denotes the convolution in the x-y-space.

Degraded images can be restored using Wiener filtering in the Fourier space (Banham and Katsaggelos, 1997; Podilchuk, 1998). The convolution of the original image and PSF in the x-y-space (4.13) can be performed by filtering in the 2D Fourier space

$$G = P \cdot F, \tag{4.14}$$

where G, P and F are the Fourier images of $g(x, y)$, $p(x, y)$ and $f(x, y)$, respectively. The original image is restored by the inverse filtering

$$\hat{F} = G \cdot W, \tag{4.15}$$

where W is the Wiener filter (Podilchuk, 1998)

$$W = \frac{P^* F}{|P|^2 F + S_{\mathrm{n}}}. \tag{4.16}$$

Here the superscript * denotes the complex conjugate, and S_{n} is the noise spectrum. Inverse Fourier transform of the filtered spectrum $\hat{F}$ returns the corrected image $\hat{f}(x, y)$.

The point spread function (PSF) of the atmosphere as a function of aerosol optical thickness was estimated by Liang et al. (2001) in numerical simulations,

$$p(s) = f_1(\tau)\exp(-q_1 s) + f_2(\tau)\exp(-q_2 s), \tag{4.17}$$

where

$$\begin{aligned} f_1(\tau) &= 0.03\tau \\ f_2(\tau) &= 0.071\tau^3 - 0.061\tau^2 - 0.439\tau + 0.996. \end{aligned} \tag{4.18}$$

Here s is the radial distance from the pixel (km) and τ is the optical thickness of the atmospheric aerosol.

The scale parameters q_1 and q_2 of the PSF were estimated from the requirement that the corrected red and NIR reflectance of a narrow lake in the scene remains in the range 0–1% as reported in measurements of natural water bodies (Froidefond et al., 2002; Novo et al., 2004; Feng et al., 2005; Cannizzaro and Carder, 2006).

$$\begin{aligned} q_1 &= 5.70 \text{ km}^{-1}, \\ q_2 &= 5.17 \times 10^4 \text{ km}^{-1}. \end{aligned} \tag{4.19}$$

The noise spectrum S_n was supposed to be increasing exponentially with spatial frequency; magnitude of it was estimated using signal/noise ratio from CHRIS documentation (CHRIS, 2002), and the mean reflectance of every spectral image.

The relative change of the mean stand reflectance due the adjacency correction $(\rho_2(\lambda) - \rho_1(\lambda))/\rho_2(\lambda)$ is plotted in Fig. 4.3, where $\rho_1(\lambda)$ and $\rho_2(\lambda)$ are the stand reflectance before and after adjacency correction, respectively.

This procedure of adjacency correction cannot be applied in the case of oblique view (the scene 5707). Then the view path crosses atmosphere layers over different targets, and the prepositions of PSF derivation are not fulfilled. Possible errors in adjacency correction increase with increasing view angle (Reinersman and Carder, 1995).

CHRIS calibration revised

The radiometric calibration of satellite sensors is problematic. Calibration procedures of optical sensors of large satellites have been developed for years and calibration is provided by a mixture of on-ground and in-orbit measurements (Slater et al., 1996; Slater and Biggar, 1996; Thome et al., 1997; Abdou et al., 2002). Preflight calibrations are subject to change during launch and exposure to the space

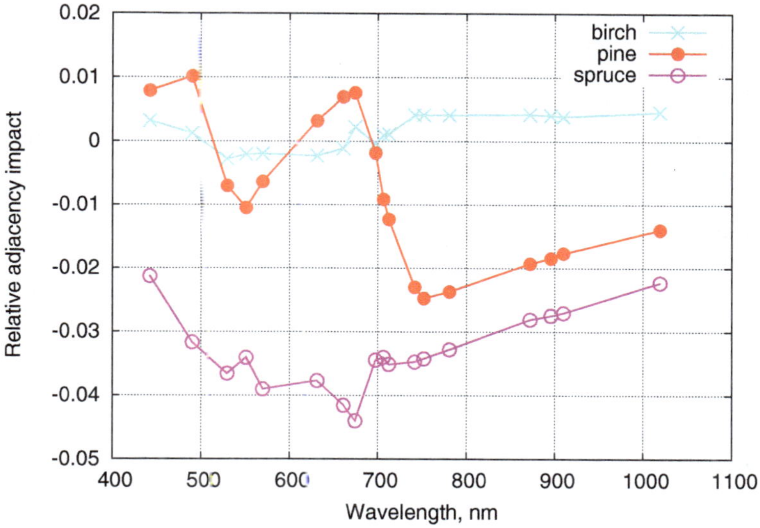

Fig. 4.3. The role of adjacency correction.

environment. With increasing time since the launch the role of in-orbit methods increase; however, small platforms offer limited scope for on-board calibration facilities. A small satellite is also limited in keeping stable conditions of its instruments (Cutter, 2004) and therefore some vicarious method should be used as an option. For the vicarious calibration of sensors onboard large operational satellites which measure continuously large areas (TM, ETM+, HRV, MISR) one can choose bright stable targets at good atmospheric conditions (Abdou et al., 2002). The small acquisition resource of the CHRIS imaging spectrometer onboard the PROBA satellite (Barnsley et al., 2004) – two or three sites of about 15×15 km per day – limits opportunities for the vicarious calibration of the spectrometer. In the CHRIS scene over the forest test site Järvselja we cannot find bright targets large enough for satellite measurements. Therefore, the CHRIS calibration is validated against top-of-canopy directional reflectance of large forest stands measured onboard a helicopter.

Reflectance spectra of several homogeneous stands in the CHRIS scene 5703 were compared to airborne measured data of 26 July 2006. Spectral bands of the UAVSpec were combined to the equivalent CHRIS bands. The footprint of the UAVSpec spectrometer's field-of-view from the height 100 m is 9.5 m^2, which is significantly less than the CHRIS pixel; however, altogether 1302 recorded UAVSpec spectra over 520 CHRIS pixels over 63 homogeneous stands were involved in the comparison which represent all dominating species at the test site. Difference in view angles was accounted for by numerical simulations of angular dependence of stand reflectance for these targets with the FRT forest reflectance model (Kuusk and Nilson, 2000). The comparison gave us correction factors for the CHRIS calibration coefficients (Table 4.3 and Fig. 4.4). Correction factors were calculated as the ratio of the mean stand directional reflectance from helicopter measurements to the mean top-of-canopy stand reflectance from CHRIS measurements separately

Table 4.8. Mode 3 spectral bands (nm) and the corresponding correction factors for CHRIS calibration coefficients

Band	Low	High	Middle	C_λ	STD
Band 1	437.3	447.8	442.4	0.6373	0.0684
Band 2	484.5	496.1	490.2	0.7658	0.0848
Band 3	524.4	535.9	530.0	0.9687	0.1018
Band 4	545.0	557.9	551.3	0.8756	0.0920
Band 5	564.7	575.4	570.0	1.0063	0.1115
Band 6	624.5	638.6	631.4	1.0439	0.1267
Band 7	653.5	669.2	661.2	1.0587	0.1446
Band 8	669.2	680.2	674.6	1.0720	0.1513
Band 9	691.6	703.4	697.5	1.0281	0.1171
Band 10	703.4	709.6	706.5	1.0973	0.1172
Band 11	709.6	715.7	712.6	1.1456	0.1214
Band 12	735.1	748.6	741.8	0.9907	0.1038
Band 13	748.6	755.6	752.1	1.0294	0.1069
Band 14	770.0	792.5	781.1	1.0116	0.1025
Band 15	858.8	886.2	872.3	1.0028	0.0962
Band 16	886.2	905.2	895.7	1.0208	0.0956
Band 17	905.2	914.9	910.0	1.0021	0.0934
Band 18	997.5	1041.3	1019.0	0.8978	0.0848

C_λ – correction factor.
STD – standard deviation of the correction factor.

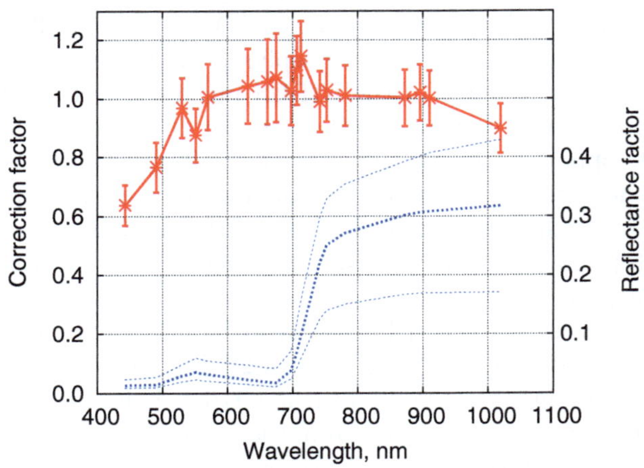

Fig. 4.4. Correction factors for CHRIS calibration coefficients (the upper curve with error bars and left axis), and mean stand reflectance and the range of stand reflectances involved in the comparison (the right axis).

for every stand. Correction factors C_λ and their standard deviations in Table 4.8 are the mean value and standard deviation over 63 stands. All three sets of spectral images were radiometrically rescaled using these correction factors.

4.6 Data

4.6.1 Illumination conditions

Diffuse-to-total spectral flux ratio D_λ/Q_λ was measured at the test site during PROBA overpass and helicopter measurements. The fish-eye sky image (Fig. 4.5) shows the free horizon at the place of incident flux measurements. During PROBA overpass the aerosol optical depth τ_{550} at AERONET Tõravere station was very low, $\tau_{550} = 0.08$, which is about three times less than in the profiles by Elterman (1968) which are often used in simulation studies. Therefore the D_λ/Q_λ ratio was very low, too. During helicopter measurements in 2006 and 2008 the atmosphere transparency was even better than during CHRIS measurements, only in August 2008 the sky flux was during helicopter measurements remarkably higher than during other measurements (Fig. 4.6). Sky spectral radiance in four spectral bands was measured at Tõravere, 45 km from the test site by an AERONET sun photometer. AERONET sun photometer measures sky radiance at almucantar and in the principal plane (Holben et al., 1998). Data are available for dates of satellite and airborne measurements except of 8 August 2007. Sky radiance profiles for forest observation dates are in Figs. 4.7–4.14. Data for 9 August 2007 characterize sky radiance during helicopter measurements on 8 August 2007. In Figs. 4.7–4.10 the azimuth is relative to the sun azimuth, thus the azimuth 0° corresponds to the Sun direction. As the sun zenith angle was varying, the aureole peaks in Figs. 4.11–4.14 do not overlap.

Fig. 4.5. Fish-eye image at the test site where incident spectral fluxes were measured.

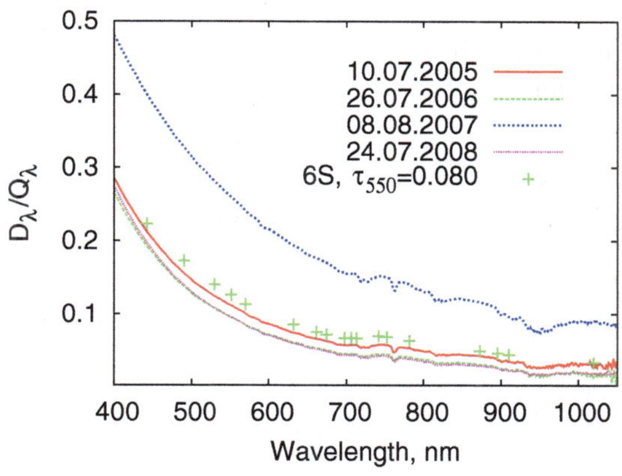

Fig. 4.6. Diffuse-to-total flux ratio D_λ/Q_λ from FieldSpec measurements during PROBA overpass (10.07.2005), helicopter measurements (26.07.2006, 08.08.2007, 24.07.2008), and simulated with 6S for CHRIS spectral bands using AERONET level 2 atmosphere data during CHRIS acquisition (Also see Fig. 4.15.).

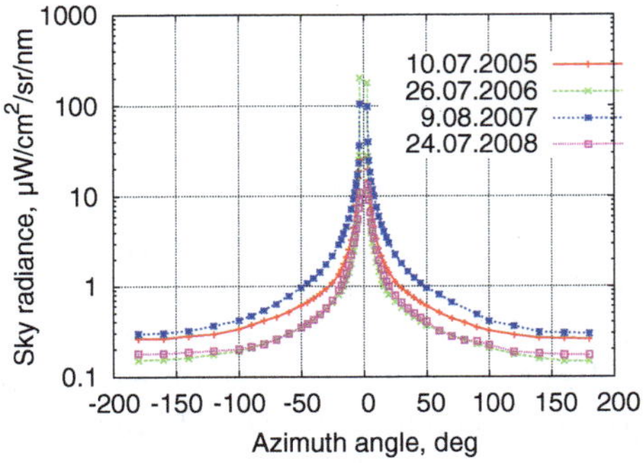

Fig. 4.7. Almucantar sky radiance, 1020 nm.

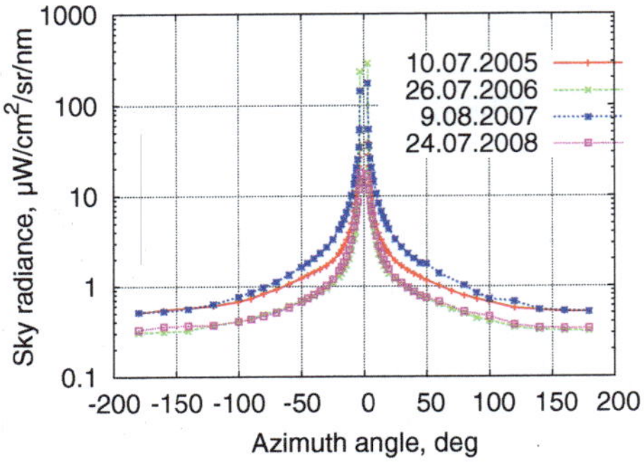

Fig. 4.8. Almucantar sky radiance, 870 nm.

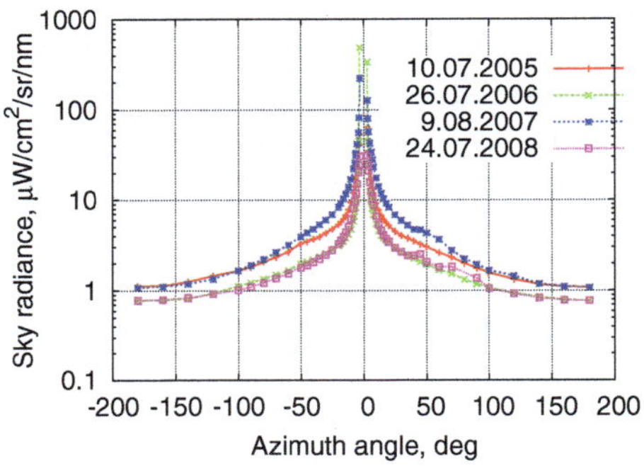

Fig. 4.9. Almucantar sky radiance, 675 nm.

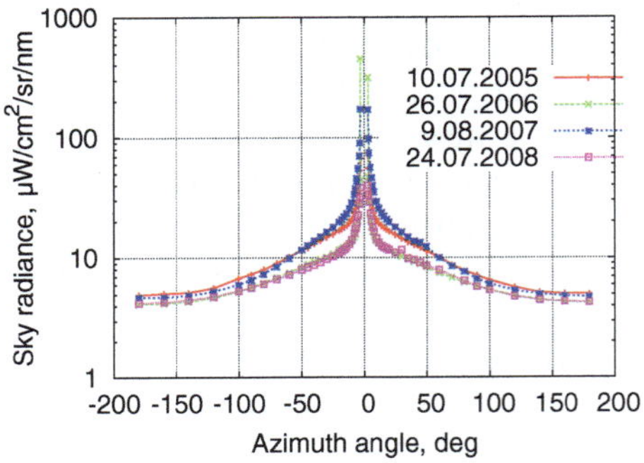

Fig. 4.10. Almucantar sky radiance, 440 nm.

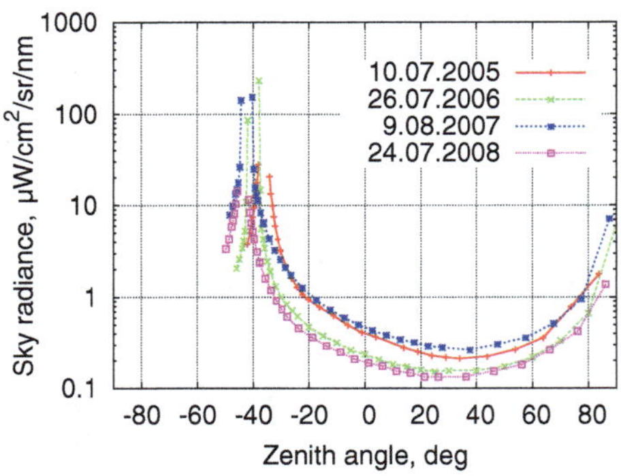

Fig. 4.11. Sky radiance in the principal plane at Tõravere, 1020 nm.

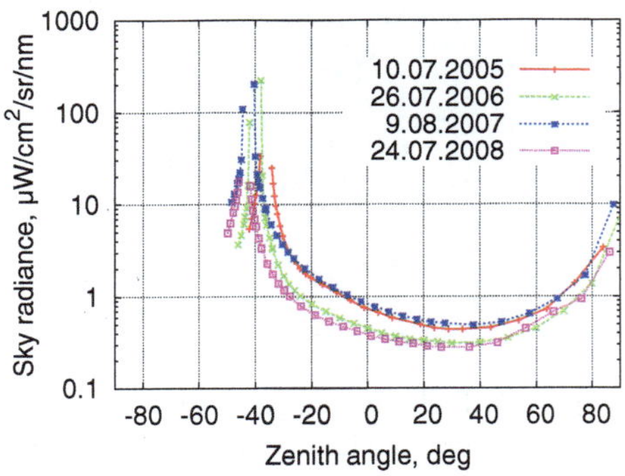

Fig. 4.12. Sky radiance in the principal plane at Tõravere, 870 nm.

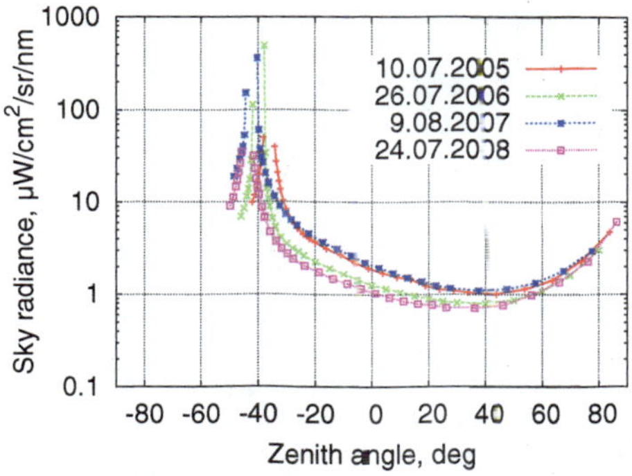

Fig. 4.13. Sky radiance in the principal plane at Tõravere, 675 nm.

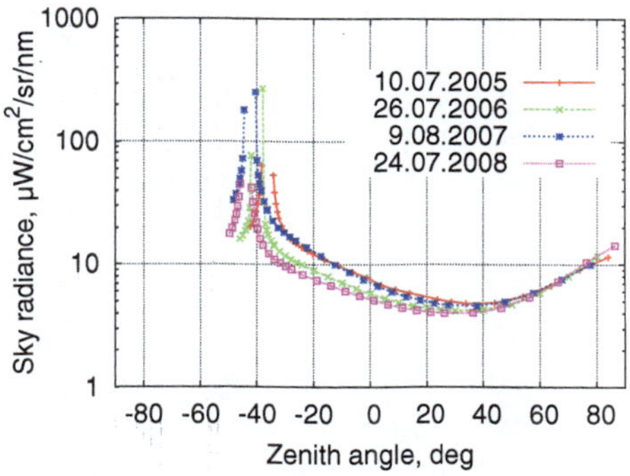

Fig. 4.14. Sky radiance in the principal plane at Tõravere, 440 nm.

4.6.2 Stands

Canopy cover and crown cover estimates, and effective and corrected LAI values for all three stands are reported in Table 4.9. The Nilson and Kuusk (2004) procedure was applied for the correction of LAI-2000 data. Two different correction procedures were applied: first using measured canopy cover and recorded LAI-2000 gap fractions; second, one iteration was performed which modified crown radius, and consequently canopy cover and gap fractions in LAI-2000 rings.

Table 4.9. Canopy parameters

Stand	LAI_{all}	LAI_{eff}	STD_{LAI}	LAI_1	LAI_2	CaC	CrC
Birch[1]	3.93	2.94/0.80	0.346/0.126	3.14	2.89	0.80	1.09
Pine	1.86	1.75	0.149	2.55	2.21	0.74	0.79
Spruce	4.36	3.76	0.617	5.03	4.32	0.90	1.25

[1] LAI_{eff}(July)/LAI_{eff}(November); STD_{LAI} – standard deviation of LAI_{eff}. LAI_{all} – allometric LAI; LAI_{eff} – effective LAI, measured with LAI-2000; CaC – canopy cover; CrC – crown cover; LAI_1 – corrected LAI, using measured LAI-2000 gap fractions (Nilson and Kuusk, 2004); LAI_2 – corrected LAI, using modified crown radii (Nilson and Kuusk, 2004).

Crown envelopes for deciduous species and Scots pine were most similar to ellipsoidal model. For the upper layer Norway spruce trees about 20% of the lower crown was close to cylinder and the upper part was characteristically cone-shaped. Norway spruces in lower tree layers had cone-shaped crowns.

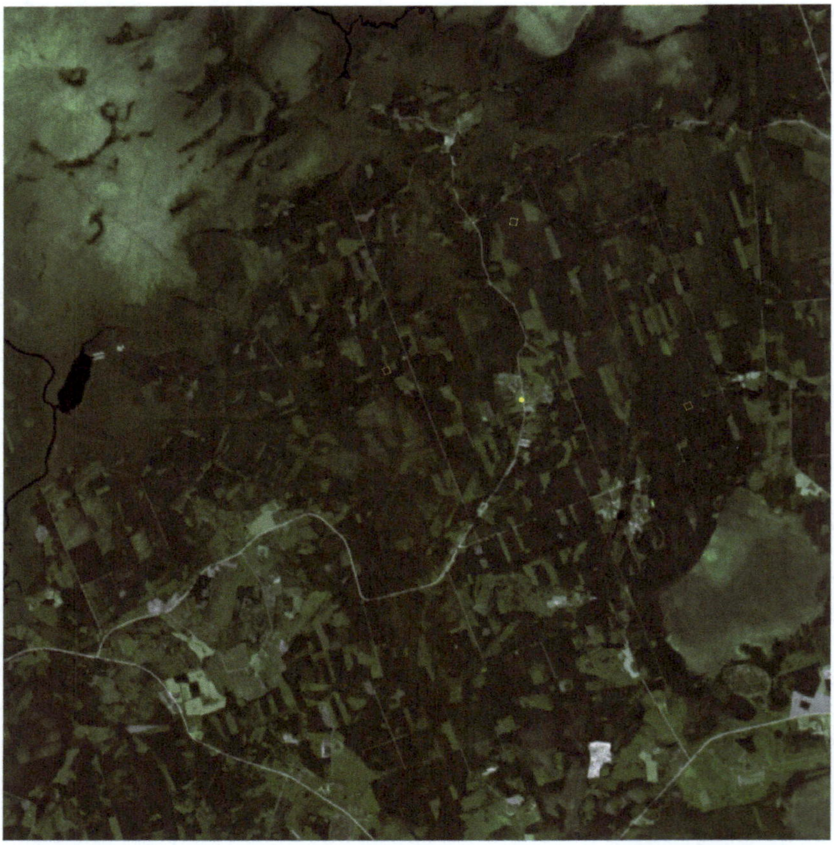

Fig. 4.15. The RGB image of the CHRIS scene 5703. Yellow squares mark the selected stands, and the yellow dot marks the place of the FieldSpec measurements of incident radiation. The image size is 12.6 × 12.6 km.

4.6.2.1 Birch stand

Structure

This deciduous stand grows on the typical brown gley-soil (*Eutri Mollic Gleysol* – FAO-UNESCO soil classification). Growth conditions are good for forest ($H_{100} = 28.7$). Stand age is 49 years. Dominating species are birch (*Betula pendula*) 57%, Common alder (*Alnus glutinosa*) 29.5% and aspen (*Populus tremula*) 11%, the total number of trees is 1031. There are two tree layers distinguishable according to the social status of the trees, the lower tree layer is mostly consisting of *Tilia cordata* and *Picea abies*. The stand was thinned in September–October 2004. Forest understorey vegetation is dominated by the mixture of several grass species. Moss layer is sparse or missing.

Coordinates of the plot center are: 58° 16′ 49.81″ N, 27° 19′ 51.53″ E; L-EST97: X:6464827.8, Y:695338.6. Azimuth of the y-coordinate of the 100 × 100 m plot is 348.7°.

The perspective and vertical view of the stand are in Figs. 4.16 and 4.17, respectively. Images are created with SVS software (McGaughey, 1997) using individual tree data. Also see Figs. 4.18 to 4.21 and Table 4.10.

Fig. 4.16. The birch stand, perspective view.

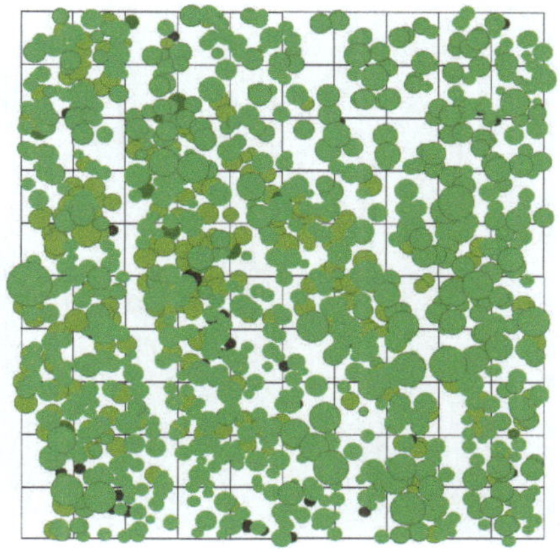

Fig. 4.17. The birch stand, vertical view.

Fig. 4.18. A sample understory photo in the birch stand, 10 August 2011.

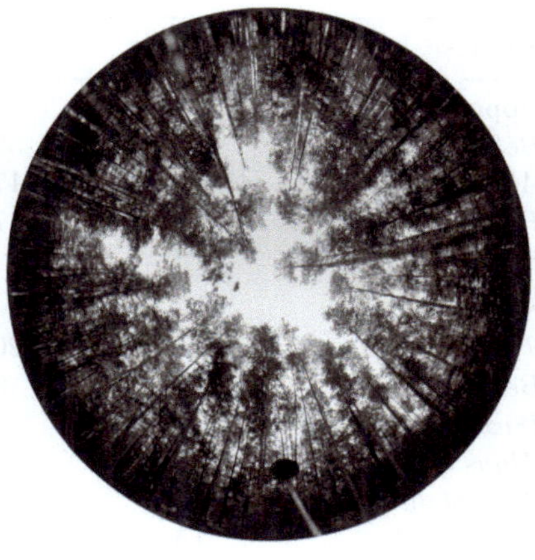

Fig. 4.19. A sample hemispherical image in the birch stand, 29 July 2007.

Fig. 4.20. The birch stand, 10 May 2007.

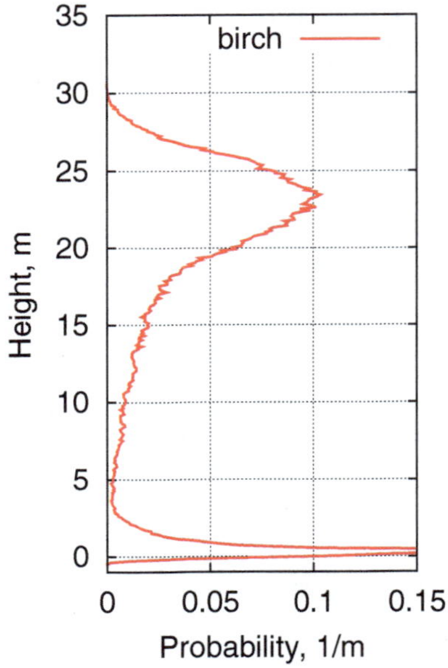

Fig. 4.21. Vertical distribution of the lidar first return – the birch stand.

Table 4.10. Mean tree parameters – the birch stand

Species	Code	N	H	$D_{1.3}$	L	R_{cr}
Upper layer						
Betula pendula	KS	399	26.5	20.7	9.2	1.6
Alnus glutinosa	LM	176	23.4	22.4	9.8	2.0
Populus tremula	HB	78	26.8	21.6	8.2	2.0
Salix ssp.	PJ	1	24.0	23.8	9.9	2.0
Second layer						
Tilia cordata	PN	205	15.9	12.8	8.1	1.9
Betula pendula	KS	66	17.9	10.5	5.6	1.0
Fraxinus excelsior	SA	30	15.4	10.9	4.0	1.6
Alnus glutinosa	LM	20	17.5	13.1	8.5	1.4
Acer platanoides	VA	16	15.7	11.3	4.3	1.9
Ulmus glabra	JA	1	15.9	11.4	8.1	1.0
Regeneration layer						
Picea abies	KU	39	8.9	8.9	4.8	1.2

N – number of trees; H – mean tree height, m; $D_{1.3}$ – mean breast-height-diameter, cm; L – mean length of live crown, m; R_{cr} – mean maximum radius of crown, m.

Optical properties

Reflectance and transmittance spectra of birch and alder leaves are plotted in Figs. 4.22 and 4.23. Spectra of aspen leaves are similar to those of birch leaves. Spectra of leaf reflectance and transmittance, understorey reflectance, and top-of-canopy-reflectance from helicopter and CHRIS measurements for CHRIS Mode 3 wavelengths in July 2008 are tabulated in Table 4.11. (Also see Figs. 4.24 to 4.30.)

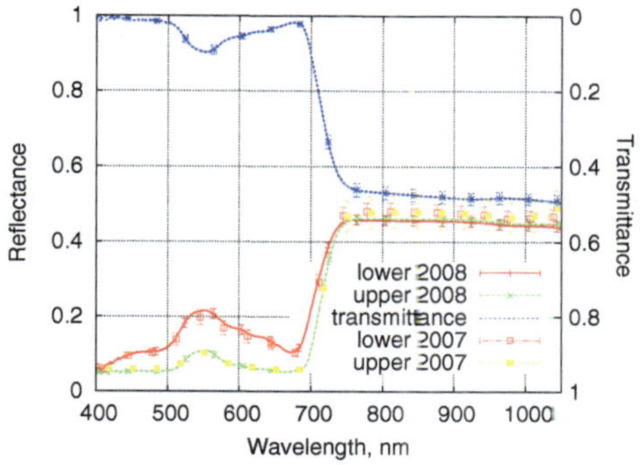

Fig. 4.22. Leaf reflectance and transmittance of *Betula pendula*.

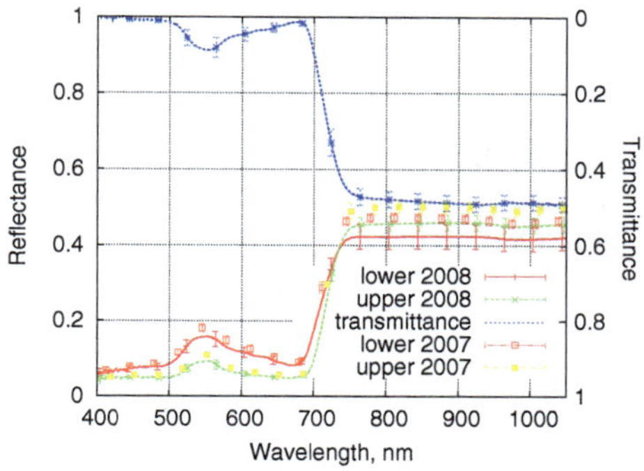

Fig. 4.23. Leaf reflectance and transmittance of *Alnus glutinosa*.

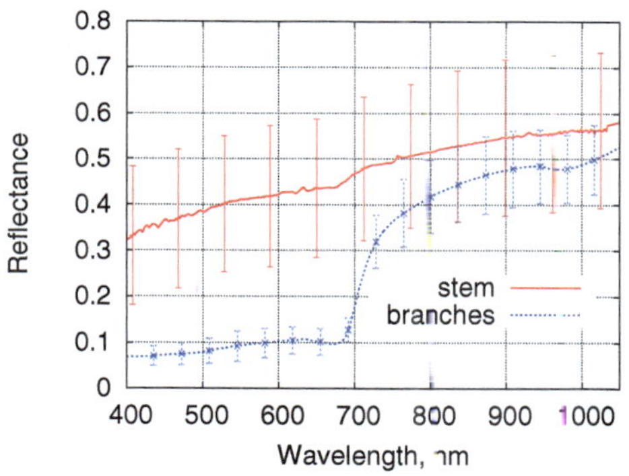

Fig. 4.24. Stem and branch bark reflectance of *Betula pendula*. The stem bark was measured with GER-2600 in July 2001.

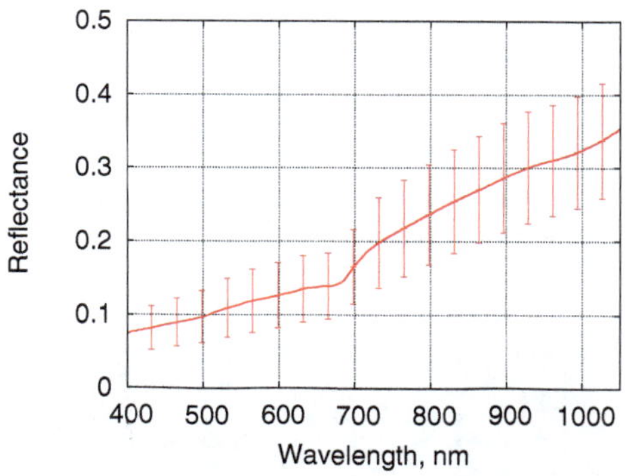

Fig. 4.25. Stem bark reflectance of *Alnus glutinosa*.

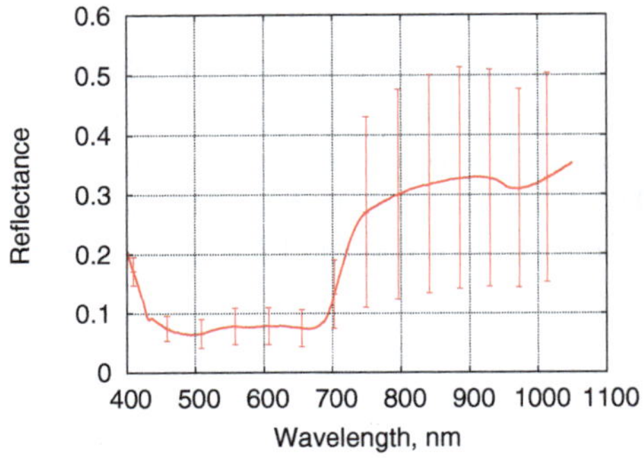

Fig. 4.26. Stem bark reflectance of *Populus tremula*, BOREAS Old Aspen, August 1994 (Newcomer et al., 2000).

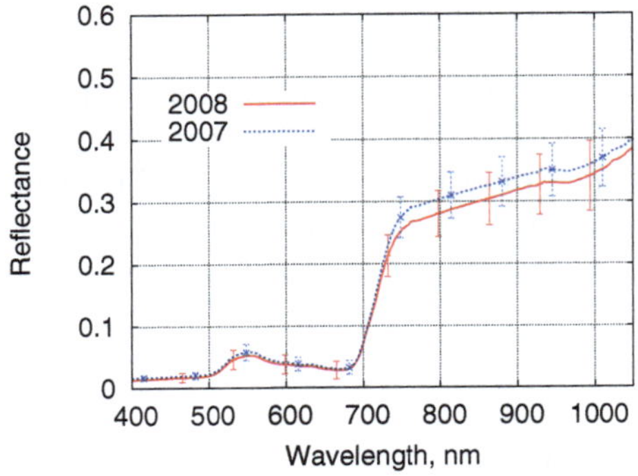

Fig. 4.27. Average reflectance of understorey vegetation in the birch stand. Error bars show the standard deviation of nine mean reflectance at LAI points L1–L9.

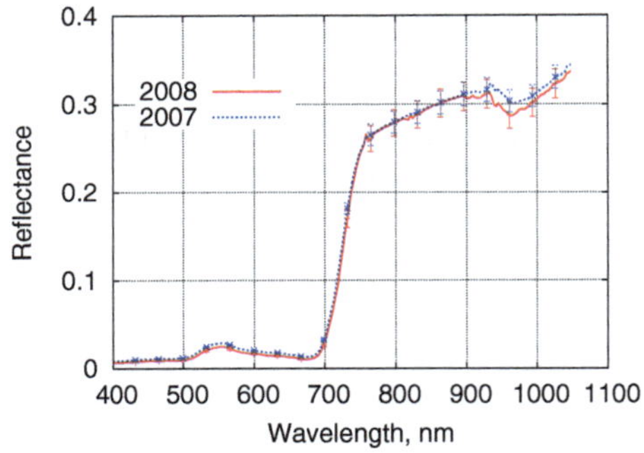

Fig. 4.28. Average top-of-canopy nadir reflectance of the birch stand. Error bars show the standard error of the mean value.

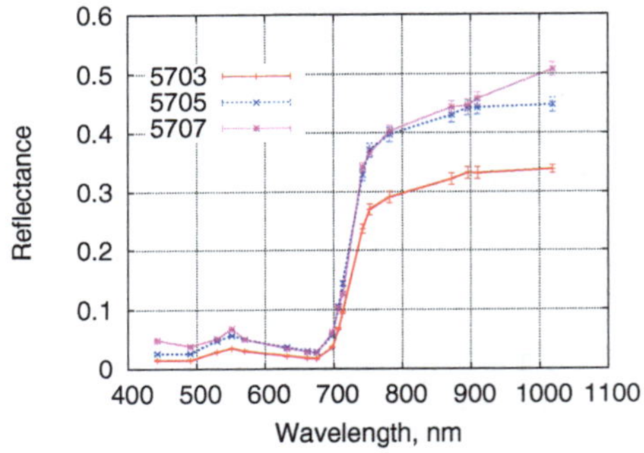

Fig. 4.29. Average top-of-canopy directional reflectance of the birch stand from CHRIS images.

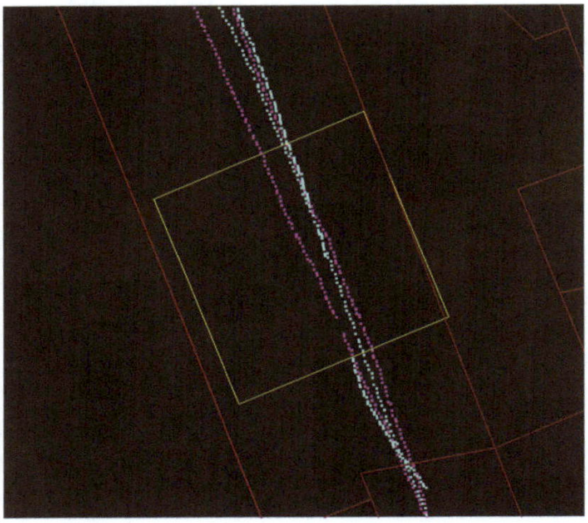

Fig. 4.30. The birch stand in the CHRIS image: red – stand boundaries, yellow – the boundary of the study area, magenta – helicopter measurements in 2007, blue – helicopter measurements in 2008.

Table 4.11. Reflectance and transmittance spectra in the birch stand

λ(nm)	$r_{B,u}$	$r_{B,l}$	t_B	$r_{A,u}$	$r_{A,l}$	t_A	$r_{P,u}$	$r_{P,l}$	t_P	r_{us}	r_{toc}	STD_{toc}	r_{CH}	STD_{CH}
442	0.0531	0.0914	0.0080	0.0502	0.0688	0.0041	0.0374	0.0784	0.0170	0.0157	0.0087	0.0005	0.0157	0.0013
490	0.0539	0.1097	0.0153	0.0498	0.0791	0.0108	0.0371	0.0900	0.0219	0.0186	0.0097	0.0007	0.0157	0.0010
530	0.0943	0.1991	0.0770	0.0809	0.1451	0.0696	0.0687	0.1703	0.0915	0.0450	0.0202	0.0016	0.0297	0.0010
551	0.1100	0.2163	0.0964	0.0932	0.1585	0.0872	0.0807	0.1838	0.1129	0.0530	0.0257	0.0022	0.0357	0.0010
570	0.0934	0.1989	0.0809	0.0794	0.1437	0.0713	0.0665	0.1675	0.0953	0.0464	0.0215	0.0019	0.0315	0.0010
631	0.0611	0.1465	0.0414	0.0547	0.1051	0.0343	0.0410	0.1216	0.0493	0.0348	0.0153	0.0014	0.0228	0.0010
661	0.0540	0.1164	0.0246	0.0495	0.0865	0.0189	0.0359	0.0955	0.0294	0.0289	0.0119	0.0010	0.0189	0.0012
675	0.0547	0.1062	0.0179	0.0511	0.0843	0.0132	0.0377	0.0877	0.0224	0.0287	0.0111	0.0009	0.0178	0.0012
698	0.0959	0.2004	0.0873	0.0836	0.1593	0.0797	0.0720	0.1725	0.1025	0.0663	0.0280	0.0027	0.0374	0.0011
707	0.1724	0.2707	0.1731	0.1484	0.2226	0.1641	0.1372	0.2316	0.1928	0.1090	0.0530	0.0049	0.0685	0.0016
713	0.2314	0.3141	0.2316	0.2024	0.2642	0.2227	0.1909	0.2692	0.2534	0.1382	0.0713	0.0056	0.0963	0.0027
742	0.4395	0.4483	0.4340	0.4254	0.4097	0.4397	0.4104	0.4012	0.4742	0.2410	0.2182	0.0126	0.2373	0.0077
752	0.4559	0.4574	0.4531	0.4472	0.4213	0.4616	0.4294	0.4107	0.4946	0.2564	0.2470	0.0138	0.2705	0.0092
781	0.4614	0.4584	0.4657	0.4565	0.4237	0.4749	0.4353	0.4110	0.5067	0.2740	0.2726	0.0148	0.2912	0.0100
872	0.4618	0.4565	0.4792	0.4606	0.4253	0.4857	0.4355	0.4089	0.5165	0.3079	0.3042	0.0160	0.3219	0.0099
896	0.4610	0.4554	0.4818	0.4605	0.4251	0.4885	0.4346	0.4082	0.5194	0.3168	0.3088	0.0161	0.3322	0.0103
910	0.4608	0.4550	0.4830	0.4606	0.4252	0.4891	0.4346	0.4077	0.5206	0.3214	0.3096	0.0161	0.3321	0.0099
1019	0.4491	0.4440	0.4879	0.4539	0.4195	0.4894	0.4301	0.4017	0.5262	0.3578	0.3182	0.0164	0.3386	0.0070

λ – wavelength; r – reflectance; t – transmittance; STD – standard deviation.
Subscripts: B – *Betula pendula*; A – *Alnus glutinosa*; P – *Populus tremula*; u – upper side; l – lower side; us – understorey; toc – top-of-canopy; CH – CHRIS scene 5703.

4.6.2.2 Pine stand

Structure

Pine (*Pinus sylvestris*) stand grows on the transitional bog (see Fig. 4.31). The soil is deep (>1.3 m) *Sphagnum* peat. The growth conditions are poor, as indicated by $H_{100} = 10.8$ (stand height at age of 100 years) in the forest inventory database. Now the stand height is 15.6 meters, the stand is 124 years old, and stand density is 1122 trees per hectare. Forest understorey vegetation is composed of *Ledum palustre*, sparse *Eriophorum vaginatum*, and continuous *Sphagnum ssp.* moss layer.

Coordinates of the stand center are: 58° 18′ 41.19″ N, 27° 17′ 48.63″ E; L-EST97: X:6468170.5, Y:693169.0. Azimuth of the y-coordinate is 16.3°.

The perspective and vertical view of the stand are in Figs. 4.33 and 4.34, respectively. Also see Figs. 4.32, 4.35, and 4.36 and Table 4.12.

Table 4.12. Mean tree parameters, the pine stand

Species	Code	N	H	$D_{1.3}$	L	R_{cr}
Upper layer						
Pinus sylvestris	MA	1115	15.9	18.0	4.2	1.5
Understorey						
Betula pendula	KS	6	4.1	5.5	2.9	0.8
Picea abies	KU	1	5.0	5.6	3.6	1.1

Notation: see in Table 4.10.

Fig. 4.31. The pine stand, 10 May 2007.

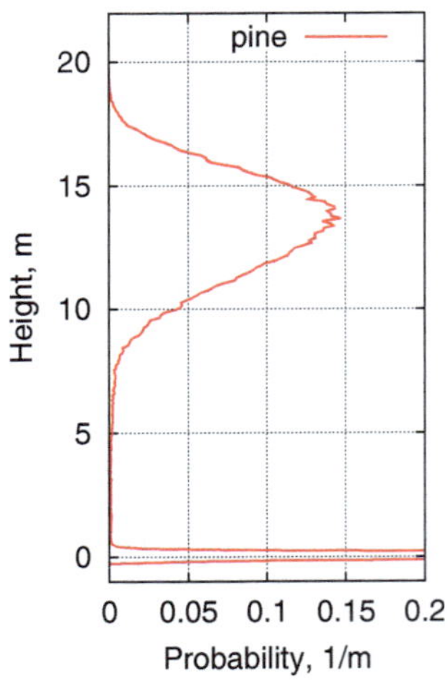

Fig. 4.32. Vertical distribution of the lidar first return – the pine stand.

Fig. 4.33. The pine stand – perspective view.

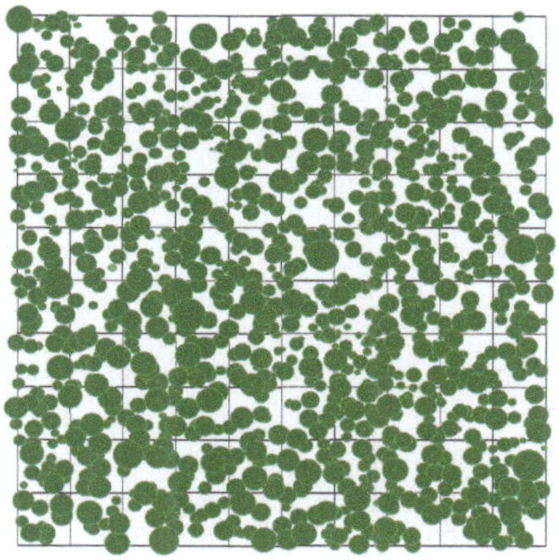

Fig. 4.34. The pine stand – vertical view.

Fig. 4.35. A sample understorey photo in the pine stand, 10 August 2011.

Fig. 4.36. A sample hemispherical image in the pine stand, 29 July 2007.

Optical properties

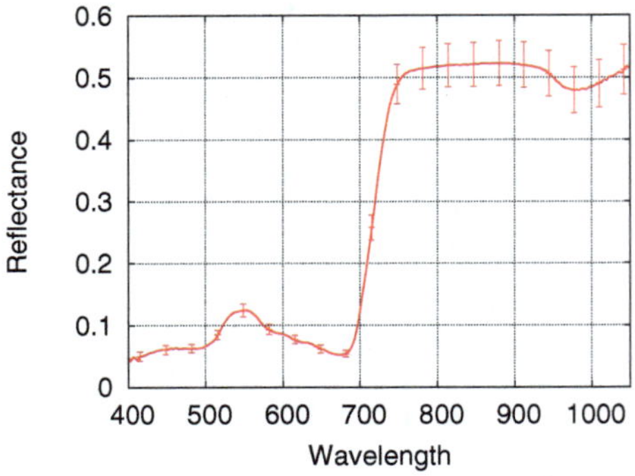

Fig. 4.37. Reflectance of a bunch of needles of *Pinus silvestris*.

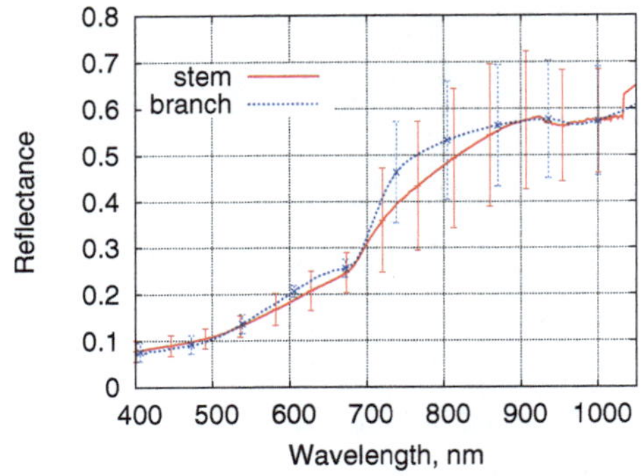

Fig. 4.38. Reflectance of stem and branch bark of *Pinus silvestris*. The stem bark was measured with GER-2600 in July 2001.

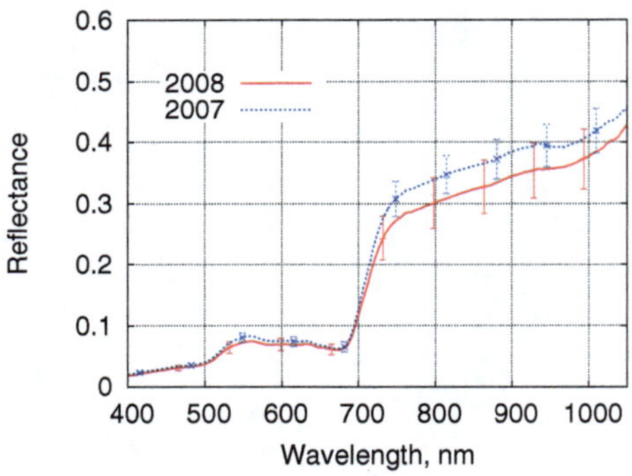

Fig. 4.39. Average reflectance of understorey vegetation in the pine stand. Error bars show the standard deviation of nine mean reflectance at LAI points L1–L9.

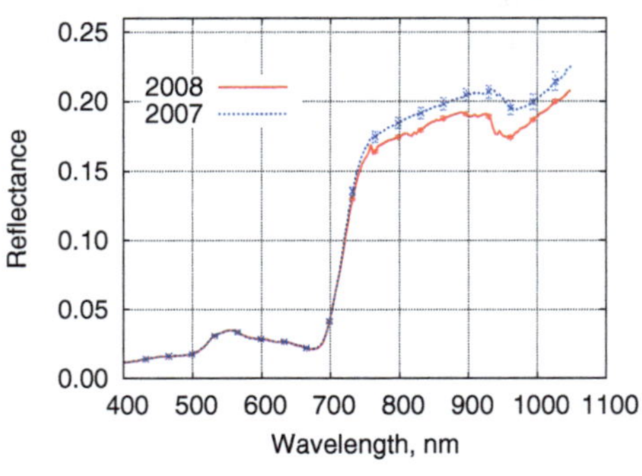

Fig. 4.40. Average top-of-canopy nadir reflectance of the pine stand. Error bars show the standard error of the mean value.

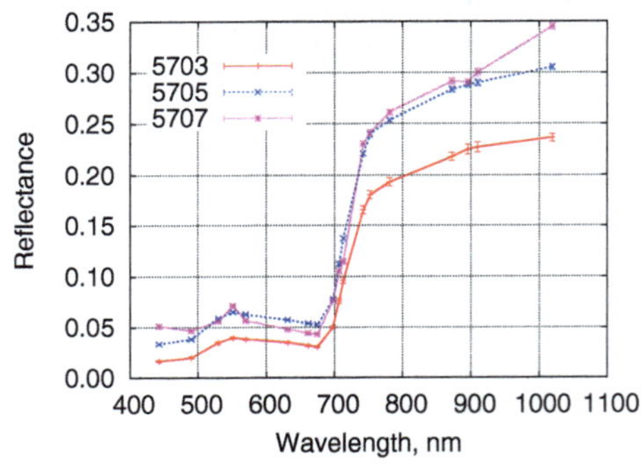

Fig. 4.41. Average top-of-canopy directional reflectance of the pine stand from CHRIS images.

Table 4.13. Reflectance and transmittance spectra in the pine stand

λ (nm)	r_l	r_{st}	r_{us}	r_{toc}	STD_{toc}	r_{CH}	STD_{CH}
442	0.0611	0.0900	0.0276	0.0150	0.0001	0.0168	0.0009
490	0.0640	0.1056	0.0350	0.0170	0.0001	0.0200	0.0009
530	0.1151	0.1289	0.0641	0.0309	0.0002	0.0349	0.0012
551	0.1254	0.1435	0.0740	0.0344	0.0003	0.0399	0.0011
570	0.1058	0.1587	0.0722	0.0318	0.0003	0.0385	0.0009
631	0.0725	0.2099	0.0702	0.0266	0.0002	0.0350	0.0009
661	0.0571	0.2338	0.0631	0.0224	0.0002	0.0319	0.0009
675	0.0532	0.2482	0.0511	0.0212	0.0002	0.0306	0.0009
698	0.1142	0.3048	0.1100	0.0413	0.0004	0.0505	0.0014
707	0.1952	0.3296	0.1450	0.0600	0.0005	0.0762	0.0025
713	0.2300	0.3449	0.1720	0.0780	0.0006	0.0965	0.0028
742	0.4683	0.3990	0.2640	0.1499	0.0010	0.1656	0.0038
752	0.4969	0.4142	0.2754	0.1607	0.0011	0.1806	0.0038
781	0.5150	0.4546	0.2932	0.1715	0.0011	0.1931	0.0039
872	0.5231	0.5528	0.3300	0.1898	0.0012	0.2179	0.0041
896	0.5222	0.5695	0.3429	0.1911	0.0012	0.2251	0.0049
910	0.5206	0.5757	0.3482	0.1907	0.0012	0.2271	0.0047
1019	0.4962	0.5797	0.3919	0.1968	0.0012	0.2361	0.0036

λ – wavelength; r_l – shoot reflectance; r_{st} – stem bark reflectance; r_{us} – understorey reflectance; STD – standard deviation.
Subscripts: toc – top-of-canopy; CH – CHRIS scene 5703.

Fig. 4.42. The pine stand in the CHRIS image: red – stand boundaries, yellow – the boundary of the study area, magenta – helicopter measurements in 2007, blue – helicopter measurements in 2008.

4.6.2.3 Spruce stand

Structure

Spruce (*Picea abies*) stand grows on a *Gleyi Ferric Podzol* site (see Fig. 4.43). However, the growth conditions are rather good ($H_{100} = 29.1$) because of drainage.

Table 4.14. Mean tree parameters, the spruce stand

Species	Code	N	H	$D_{1.3}$	L	R_{cr}
Upper layer						
Picea abies	KU	624	23.2	23.5	10.8	1.8
Betula pendula	KS	143	24.5	17.9	8.5	1.5
Alnus glutinosa	LM	3	22.4	20.3	9.6	2.1
Populus tremula	HB	2	25.1	18.3	6.9	1.5
Pinus sylvestris	MA	2	24.5	26.4	10.4	2.1
Second layer						
Betula pendula	KS	152	17.5	9.3	4.5	0.9
Picea abies	KU	517	13.8	11.1	6.3	1.2
Regeneration layer						
Picea abies	KU	157	8.0	6.9	4.4	1.1
Picea abies	KU	89	5.3	5.2	3.7	1.1

Notation: see in Table 4.10.

Fig. 4.43. The spruce stand, 10 May 2007.

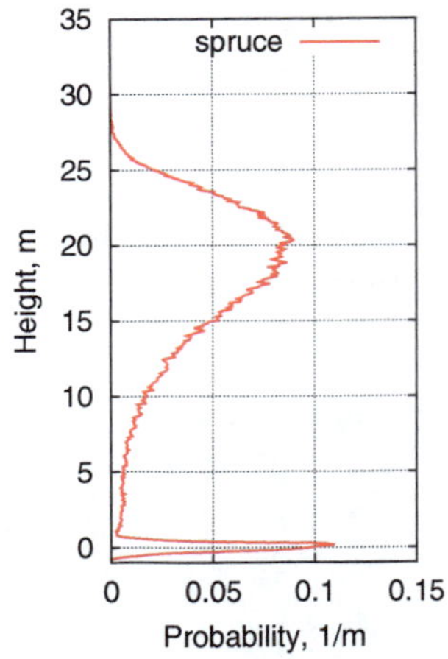

Fig. 4.44. Vertical distribution of the lidar first return – the spruce stand.

Stand age is 59 years. There are two tree layers distinguishable according to the social status of the trees. Average height in the first (upper) layer is 23.2 meters and there are 774 trees per hectare. Stand density in the second (lower) layer is 915 trees per hectare and the height of trees ranges from 3.5 to 20 meters. Canopy cover is high (0.89) and, therefore, forest understorey vegetation is either partially missing or consists only of mosses such as *Hylocomium splendens* or *Pleurozium schreberi*.

Coordinates of the stand center are: 58° 17' 43.0" N, 27° 15' 22.0" E; L-EST97: X:6466255.7, Y:690873.2. Azimuth of the y-coordinate is 349.8°.

The perspective and vertical view of the stand are in Figs. 4.45 and 4.46, respectively. Also see Figs. 4.44, 4.47, and 4.48 and Table 4.14.

Fig. 4.45. The spruce stand, perspective view.

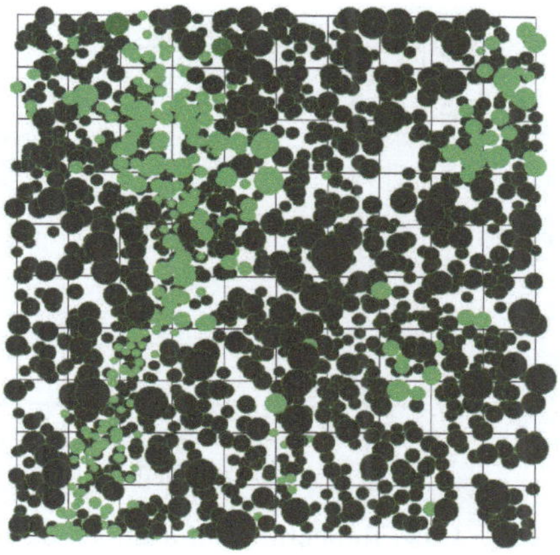

Fig. 4.46. The spruce stand, vertical view.

Fig. 4.47. A sample understory photo in the spruce stand, 10 August 2011.

Fig. 4.48. A sample hemispherical image in the spruce stand, 22 July 2007.

Optical properties

See Figs. 4.49 to 4.54 and Table 4.15.

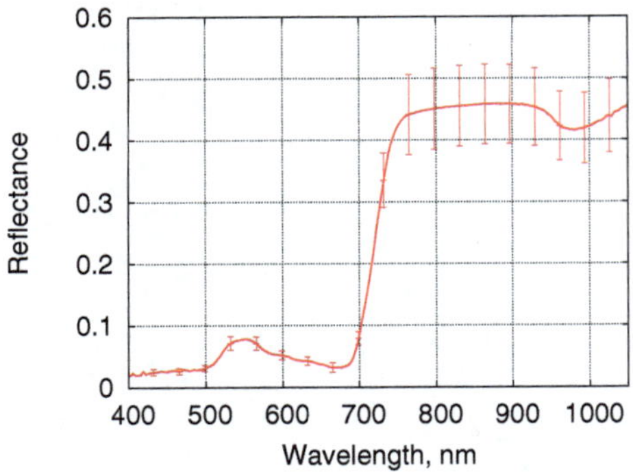

Fig. 4.49. Reflectance of a bunch of needles of *Picea abies*.

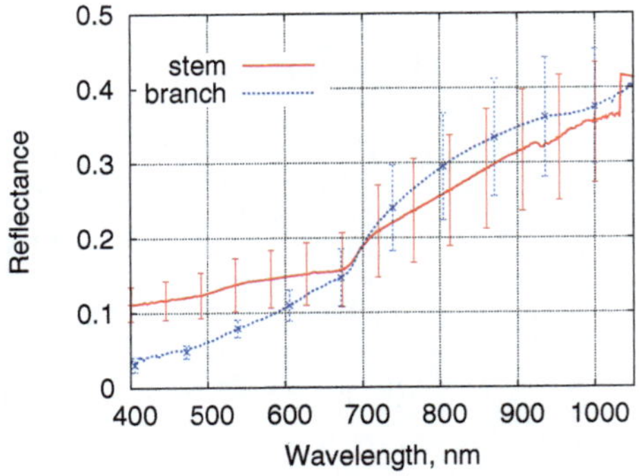

Fig. 4.50. Reflectance of stem and branch bark of *Picea abies*. The stem bark was measured with GER-2600 in July 2001.

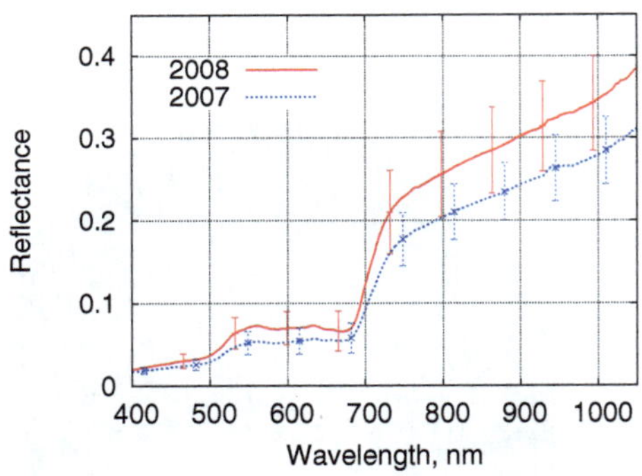

Fig. 4.51. Average reflectance of understorey vegetation in the spruce stand. Error bars show the standard deviation of nine mean reflectance at LAI points L1–L9.

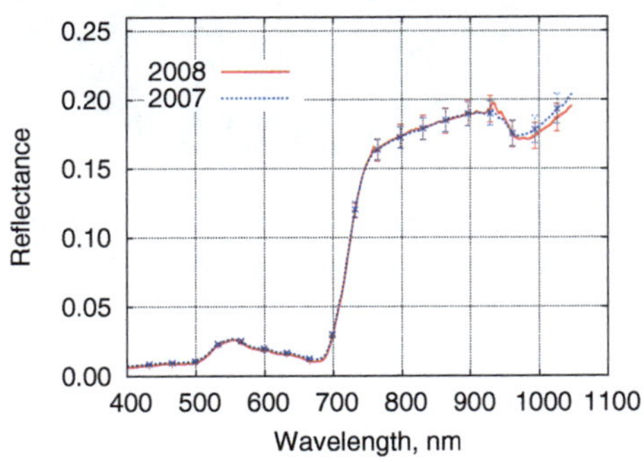

Fig. 4.52. Average top-of-canopy nadir reflectance of the spruce stand. Error bars show the standard error of the mean value.

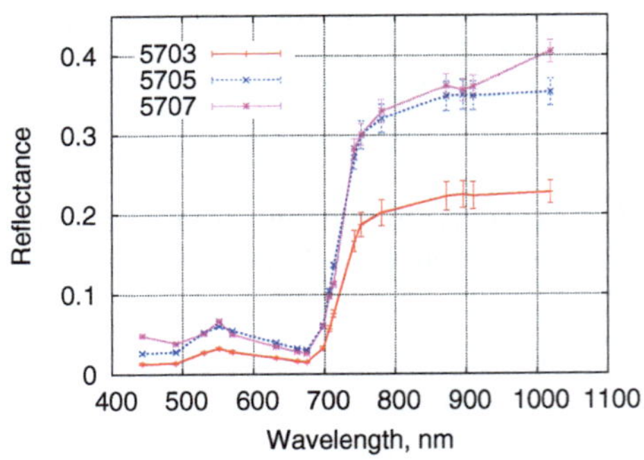

Fig. 4.53. Average top-of-canopy directional reflectance of the spruce stand from CHRIS images.

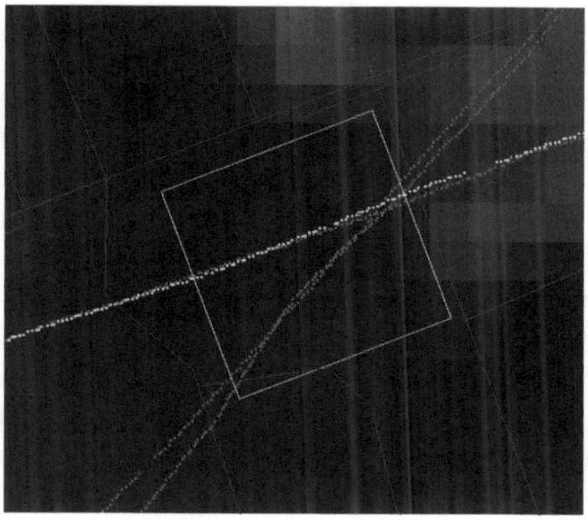

Fig. 4.54. The spruce stand in the CHRIS image: red – stand boundaries, yellow – the boundary of the study area, magenta – helicopter measurements in 2007, blue – helicopter measurements in 2008.

Table 4.15. Reflectance and transmittance spectra in the spruce stand

λ (nm)	r_l	r_{st}	r_{us}	r_{toc}	STD_{toc}	r_{CH}	STD_{CH}
442	0.0266	0.1166	0.0265	0.0083	0.0003	0.0135	0.0014
490	0.0289	0.1235	0.0341	0.0093	0.0003	0.0142	0.0012
530	0.0719	0.1357	0.0631	0.0225	0.0009	0.0273	0.0016
551	0.0791	0.1412	0.0710	0.0262	0.0010	0.0326	0.0015
570	0.0691	0.1438	0.0711	0.0230	0.0009	0.0282	0.0020
631	0.0430	0.1521	0.0733	0.0159	0.0006	0.0205	0.0015
661	0.0333	0.1546	0.0674	0.0118	0.0004	0.0164	0.0015
675	0.0323	0.1579	0.0667	0.0106	0.0004	0.0151	0.0014
698	0.0794	0.1870	0.1172	0.0286	0.0010	0.0326	0.0022
707	0.1301	0.1979	0.1510	0.0480	0.0014	0.0571	0.0035
713	0.1800	0.2037	0.1670	0.0584	0.0020	0.0763	0.0046
742	0.3946	0.2219	0.2225	0.1431	0.0067	0.1667	0.0132
752	0.4260	0.2274	0.2305	0.1569	0.0076	0.1875	0.0151
781	0.4471	0.2439	0.2476	0.1695	0.0083	0.2021	0.0166
872	0.4589	0.2986	0.2889	0.1863	0.0092	0.2229	0.0181
896	0.4584	0.3113	0.3002	0.1904	0.0094	0.2252	0.0168
910	0.4574	0.3187	0.3063	0.1915	0.0095	0.2238	0.0174
1019	0.4344	0.3616	0.3603	0.1836	0.0097	0.2283	0.0146

λ – wavelength; r_l – shoot reflectance; r_{st} – stem bark reflectance; r_{us} – understorey reflectance; STD – standard deviation.
Subscripts: toc – top-of-canopy; CH – CHRIS scene 5703.

4.7 Concluding remarks

Extensive ground based measurements of the structural and spectral properties have been performed at three forest stands near Järvselja, Estonia. This dataset is complemented by high-resolution airborne measurements and multi-view hyperspectral CHRIS data.

The obtained tree location coordinates are in the radius of 5–20 cm of their true position. Standard errors of regression models for tree height, crown length and crown radius are 1.6 m, 1.7 m and 0.28 m, respectively. The compiled dataset of tree inventory parameters makes it possible to approximate individual crown architecture or calculate average estimates at stand or stand element level.

Low aerosol optical depth and water content of the atmosphere at such a high geographical latitude, and good weather conditions during acquisition made atmospheric correction of satellite data less sensitive to atmosphere parameters. Top-of-canopy helicopter measurements were used for the radiometric correction of CHRIS data. The radiometric quality of helicopter measurements is according to equation (4.11) determined by the gain and offset errors of spectrometers, and errors in the reference reflectance. The sensitivity of silicon-based sensors and gain coefficients do not change in the range of temperatures encountered. The applied stray light and aliasing correction removed artifacts in reflectance spectra and allow to extend the reliable spectral domain of the Zeiss MMS-1 spectrometer module. For the calibration we used a gray Spectralon panel which has certified reflectance spectrum with 0.5% variance. The footprint of the FOV of UAVSpec looking vertically on it was about 3 cm during calibration, therefore both the horizontal variance and deviation from Lambertian reflection of the reference panel may cause some systematic errors in the calibration of UAVSpec.

Helicopter measurements which were used for the vicarious calibration of CHRIS sensor were done one year later than CHRIS acquisition in the same phase of vegetation growth and in very similar illumination conditions. Several studies confirm that there is almost no change of reflectance of hemi-boreal mature stands in the age range of 40–60 years. Some changes of forest spectral reflectance in different years are possible, caused by changes in moisture conditions. The comparative study of forest hyperspectral reflectance at the test site in dry and normal summer revealed changes up to 10% in some wavelengths (Kuusk et al., 2010). As the summer 2006 of helicopter measurements was dry while in 2005 of CHRIS measurements we had normal amount of precipitation, the suggested correction factors may be systematically overestimated in all spectral bands but red bands of chlorophyll absorption.

The impact of adjacency correction depends on the reflectance pattern of neighboring stands. The birch stand is surrounded by other broadleaved stands, so the adjacency correction had almost no effect. As the wavelength-dependence of pine stand reflectance differs from that of its neighbors, the relative impact varies with wavelength. There are recent clearcuts close to the spruce stand which have higher reflectance than the spruce stand throughout the whole VNIR spectrum, thus the adjacency correction decreased the stand reflectance by 2–4%.

CHRIS acquisition and helicopter measurements were at slightly different zenith angles. The leveling of the UAVSpec spectrometer during flight was controlled and adjusted using a bubble level. Systematic error in view direction was suppressed flying over test plots in forth and back directions. Swaying of the helicopter increases the effective field-of-view of the UAVSpec to some extent. The reflectance change due to different view angles was corrected by numerical simulations with the reflectance model FRT (Kuusk and Nilson, 2000). As stands of different structures have different angular dependence of directional reflectance (Rautiainen et al.,

2004), some random error is added by this correction, the magnitude of which can be estimated only roughly.

Stem reflectance spectra measured with GER-2600 in (Lang et al., 2002) have no correction of spectral aliasing. The decrease of birch, pine and spruce stem reflectance between 920 and 1040 nm in Figs. 4.24, 4.38 and 4.50 is partly an artifact caused by the spectral aliasing.

Both in helicopter measurements and in CHRIS images the signal variance due to noise is in most spectral bands significantly lower than the variance of target reflectance at such high spatial resolution. Thus the variance of reported spectra is mainly due to the target variance and not the instrumental noise.

Acknowledgments

The CHRIS image data have been provided by the European Space Agency, using the ESA PROBA platform and the Surrey Satellite Technology Ltd CHRIS instrument. The sun-photometer data are provided by the International AERONET Federation, we thank Drs O. Kärner and M. Sulev for their effort in establishing and maintaining the Tõravere AERONET site. The European Commission's DG Joint Research Center (Ispra, Italy) supported the collection of field data. We acknowledge the contribution by Drs Miina Rautiainen, Matti Mõttus and Tiit Nilson in collecting field data and discussing results. The study has been funded by Estonian target financed projects SF0062466s03 and SF0060115s08, and by Estonian Science Foundation, Grants no. 6100, 6812 and 6815.

References

Abdou, W. A., C. J. Bruegge, M. C. Helmlinger, J. E. Conel, S. H. Pilorz, W. Ledeboer, B. J. Gaitley, and K. J. Thome (2002). Vicarious calibration experiment in support of the Multi-angle Imaging SpectroRadiometer. *IEEE Transactions on Geoscience and Remote Sensing*, **40**(7):1500–1511.

Banham, M. R., and A. K. Katsaggelos (1997). Digital image restoration. *IEEE Signal Processing Magazine*, **14**(2):24–41.

Barnsley, M. J., J. J. Settle, M. A. Cutter, D. R. Lobb, and F. Teston (2004). The PROBA/CHRIS mission: A low-cost smallsat for hyperspectral multiangle observations of the earth surface and atmosphere. *IEEE Transactions on Geoscience and Remote Sensing*, **42**(7):1512–1520.

Cannizzaro, J. P., and K. L. Carder (2006). Estimating chlorophyll a concentrations from remote-sensing reflectance in optically shallow waters. *Remote Sensing of Environment*, **101**(1):13–24.

CHRIS (2002). The CHRIS Instrument. Web-page http://www.chris-proba.org.uk/frames/index2.html.

Cutter, M. (2004). Review of aspects associated with the CHRIS calibration. In *Proc. of the 2nd ESA CHRIS/Proba Workshop, 28–30 April, ESRIN, Frascati, Italy*, number SP-578, pages 1–5. ESA. CD-ROM.

Deschamps, P. Y., F. M. Breon, M. Leroy, A. Podaire, A. Bricaud, J. C. Buriez, and G. Seze (1994). The POLDER mission – instrument characteristics and scientific objectives. *IEEE Transactions on Geoscience and Remote Sensing*, **32**(3):598–615.

Elterman, L. (1968). *UV, Visible, and IR Attenuation for Altitudes to 50 km. Rep. AFCRL-68-0153*. U.S. Air Force Cambridge Research Laboratory, Bedford, Mass.

ESRI (2007). ESRI world basemap data. Web-page http://www.esri.com/data/download/basemap/index.html.

Feng, H., J. W. Campbell, M. D. Dowell, and T. S. Moore (2005). Modeling spectral reflectance of optically complex waters using bio-optical measurements from Tokyo Bay. *Remote Sensing of Environment*, **99**(3):232–243.

Froidefond, J. M., L. Gardel, D. Guiral, M. Parra, and J. F. Ternon (2002). Spectral remote sensing reflectances of coastal waters in French Guiana under the Amazon influence. *Remote Sensing of Environment*, **80**(2):225–232.

Hadjimitsis, D. G., C. R. I. Clayton, and V. S. Hope (2004). An assessment of the effectiveness of atmospheric correction algorithms through the remote sensing of some reservoirs. *International Journal of Remote Sensing*, **25**(18):3651–3674.

Hoffmann, C. W., and V. A. Usoltsev (2002). Tree-crown biomass estimation in forest species of the Ural and of Kazakhstan. *Forest Ecology and Management*, **158**(1–3):59–69.

Holben, B. N., T. F. Eck, I. Slutsker, D. Tanre, J. P. Buis, A. Setzer, E. Vermote, J. A. Reagan, Y. J. Kaufman, T. Nakajima, F. Lavenu, I. Jankowiak, and A. Smirnov (1998). AERONET – A federated instrument network and data archive for aerosol characterization. *Remote Sensing of Environment*, **66**(1):1–16.

HYSPLIT (2007). HYSPLIT: HYbrid Single-Particle Lagrangian Integrated Trajectory Model Version 4.8. Web-page http://www.arl.noaa.gov/ready/hysp_info.html.

Järvselja (2007). Järvselja experimental and training forest district. Web page http://www.jarvselja.ee/.

Jennings, S. B., N. D. Brown, and D. Sheil (1999). Assessing forest canopies and understorey illumination: canopy closure, canopy cover and other measures. *Forestry*, **72**(1):59–74.

Johansson, T. (1999). Dry matter amount and increment in 21- to 91-year-old common alder and gray alder and some practical implications. *Canadian Journal of Forest Research*, **29**(11):1679–1690.

Korhonen, L., T. K. Korhonen, M. Rautiainen, and P. Stenberg (2006). Estimation of forest canopy cover: a comparison of field measurement techniques. *Silva Fennica*, **40**(4):577–588.

Kostkowski, H. J. (1997). *Reliable Spectroradiometry*. Spectroradiometry Consulting, La Plata, MD.

Kuusk, A. (1994). A multispectral canopy reflectance model. *Remote Sensing of Environment*, **50**:75–82.

Kuusk, A. (1998). Monitoring of vegetation parameters on large areas by the inversion of a canopy reflectance model. *International Journal of Remote Sensing*, **19**(15):2893–2905.

Kuusk, A., and T. Nilson (2000). A directional multispectral forest reflectance model. *Remote Sensing of Environment*, **72**(2):244–252.

Kuusk, J. (2011). *Measurement of forest reflectance. Top-of-canopy spectral reflectance of forests for developing vegetation radiative transfer models*. Lambert Academic Publishing, Saarbrücken, Germany. 120 pp.

Kuusk, J., A. Kuusk, M. Lang, and A. Kallis (2010). Hyperspectral reflectance of boreonemoral forests in a dry and normal summer. *International Journal of Remote Sensing*, **31**(1):159–175.

Lang, M., A. Kuusk, T. Nilson, T. Lükk, M. Pehk, and G. Alm (2002). Reflectance spectra of ground vegetation in sub-boreal forests. Web page http://www.aai.ee/bgf/ger2600/.

Lang, M., M. Jürjo, V. Adermann, and H. Korjus (2006). Integrated approach for quantitative assessment of illegal forest fellings in Estonia. *Baltic Forestry*, **12**(1):103–109.

Liang, S. (2004). *Quantitative Remote Sensing of Land Surfaces*. Wiley-Interscience.

Liang, S. L., H. L. Fang, and M. Z. Chen (2001). Atmospheric correction of Landsat ETM+ land surface imagery – Part I: Methods. *IEEE Transactions on Geoscience and Remote Sensing*, **39**(11):2490–2498.

Marklund, L. G. (1988). *Biomass Functions for Pine, Spruce and Birch in Sweden*. Swedish University of Agricultural Sciences, Umeå.

Martin, J. G., B. D. Kloeppel, T. L. Schaefer, D. L. Kimbler, and S. G. McNulty (1998). Aboveground biomass and nitrogen allocation of ten deciduous southern Appalachian tree species. *Canadian Journal of Forest Research*, **28**:1648–1659.

McGaughey, R. J. (1997). Visualizing forest stand dynamics by using the stand visualization system. In *Proceedings ACSM/ASPRS Annual Convention and Exposition Seattle, WA. Betsheda*, Vol. 4, pages 248–257. MD: American Society for Photogrammetry and Remote Sensing.

McPeters, R. (2007). Total Ozone Mapping Spectrometer. Web-page http://jwocky.gsfc.nasa.gov/.

MODIS (2005). MODIS Atmosphere Terra Images. Web-page http://modis-atmos.gsfc.nasa.gov/IMAGES/.

Newcomer, J., D. Landis, S. Conrad, S. Curd, K. Huemmrich, D. Knapp, A. Morrell, J. Nickeson, A. Papagno, D. Rinker, R. Strub, T. Twine, F. Hall, and P. Sellers, editors (2000). *Collected Data of The Boreal Ecosystem-Atmosphere Study*. NASA. CD-ROM.

Niinemets, U., and K. Kull (1994). Leaf weight per area and leaf size of 85 Estonian woody species in relation to shade tolerance and light availability. *Forest Ecology and Management*, **70**:1–10.

Nilson, T., and A. Kuusk (2004). Improved algorithm for estimating canopy indices from gap fraction data in forest canopies. *Agricultural and Forest Meteorology*, **124**(3–4):157–169.

Novo, E. M. L. M., W. Pereira, and J. M. Melack (2004). Assessing the utility of spectral band operators to reduce the influence of total suspended solids on the relationship between chlorophyll concentration and the bidirectional reflectance factor in Amazon waters. *International Journal of Remote Sensing*, **25**(22):5105–5116.

Pensa, M., and A. Sellin (2002). Needle longevity of Scots pine in relation to foliar nitrogen content, specific leaf area, and shoot growth in different forest types. *Canadian Journal of Forest Research*, **32**(7):1225–1231.

Podilchuk, C. (1998). Signal recovery from partial information. In V. K. Madisetti and D. B. Williams, editors, *Signal Processing Handbook*, pages 25–1 to 25–21. CRC Press LLC, Boca Raton, Florida 33431, USA.

R-project (2007). The R Project for Statistical Computing. Web-page http://www.r-project.org/.

Rabiner, L. R., and B. Gold (1975). *Theory and Application of Digital Signal Processing*. Prentice-Hall, Englewood Cliffs.

Rautiainen, M., P. Stenberg, T. Nilson, and A. Kuusk (2004). The effect of crown shape on the reflectance of coniferous stands. *Remote Sensing of Environment*, **89**(1):41–52.

Reinersman, P. N., and K. L. Carder (1995). Monte Carlo simulation of the atmospheric point-spread function with an application to correction for the adjacency effect. *Applied Optics*, **34**(21):4453–4471.

Schaepman, M., and S. Dangel (2000). Solid laboratory calibration of a nonimaging spectroradiometer. *Applied Optics*, **39**:3754–3764.

Sellin, A. (2000). Estimating the needle area from geometric measurements: Application of different calculation methods to Norway spruce. *Trees*, **14**(4):215–222, 819.

Slater, P. N., and S. F. Biggar (1996). Suggestions for radiometric calibration coefficient generation. *Journal of Atmospheric and Oceanic Technology*, **13**(2):376–382.

Slater, P. N., S. F. Biggar, K. J. Thome, D. I. Gellman, and P. R. Spyak (1996). Vicarious radiometric calibrations of EOS sensors. *Journal of Atmospheric and Oceanic Technology*, **13**(2):349–359.

Song, C. H., and C. E. Woodcock (2003). Monitoring forest succession with multitemporal Landsat images: Factors of uncertainty. *IEEE Transactions on Geoscience and Remote Sensing*, **41**(11):2557–2567.

Tamm, U. (2000). *Aspen in Estonia.* Eesti Loodusfoto, Tartu.

Thome, K., B. Markham, J. Barker, P. Slater, and S. Biggar (1997). Radiometric calibration of Landsat. *Photogrammetric Engineering and Remote Sensing*, **63**(7):853–858.

VALERI (2005). Web page http://www.avignon.inra.fr/valeri/, accessed on March 15, 2005.

Vermote, E. F., D. Tanre, J. L. Deuze, M. Herman, and J.-J. Morcrette (1997). Second simulation of the satellite signal in the solar spectrum, 6S: An overview. *IEEE Transactions on Geoscience and Remote Sensing*, **35**(3):675–686.

Wang, C. (2006). Biomass allometric equations for 10 co-occurring tree species in Chinese temperate forests. *Forest Ecology and Management*, **222**:9–V16.

Wendlandt, W. W., and H. G. Hecht (1966). *Reflectance Spectroscopy.* Wiley.

Widlowski, J. L., M. M. Verstraete, B. Pinty, and N. Gobron (2003). *Allometric Relationships of Selected European Tree Species.* Tech. rep., EC Joint Research Centre, Ispra, Italy.

Widlowski, J. L., M. Taberner, B. Pinty, V. Bruniquel-Pinel, M. Disney, R. Fernandes, J. P. Gastellu-Etchegorry, N. Gobron, A. Kuusk, T. Lavergne, S. Leblanc, P. E. Lewis, E. Martin, M. Mottus, P. R. J. North, W. Qin, M. Robustelli, N. Rochdi, R. Ruiloba, C. Soler, R. Thompson, W. Verhoef, M. M. Verstraete, and D Xie (2007). Third radiation transfer model intercomparison (RAMI) exercise: Documenting progress in canopy reflectance models. *Journal of Geophysical Research–Atmospheres*, **112**(D9):D09111.

Part II

Optical Properties of Snow and Natural Waters

5 Reflection properties of snow surfaces

Teruo Aoki

5.1 Introduction

The cryosphere plays an important role for the energy budget on the Earth because of the high albedo and large seasonal variations. The surface albedo in the cryosphere varies drastically not only by phase change of snow and ice but also by change from new snow to melting snow. The melting of polar snow and ice due to global warming and the resultant sea level rise are currently causing much public anxiety (Bindoff et al., 2007). The mass balance loss from the Greenland Ice Sheet increased significantly after the mid-1990s (Steffen et al., 2008). Moreover, the decrease of the Arctic sea ice extent, recorded since 1978, accelerated from 1996 to 2006 (Comiso and Nishio, 2008). The smallest ice extent, recorded in 2007, was less than any value predicted by climate models (Stroeve et al., 2007). This abrupt melting of Arctic snow and ice has not been accurately simulated by many general circulation models (GCMs). Thus, better understanding of snow optical properties is needed to accurately simulate the future climate in the cryosphere. The albedo is an important optical snow parameter and is necessary for radiation budget calculations on snow surfaces. Bidirectional reflectance is another important optical parameter and is necessary for satellite remote sensing of snow properties. These applications require modeling of the reflection properties of snow surface with both broad and high spectral resolutions.

Over the past five decades, numerous studies have focused on the modeling of the surface albedo of snow. The first attempt to calculate the spectral albedo of snow by multiple scattering radiative transfer in the visible and near-infrared regions was conducted by Dunkle and Bevans (1956), who used Schuster's two-stream approximation. Giddings and LaChapelle (1961) calculated snow albedo using a diffusion model. These two models were actually equivalent and were valid for only diffuse incidence and high albedo (Warren, 1982). They did not start from single scattering by each snow grain. In the 1970s, many multiple scattering models for snow albedo considered single scattering by each snow grain (Barkstrom, 1972; Bohren and Barkstrom, 1974; Barkstrom and Querfeld, 1975; Berger, 1979; Choudhury and Chang, 1979a,b). Bergen (1970, 1971) calculated the transmittance of a snow layer with radiative transfer. However, these models were not accurate enough for a wide range of wavelengths, grain sizes, and solar zenith angles. Wiscombe and

Warren (1980a) and Warren and Wiscombe (1980) simulated the spectral albedo of snow using the delta-Eddington approximation for multiple scattering and Mie theory for single scattering based on realistic physical parameters. They demonstrated the dependence of the spectral albedo of snow surfaces on solar zenith angle, snow grain size, illuminating conditions, and snow impurities, and compared them with observations. They also discussed the effect of close packing and nonsphericity of each snow grain. Choudhury and Chang (1981a,b) and Choudhury (1981) developed a two-stream model considering the effects of atmospheric gaseous absorption and surface reflection.

Mellor (1977) reviewed various snow studies in the field of dynamics, chemistry, and optics; and Warren (1982) reviewed the detailed optical properties of snow albedo models and observations. Extending the model of Wiscombe and Warren (1980a), Chýlek et al. (1983) calculated the albedo of soot-contaminated snow. Carroll (1982) and Wendler and Kelley (1988) calculated the effects of snow surface striations and sastrugi (a wavelike structure on snow surfaces formed by strong wind) on snow albedo. Blanchet and List (1987) examined the effect of anthropogenic aerosols in Arctic haze and snow on the radiation budget, using the atmospheric radiative transfer model with an interactive snow layer based on the delta-Eddington approximation. Warren et al. (1990) extended their model for calculating the spectral albedo and emissivity of CO_2 in Martian polar caps. Grenfell et al. (1994) used a two-layer snow model extended from the model of Wiscombe and Warren (1980a) and obtained a good agreement of spectral albedo with observation under cloudy conditions in Antarctica. However, some discrepancies remained at near-infrared wavelengths under clear sky. They stated three possible reasons for these discrepancies: the underestimate of snow grain size, inaccurate correction of the instrument, and a fault in the model. Aoki et al. (1998, 2000) demonstrated that the optically equivalent snow grain size has a dimension of the branch width of dendrites or the width of the narrower portion of broken crystals, based on the spectral albedos measured on the snowfield and those theoretically calculated. The effects of snow impurities and snow grain size on broadband albedos were identified from long-term radiation budget measurements on the snow surface and snow pit work in a dry snow area of Japan (Aoki et al., 2003, 2006, 2007a), and in a wet snow area of Japan (Motoyoshi et al., 2005).

The model of Wiscombe and Warren (1980a) has been used operationally since the 1980s (Chýlek et al., 1983; Marshall and Warren, 1986). Since this model is based on the delta-Eddington approximation, it is not applicable for large solar zenith angles and cannot be extended to the radiance model. With regard to the atmospheric effects, Wiscombe and Warren (1980b) calculated the spectral albedo at snow surface and top of the atmosphere (TOA), and the spectrally integrated planetary albedos with summer atmospheric conditions of the Antarctic Plateau. They reported that the differences in albedo between the snow surface and TOA were due to Rayleigh scattering and gaseous absorption, and varied with the solar zenith angle. Choudhury and Chang (1981a,b) and Choudhury (1981) calculated the spectral albedo, taking into account the atmospheric effects by parameterizing the effect of aerosols, cloud cover, and atmospheric gases, but they did not incorporate the radiative interaction between the atmosphere and snow. Blanchet and List (1987) calculated the effect of aerosols on the radiation budget considering this interaction and the atmospheric absorption, but they also used the delta-Eddington approximation. Aoki et al. (1999) examined the atmospheric effects (e.g., atmospheric molecules, absorptive gases, aerosols, and clouds) on the spectral and spectrally integrated snow albedos at the surface and TOA. For the same snow conditions, the broadband snow albedo is higher under a cloudy sky than under a clear sky (e.g., Yamanouchi, 1983) because of the differences in the spectral distribution of the downward solar flux between clear and cloudy skies (Liljequist, 1956).

The first bidirectional reflectance distribution function (BRDF) model of snow surface was developed by Li (1982), using Mie theory and the doubling method, which indicated the anisotropic reflection property of snow. BRDF is only defined for direct solar beam in a strict sense. When the incident radiation includes diffuse component, thus in the atmosphere, it is known as the hemispherical directional reflectance distribution function (HDRDF). The detailed definitions are presented in section 5.2. Here we use the same terminology as in the references. Han (1996) developed a snow BRDF model using the discrete ordinate method and Mie theory to retrieve the surface albedo from satellite measurements in the Arctic. Leroux and Fily (1998) developed a BRDF model including the effect of sastrugi with regularly spaced identical rectangular protrusions. Leroux et al. (1998) and Leroux et al. (1999) developed a polarized BRDF model using doubling and adding method along with Mie theory and ray tracing technique, and compared the theoretical values with measurements in the principal plane at wavelength $\lambda = 1.65\,\mu\text{m}$. They demonstrated that the snow grain shape strongly affects the BRDF in the near-infrared region and that hexagonal particles, rather than spherical particles, give better agreement with measurements. Aoki et al. (2000) compared the theoretically calculated snow HDRF at selected wavelengths of 0.52 to 2.21 μm with those measured on the snowfield in eastern Hokkaido, Japan. They demonstrated that the anisotropy of HDRF is very remarkable at $\lambda = 1.65\,\mu\text{m}$ and 2.21 μm, and that the shape of the scattering phase function is important for BRDF simulation. Comparison between in situ measurements of bidirectional (or hemispherical directional) reflectance and the theoretical simulations were conducted to investigate the effect of nonspherical snow particles (Painter and Dozier, 2004; Tanikawa et al., 2006) or to validate an approximate asymptotic theory for snow optical properties (Kokhanovsky et al., 2005).

In this review, section 5.2 presents basic definitions and terms, and section 5.3 describes the feedback effect between albedo and snow physical parameters. Section 5.4 discusses the atmospheric effects on snow albedo. Sections 5.5 and 5.6 present the effects of snow physical parameters on spectral albedo, bidirectional reflectance, and broadband albedo. Section 5.7 briefly discusses satellite remote sensing of snow physical parameters and broadband snow albedo models used in GCMs as applications of radiative transfer modeling for the atmosphere–snow system.

5.2 Basic definitions and terminologies

Various articles (e.g., Warren, 1982; Hapke, 1993; Leroux et al., 1999) introduce basic definitions and terms for reflection properties at or over snow surfaces. Here, we confirm the definitions of the parameters frequently used in snow studies and the relationships among them. We assume that an incident solar beam is coming from the direction of zenith angle θ_0 and azimuth angle ϕ_0, and is reflected in the direction of zenith angle θ' and azimuth angle ϕ' in spherical polar coordinates (Fig. 5.1).

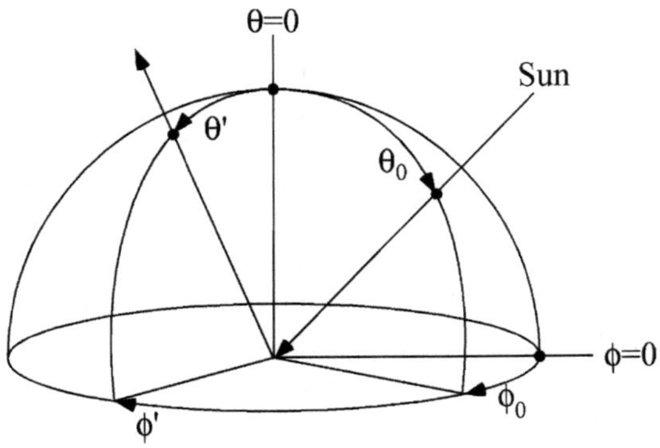

Fig. 5.1. Relationship between incident and reflected solar radiation in spherical polar coordinates.

When there is no diffuse component in incident solar radiation, BRDF with the unit of [sr^{-1}] is calculated by

$$BRDF(\theta_0, \phi_0, \theta', \phi') = \frac{I^{\uparrow}(\theta', \phi')}{\mu_0 F_{\odot}^{\downarrow}(\theta_0, \phi_0)}, \tag{5.1}$$

where $\mu_0 = \cos\theta_0$, $F_{\odot}^{\downarrow}(\theta_0, \phi_0)$ is the incident solar flux on a surface normal to the direct beam, and $I^{\uparrow}(\theta', \phi')$ is the reflected radiance. The dimensionless parameter known as bidirectional reflectance factor (BRF)[1] is given by

$$BRF(\theta_0, \phi_0, \theta', \phi') = \pi BRDF(\theta_0, \phi_0, \theta', \phi'). \tag{5.2}$$

When the incident solar radiation includes a diffuse component, the HDRDF with the unit of [sr^{-1}] is calculated by

$$HDRDF(\theta_0, \phi_0, \theta', \phi') = \frac{I^{\uparrow}(\theta', \phi')}{\mu_0 F_{\odot}^{\downarrow}(\theta_0, \phi_0) + F_{diff}^{\downarrow}}, \tag{5.3}$$

where $F_{diff}^{\downarrow}$ is the diffuse component in incident solar flux, and total incident flux density or irradiance is $F^{\downarrow}(\theta_0, \phi_0) = \mu_0 F_{\odot}^{\downarrow}(\theta_0, \phi_0) + F_{diff}^{\downarrow}$. The dimensionless parameter of HDRDF known as the hemispherical-directional reflectance factor (HDRF) is given by

$$HDRF(\theta_0, \phi_0, \theta', \phi') = \pi HDRDF(\theta_0, \phi_0, \theta', \phi'). \tag{5.4}$$

[1]Sometimes BRF is called the reflection function (Kokhanovsky et al., 2005).

The dimensionless parameter, the anisotropic reflectance factor f, is the ratio of BRF or HDRF to albedo $\alpha(\theta_0, \phi_0)$:

$$f(\theta_0, \phi_0, \theta', \phi') = \frac{\pi BRDF(\theta_0, \phi_0, \theta', \phi')}{\alpha(\theta_0, \phi_0)} \quad \text{or} \quad \frac{\pi HDRDF(\theta_0, \phi_0, \theta', \phi')}{\alpha(\theta_0, \phi_0)}. \tag{5.5}$$

Albedo α is the ratio of the reflected (upward) flux $F^{\uparrow}$ to the incident (downward) solar flux $F^{\downarrow}(\theta_0, \phi_0)$:

$$\alpha(\theta_0, \phi_0) = \frac{F^{\uparrow}}{F^{\downarrow}(\theta_0, \phi_0)}. \tag{5.6}$$

Upward flux $F^{\uparrow}$ can be expressed by integrating $I^{\uparrow}(\theta', \phi')$ over the hemisphere as:

$$F^{\uparrow} = \int_0^{2\pi} \int_0^{\pi/2} I^{\uparrow}(\theta', \phi') \cos\theta' \sin\theta' \, d\theta' \, d\phi'. \tag{5.7}$$

From Eqs. (5.1) or (5.3), (5.6), and (5.7), the relationship between α and (*BRDF* or *HDRDF*) is expressed as follows:

$$\alpha(\theta_0, \phi_0) = \int_0^{2\pi} \int_0^{\pi/2} BRDF \text{ or } HDRDF(\theta_0, \phi_0, \theta', \phi') \cos\theta' \sin\theta' \, d\theta' \, d\phi'. \tag{5.8}$$

5.3 Feedback effect between snow physical parameters and albedo

Studies on the basic behavior of the spectral albedo of snow surface indicate that the snow albedo is essentially determined by light scattering by snow grains and radiative interaction between the snowpack and the atmosphere. More specifically, the snow albedo depends on (1) snow physical parameters and (2) external parameters (e.g., atmospheric conditions and solar zenith angle). Snow physical parameters furthermore consist of those related to single scattering (snow grain size, morphology, and impurities) and multiple scattering (snow depth, layer structure, density, water contents, surface condition, and impurities). External parameters include atmospheric conditions (cloud cover, aerosols, air pressure, and atmospheric gases) and solar zenith angle.

Snow grain size and snow impurities (light-absorbing aerosols contained in snowpack) are very important snow physical parameters related to the albedo. Fig. 5.2 depicts spectral albedos depending on snow grain size and mass concentration of black carbon (BC) in snowpack, theoretically calculated using a radiative transfer model for the atmosphere- snow system (Aoki et al., 1999, 2000). We employed the refractive index of ice revised by Warren and Brandt (2008), in which the imaginary part at wavelengths $0.2 < \lambda < 0.5\,\mu$m is much less than that in the previous data set compiled by Warren (1984). With this change, the spectral albedos calculated for pure snow with a semi-infinite snow depth are close to 1.0 for $0.2 < \lambda < 0.5\,\mu$m (Fig. 5.2(a)). The spectral albedos in the near-infrared region strongly depend on snow grain size (Fig. 5.2(a)). On the other hand, the spectral albedos (mainly in

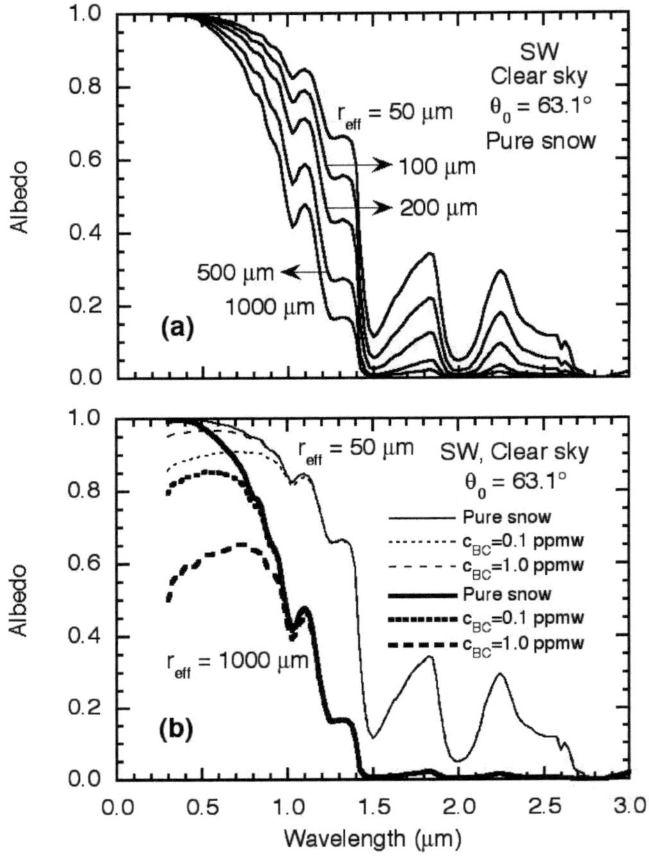

Fig. 5.2. Theoretically calculated spectral snow albedos depending on (a) snow grain size in pure snow and (b) mass concentration of snow impurities. r_{eff} is the effective snow grain radius, θ_0 is the solar zenith angle, and c_{BC} is the mass concentration of black carbon (BC) in snow. Snow depth is assumed to be a semi-infinite, and the model atmosphere employed is subarctic winter (SW).

the visible region) decrease with an increase of BC concentration (Fig. 5.2(b)). The albedo reduction rate for the same BC concentration is enhanced by larger snow grains. Therefore, positive feedback for snow albedo reduction is triggered by snow pollution with light absorbing aerosols (Fig. 5.3). Anthropogenic light-absorbing snow impurities have also the effect to accelerate snow and ice melting, together with snow grain growth associated with an air-temperature increase by greenhouse gases through this feedback mechanism.

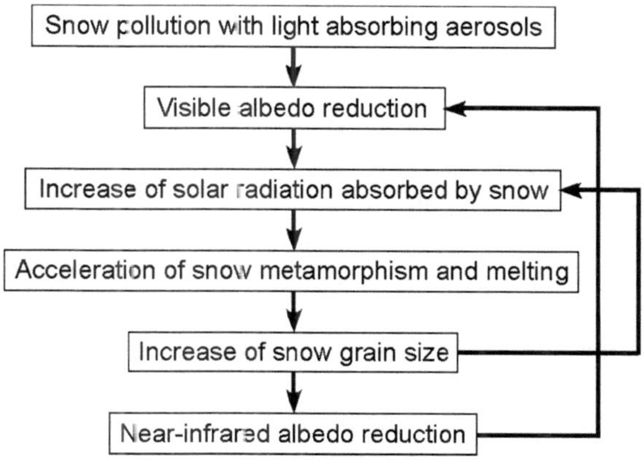

Fig. 5.3. Positive feedback effect for snow albedo reduction triggered by snow pollution with light absorbing aerosols.

5.4 Atmospheric effects on snow albedo

5.4.1 Radiative transfer model for the atmosphere–snow system

This section discusses the atmospheric effects on snow albedo based on Aoki et al. (1999). It should be noted that the atmospheric effects on snow albedo are related to the snow physical parameters as well. To consider the radiative interaction between the atmosphere and snowpack (hereafter, 'snow'), one snow layer is added below the atmospheric layers. Natural snow grains are nonspherical and packed closely together. Wiscombe and Warren (1980a) noted that possible adjustment for near-field effects and nonsphericity of snow grains are a few percent reduction of albedo for all wavelengths. In our model, snow grains are assumed to be mutually independent spherical ice particles, and radiative transfer in snow is treated as in a multiple scattering model in the atmosphere containing aerosols or cloud particles. The detailed processes and parameters used in the model are presented in Table 1 in Aoki et al. (1999). We used Mie theory for single scattering calculation, and doubling and adding method for multiple scattering calculation without polarization. The snow grain is so large that the Mie phase function has a very sharp forward peak, which induces large errors in the direct calculation of multiple scattering. Aoki et al. (1997) examined four kinds of approximations for phase function: Hansen's renormalization (Hansen, 1971), Grant's renormalization (Wiscombe, 1976), the delta-M method (Wiscombe, 1977), and truncation method (Hansen, 1969; Potter, 1970). Aoki et al. (1997) found Grant's renormalization and truncation method for snow particles to be operationally useful. With the truncation method, the result is not sensitive to the choice of truncation angle between 5° and 20°. In this study, we used the truncation method for the snow phase function with a truncation angle of 10° and the delta-M method for cloud and aerosol phase functions.

Atmospheric transmittance due to gaseous absorption was calculated for water vapor, carbon dioxide, oxygen, and ozone. For the first three gases, we used the extended exponential-sum fitting of transmissions developed by Asano and Uchiyama (1987), with the spectral absorption coefficients calculated with the line-by-line algorithm developed by Uchiyama (1992). Three kinds of model atmosphere (Anderson et al., 1986) were used for the atmosphere: midlatitude winter (MW), subarctic winter (SW), and SW that does not include a lower layer less than 2 km. The last one was adopted to simulate the atmosphere over the Antarctic Plateau, and we refer to it as Antarctic summer (AS) in this study. The effects of aerosols and cloud cover were examined for only SW because we intend to simulate AS coastal aerosols and high-latitudinal middle clouds. We calculates the spectral surface albedo (α^s) and the spectral planetary albedo (α^p) of snow for three values of effective grain radii: $r_{\text{eff}} = 50\,\mu\text{m}$ (new snow), $r_{\text{eff}} = 200\,\mu\text{m}$ (fine-grained older snow), and $r_{\text{eff}} = 1000\,\mu\text{m}$ (old snow for the melting point) (Wiscombe and Warren, 1980a). The snow layer is homogeneous with semi-infinite snow depth. The effect of atmospheric aerosols was examined for three values of optical thicknesses, $\tau_a = 0.02$, 0.1, and 0.3, at $\lambda = 0.5\,\mu\text{m}$ in SW. We assumed three optical thicknesses $\tau_c = 2.5$, 5, and 10 at $\lambda = 0.5\,\mu\text{m}$ for cloud effects. The value of $\tau_a = 0.02$ is the background level in the Antarctic coast in summer (Shaw, 1982), and high values of $\tau_a = 0.1$–0.3 were observed in Antarctica after major volcanic eruptions (Herber et. al., 1996).

5.4.2 Aerosol and cloud effects on spectral surface albedo

Monochromatic snow surface albedo has the θ_0 dependence under clear sky, where it is higher at large θ_0 than that at small θ_0. Warren (1982) explained that for this reason, a photon on average undergoes its first scattering event closer to the surface if it entered to the snow at a grazing angle. If the scattering event sends it in an upward direction, its chance of escaping the snow without being absorbed is greater than it would be if it were scattered from deeper in the snow. The phenomenon is greatly enhanced by the extreme asymmetry of scattering phase function. The snow surface is generally illuminated by direct and diffuse solar radiation. The θ_0 dependence of surface albedo is also related to these factors; thus, it is modified by the diffuse component under the atmosphere. Under a clear sky for small θ_0, the surface albedo is increased by the additional contribution of diffuse components from larger zenith angles, and vice versa for large θ_0. With an overcast sky, the direct solar beam weakens or disappears; the snow surface is illuminated mainly by diffuse radiation; and the surface albedo becomes constant, being independent of θ_0. The atmospheric effects due to Rayleigh scattering and gaseous absorption on spectral surface albedo of snow under clear sky are presented in Figs. 3 and 4 in Aoki et al. (1999).

Atmosphere aerosols also increase the diffuse component in downward solar flux and can change the θ_0 dependence of surface albedo on snow. Fig. 5.4(a) plots the monochromatic surface albedos as a function of θ_0 for the atmosphere containing aerosols (α^s_{4g+a}) and for aerosol-free case (α^s_{4g}), where the subscript 4g indicates that the atmosphere contains the four absorptive gases mentioned in section 5.4.1. The θ_0 dependence of α^s_{4g+a} becomes weak with increasing optical thickness of aerosols τ_a. The difference between α^s_{4g+a} and α^s_{4g} is greater for large θ_0 than for small θ_0. Since the θ_0 dependence of each albedo varies with the wavelength,

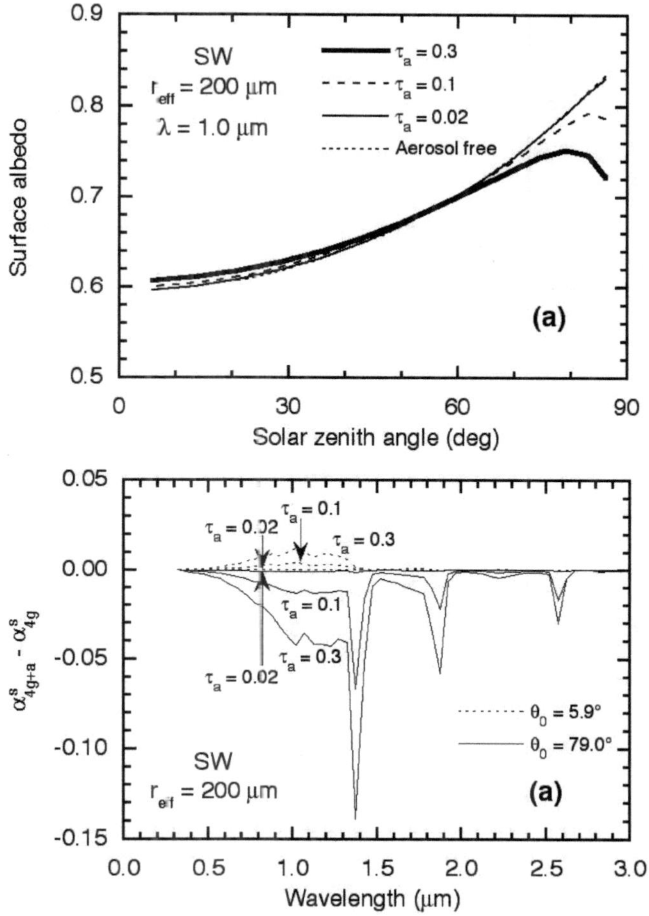

Fig. 5.4. (a) Monochromatic surface albedos of snow with $r_{\mathrm{eff}} = 200\,\mu$m at $\lambda = 1.0\,\mu$m as a function of solar zenith angle for the model atmosphere SW containing aerosols (α^s_{4g+a}) and the aerosol-free atmosphere (α^s_{4g}) at $\lambda = 0.5\,\mu$m. (b) Differences in spectral surface albedo between α^s_{4g+a} and α^s_{4g} at $\theta_0 = 5.9°$ (dashed curves) and 79.0° (solid curves). τ_a is the aerosol optical thickness at $\lambda = 0.5\,\mu$m (Aoki et al., 1999).

the difference between α^s_{4g+a} and α^s_{4g} also depends on the wavelength. Fig. 5.4(b) plots the spectral variation of $\alpha^s_{4g+a} - \alpha^s_{4g}$ at $\theta_0 = 5.9°$ and 79.0°. The effect of background aerosols ($\tau_a = 0.02$) on the surface albedo is very small. However, the difference $|\alpha^s_{4g+a} - \alpha^s_{4g}|$ rises up to 0.14 in the absorption band ($\lambda = 1.375\,\mu$m) for $\tau_a = 0.3$ and $\theta_0 = 79.0°$ because the aerosols increase the fraction of diffuse component at large θ_0 in the absorption band. At small θ_0 the diffuse component is somewhat more reduced in the absorption bands than in the non-absorbing region, resulting in very small values of $|\alpha^s_{4g+a} - \alpha^s_{4g}|$.

Fig. 5.5(a) depicts the monochromatic surface albedos as a function of θ_0 for a cloudy atmosphere (α^s_{4g+c}) and clear case (α^s_{4g}). The θ_0 dependence of α^s_{4g+c} is completely lost for $\theta_c = 10$, where the snow surface is illuminated by only diffuse radiation. Fig. 5.5(b) presents the spectral variation of $\alpha^s_{4g+c} - \alpha^s_{4g}$ at $\theta_0 = 5.9°$ and

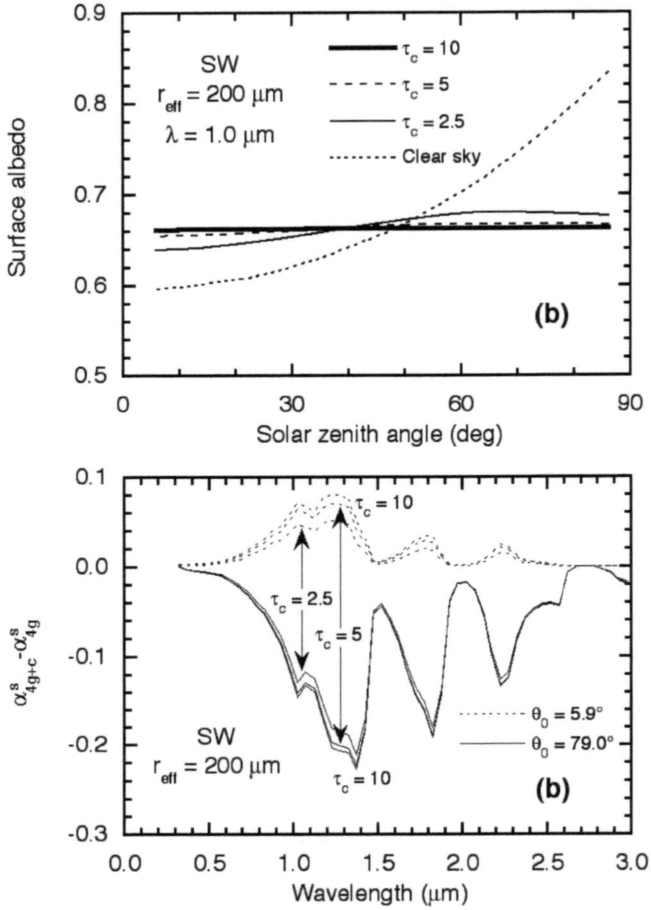

Fig. 5.5. (a) Same as Fig. 5.4(a), but for a cloudy (α^s_{4g+c}) and clear sky (α^s_{4g}). (b) Same as Fig. 5.4(b), but for a cloudy sky ($\alpha^s_{4g+c} - \alpha^s_{4g}$) at $\theta_0 = 5.9°$ (dashed curves) and 79.0° (solid curves). τ_c is the cloud optical thickness at $\lambda = 0.5\,\mu$m (Aoki et al., 1999).

79.0°. The value of $|\alpha^s_{4g+c} - \alpha^s_{4g}|$ is small at around $\lambda = 0.5$, 1.5, 2.0, and 2.8 μm, where α^s_{4g} is close to unity or zero (see Fig. 5.2(a)). When α^s_{4g} is close to unity, light absorption by ice is so weak that most incident photons escape from the snow surface, almost independently of incident angle. When the value of α^s_{4g} is close to zero, light absorption by ice is so strong that most incident photons are absorbed in snow, almost independently of incident angle. Thus, the θ_0 dependence of surface albedo is small even for a clear sky at such wavelengths, and $|\alpha^s_{4g+c} - \alpha^s_{4g}|$ becomes small. Except for those wavelengths, the effect of cloud cover on spectral surface albedo is quite large even for $\tau_c = 2.5$ at any value of θ_0.

5.4.3 Effect of the difference in atmospheric type on spectrally integrated albedo

Fig. 5.6(a) plots the spectrally integrated albedo ($\overline{\alpha}^s$) of snow as a function of θ_0 for three model atmospheres. The effect of differences in atmosphere type on the spectral surface albedo is very small, except for gaseous absorption bands at large θ_0 (Aoki et al., 1999). However, the value of $\overline{\alpha}^s$ varies depending on model atmospheres, and $\overline{\alpha}^s$ for MW ($\overline{\alpha}^s_{MW}$) is higher than that for AS ($\overline{\alpha}^s_{AS}$) at any θ_0. Namely, the cause of differences in $\overline{\alpha}^s$ between model atmospheres is not attributed to the differences in spectral surface albedo in the gaseous absorption bands.

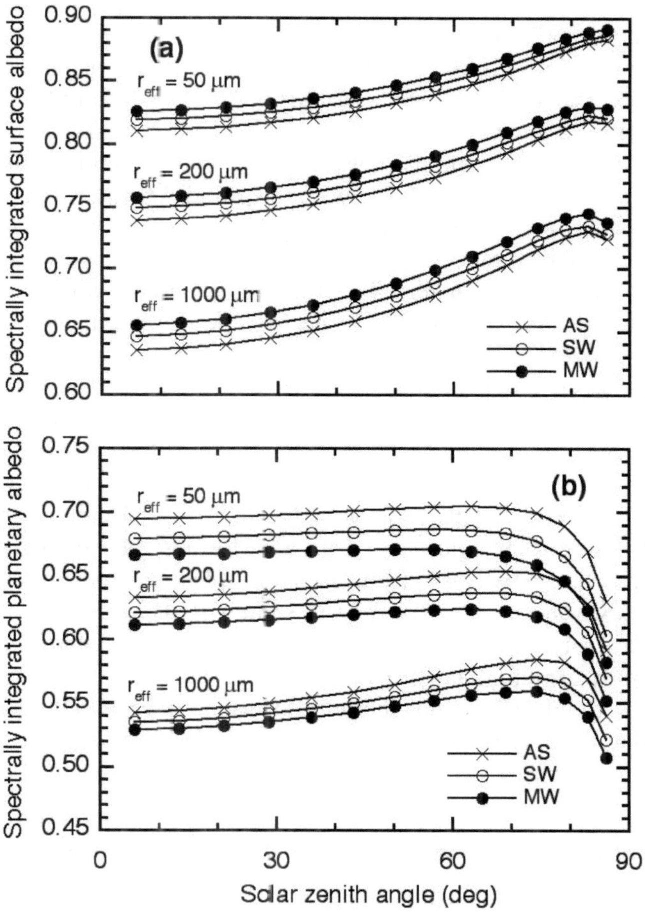

Fig. 5.6. Spectrally integrated albedos as a function of θ_0 at (a) snow surface ($\overline{\alpha}^s$) and (b) TOA ($\overline{\alpha}^p$) over the snow surface with $r_{eff} = 50$, 200, and 1000 μm for the clear condition of model atmospheres MW, SW, and AS (Aoki et al., 1999).

The spectrally integrated surface albedo $\overline{\alpha}^s$ is given by

$$\overline{\alpha}^s = \frac{\int_0^\infty \alpha^s(\lambda) F^{\downarrow}(\lambda)\, d\lambda}{\int_0^\infty F^{\downarrow}(\lambda)\, d\lambda}, \tag{5.9}$$

where $F^{\downarrow}(\lambda)$ is the spectral downward solar flux at the snow surface, and $\alpha^s(\lambda)$ is the spectral surface albedo. The value of $F^{\downarrow}(\lambda)$ varies, depending on atmospheric conditions, especially water vapor amount under clear sky. According to Eq. (5.9), $\overline{\alpha}^s$ is a weighted mean of $\alpha^s(\lambda)$ with the weight of normalized downward flux. Even if $\alpha^s(\lambda)$ does not change with the atmospheric condition, $\overline{\alpha}^s$ could change, depending on the atmosphere through the difference in $F^{\downarrow}(\lambda)$. Major water vapor bands are located in the near-infrared region. Hence, $F^{\downarrow}(\lambda)$ for MW (high water vapor amount) is smaller than for AS (low water vapor amount) in the near-infrared region, but the two are almost the same in the visible region. On the other hand, $\alpha^s(\lambda)$ is high in the visible region and low in the near-infrared region. Since the weight ($F^{\downarrow}(\lambda)$) for a low value of $\alpha^s(\lambda)$ in the near-infrared region is small for MW and large for AS, $\overline{\alpha}^s_{MW}$ becomes higher than $\overline{\alpha}^s_{AS}$. The maximum difference between $\overline{\alpha}^s_{MW}$ and $\overline{\alpha}^s_{AS}$ amounts to 0.015 for $r_{\text{eff}} = 50\,\mu$m and 0.021 for $r_{\text{eff}} = 1000\,\mu$m at $\theta_0 = 5.9°$. Thus, a few percent of spectrally integrated surface albedo could vary with the water vapor amount even for the same spectral surface albedo.

Fig. 5.6(b) presents the spectrally integrated planetary albedo ($\overline{\alpha}^p$) as a function of θ_0. In this case $\overline{\alpha}^p$ has the same form as Eq. (5.9), replacing $\alpha^s(\lambda)$ by $\alpha^p(\lambda)$ and $F^{\downarrow}(\lambda)$ by the downward solar flux at TOA. Since the downward solar flux at TOA does not depend on atmospheric conditions, the behavior of $\alpha^p(\lambda)$ is directly reflected on $\overline{\alpha}^p$. Contrary to $\overline{\alpha}^s$, $\overline{\alpha}^p_{MW}$ is lower than $\overline{\alpha}^p_{AS}$. This is because $\alpha^p_{MW} < \alpha^p_{AS}$ mainly in the water vapor bands. The difference between $\overline{\alpha}^p_{MW}$ and $\overline{\alpha}^p_{AS}$ increases with θ_0 and amounts to as much as 0.043 for $r_{\text{eff}} = 50\,\mu$m and 0.028 for $r_{\text{eff}} = 1000\,\mu$m at $\theta_0 = 79.0°$. Although $\overline{\alpha}^s$ gradually increases with θ_0, $\overline{\alpha}^p$ rapidly declines at around $\theta_0 = 75°$. The former reflects the property of θ_0 dependence of $\alpha^s(\lambda)$. The reason for the latter phenomenon is that the absolute value of the downward solar flux on the snow surface is reduced by the strong absorption of direct solar beam propagating along the long slant path at large θ_0, and the resultant upward solar flux at TOA is also small. The value of $\overline{\alpha}^p$ is lower than that of $\overline{\alpha}^s$ due to the atmospheric absorption, and the difference between $\overline{\alpha}^s$ and $\overline{\alpha}^p$ becomes roughly 0.1 to 0.3, although it depends on r_{eff}, θ_0, and the model atmosphere.

5.4.4 Aerosol and cloud effects on spectrally integrated albedo

The θ_0 dependence of spectrally integrated surface albedo for the atmosphere containing aerosols ($\overline{\alpha}^s_{4g+a}$) is presented in Fig. 5.7(a). For $\tau_a = 0.02$, the curves of $\overline{\alpha}^s_{4g+a}$ almost overlap with those for the aerosol-free atmosphere ($\overline{\alpha}^s_{4g}$). When the aerosol optical thickness τ_a increases, the θ_0 dependence of $\overline{\alpha}^s$ weakens because of the θ_0 dependence of the monochromatic surface albedo (Fig. 5.4(a)). The curve of $\overline{\alpha}^s_{4g+a}$ crosses that of $\overline{\alpha}^s_{4g}$ at $\theta_0 = 55°$. That is, the aerosols increase $\overline{\alpha}^s$ at small θ_0, but reduce it at large θ_0. Fig. 5.7(b) depicts the θ_0 dependence of spectrally integrated planetary albedo for the atmosphere containing aerosols ($\overline{\alpha}^p_{4g+a}$), where

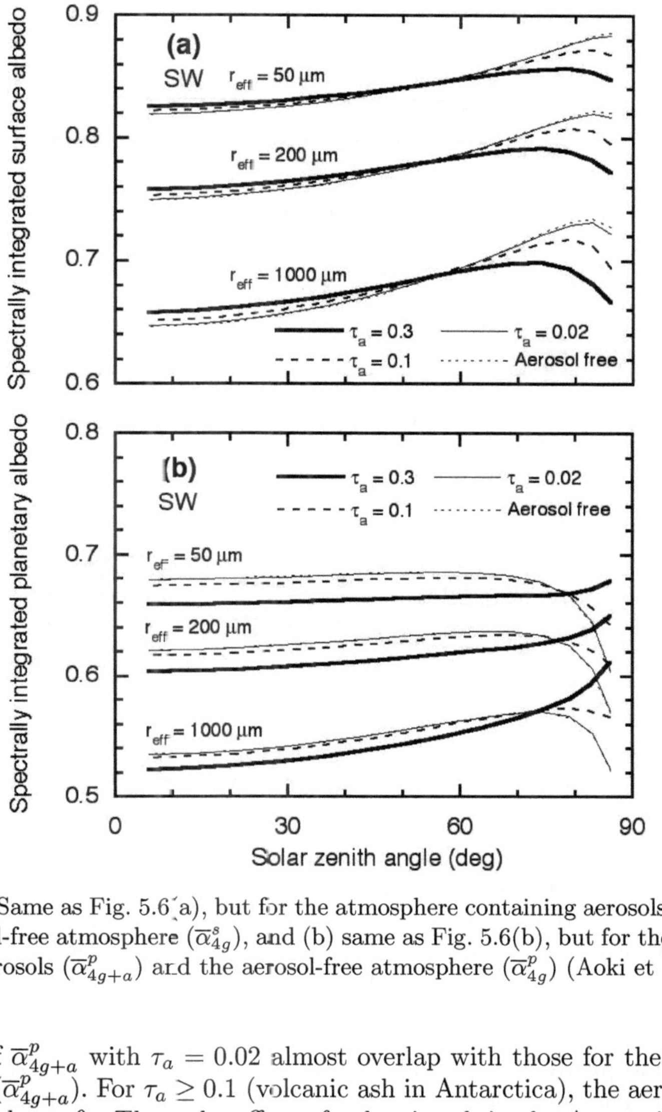

Fig. 5.7. (a) Same as Fig. 5.6(a), but for the atmosphere containing aerosols ($\overline{\alpha}^s_{4g+a}$) and for the aerosol-free atmosphere ($\overline{\alpha}^s_{4g}$), and (b) same as Fig. 5.6(b), but for the atmosphere containing aerosols ($\overline{\alpha}^p_{4g+a}$) and the aerosol-free atmosphere ($\overline{\alpha}^p_{4g}$) (Aoki et al., 1999).

the curves of $\overline{\alpha}^p_{4g+a}$ with $\tau_a = 0.02$ almost overlap with those for the aerosol-free atmosphere ($\overline{\alpha}^p_{4g+a}$). For $\tau_a \geq 0.1$ (volcanic ash in Antarctica), the aerosols reduce $\overline{\alpha}^p$ except at large θ_0. Thus, the effect of volcanic ash in the Antarctic is positive (heating) on the radiation budget in the shortwave region at TOA except at large θ_0. Blanchet and List (1987) demonstrated the reduction of $\overline{\alpha}^p$ by 0.021 caused by their Arctic aerosol model 1 with $\tau_a = 0.081$ for $\theta_0 \sim 70°$ on the snow with $r_{\text{eff}} = 200\,\mu$m. In our study, the reduction of $\overline{\alpha}^p$ is estimated to be 0.003 due to aerosols with $\tau_a = 0.1$ for $\theta_0 = 70°$ and $r_{\text{eff}} = 200\,\mu$m. However, the reduction of $\overline{\alpha}^p$ is 0.012 for $\tau_a = 0.3$. These results suggest that such thick volcanic ash aerosols over the Antarctic could have the same order of effect on the radiation budget at TOA as the Arctic haze has.

It is a well-known phenomenon that the value of $\overline{\alpha}^s$ of snow under cloudy sky is higher than that under clear sky (e.g., Liljequist, 1956; Yamanouchi, 1983). To reproduce such a situation theoretically, spectrally integrated surface albedos are calculated for a cloudy atmosphere ($\overline{\alpha}^s_{rg+c}$) for $r_{\text{eff}} = 50$ and $200\,\mu$m, and for

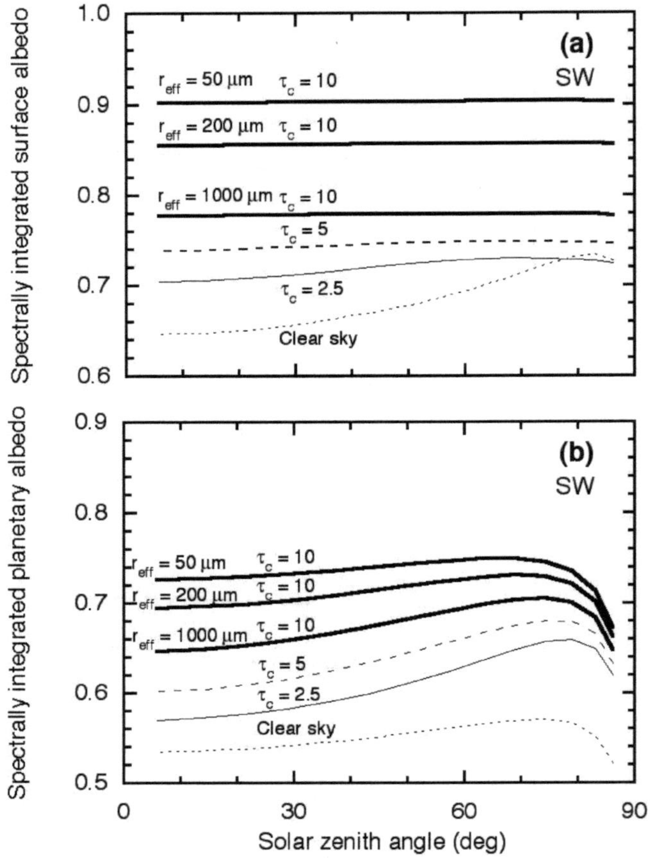

Fig. 5.8. (a) Spectrally integrated surface albedos as a function of θ_0 for a cloudy atmosphere ($\overline{\alpha}^s_{4g+c}$) with $\tau_c = 10$ for $r_{\text{eff}} = 50$ and $200\,\mu\text{m}$, and $\tau_c = 2.5$, 5, and 10 for $r_{\text{eff}} = 1000\,\mu\text{m}$, and a clear atmosphere ($\overline{\alpha}^s_{4g}$). (b) Same as (a), but for TOA in the cloudy case ($\overline{\alpha}^p_{4g+c}$) and the clear case ($\overline{\alpha}^p_{4g}$) (Aoki et al., 1999).

a cloudy atmosphere ($\overline{\alpha}^s_{4g+c}$) and a clear atmosphere ($\overline{\alpha}^s_{4g}$) for $r_{\text{eff}} = 1000\,\mu\text{m}$ (Fig. 5.8(a)). The value of $\overline{\alpha}^s_{4g+c}$ increases with τ_c and becomes higher than $\overline{\alpha}^s_{4g}$ at any θ_0 for $\tau_c \geq 5$. This phenomenon cannot be understood intuitively from the θ_0 dependence of the monochromatic surface albedo (Fig. 5.5(a)). Liljequist (1956) explained this phenomenon by the difference in spectral distribution of downward solar flux between clear and cloudy conditions. The details of this explanation could be given using Eq. (5.9). Under a cloudy atmosphere, the downward solar flux $F^{\downarrow}(\lambda)$ is much smaller than that under a clear atmosphere in the near-infrared region, whereas in the visible region both $F^{\downarrow}(\lambda)$ are not as different or $F^{\downarrow}(\lambda)$ under a cloudy atmosphere is even larger than that for a clear atmosphere at small θ_0 due to multiple reflection between the atmosphere and snow surface (see Fig. 15a in Aoki et al., 1999). On the other hand, spectral surface albedo $\alpha^s(\lambda)$ is high in the visible region and low in the near-infrared region under both clear and cloudy conditions. Since the weight ($F^{\downarrow}(\lambda)$) for low $\alpha^s(\lambda)$ in the near-infrared region is large for a clear atmosphere and small for a cloudy atmosphere, $\overline{\alpha}^s_{4g+c}$ becomes higher than $\overline{\alpha}^s_{4g}$. The θ_0 dependence of $\overline{\alpha}^s_{4g+c}$ becomes weaker with the increase of τ_c (Fig. 5.8(a)), as a result of the θ_0 dependence of the monochromatic surface albedo under cloudy sky (Fig. 5.5(a)). The spectrally integrated planetary albedos for a cloudy atmosphere ($\overline{\alpha}^p_{4g+c}$) as a function of θ_0 are depicted in Fig. 5.8(b), where $\overline{\alpha}^p_{4g+c}$ is higher than that for a clear atmosphere ($\overline{\alpha}^p_{4g}$) at any θ_0. This is because the cloud droplets are smaller than snow grains, and the single scattering albedo of cloud droplets is higher than that of snow grains. Thus, the cloud cover has the so-called 'albedo effect' (cooling effect) on the radiation budget in the shortwave region at TOA. These results are consistent with the study of Yamanouchi and Charlock (1995), who used the data of satellite and radiation budget observations at the snow surface.

5.5 Effects of snow physical parameters on spectral albedo and bidirectional reflectance

5.5.1 Observational condition, instrumentation, and radiative transfer model

This section discusses the effects of snow physical parameters on reflection properties by comparing the spectral albedos and HDRF observed on a snowfield in eastern Hokkaido, Japan, with those theoretically calculated using a radiative transfer model based on Aoki et al. (2000). Although Aoki et al. (2000) used the terms 'BRDF' or 'NBRDF' (BRDF normalized by the nadir radiance), in the strictest sense they were 'HDRF' or 'NHDRF' because the diffuse component was included in the incident radiation. We hereafter use HDRF or NHDRF in this review.

The spectral albedo and HDRF observations with snow pit work were performed on February 22 through 25, 1998, at three sites on a snowfield around Kitami in eastern Hokkaido, Japan. The spectral albedo data were selected under the conditions where θ_0 was close to the value at local solar noon (53°), and the snow surface was illuminated by a direct solar beam. The sky was clear except on February 22. New snowfall was observed on February 20 with a thickness of 10 cm and on February 21 with a thickness of less than 1 cm. The surface snow conditions changed from new snow to faceted crystals or granular snow during the observation period, and those in the lower part of the snow were depth hoar throughout this period. Snow grain size (radius) was measured at snow pit work using a handheld lens, which gave two types of dimensions of grain size: one-half length of the major axis of crystals or dendrites (r_1) and one-half the branch width of dendrites or one-half the dimension of the narrower portion of broken crystals (r_2). The snow impurities (water-insoluble solid particles) for the snow samples of the surface layer (0–5 cm) and of the sub-surface layer (5–10 cm) were filtered within a day using a nuclepore filter with a pore size of 0.2 μm after melting the snow samples. The concentrations of impurities were estimated by directly weighing the nuclepore filters before and after filtering using a balance. The imaginary part of the refractive index and the size distribution of the impurities were determined from the transmittance measurement of the nuclepore filter and electron microscopic analysis, respectively. Those data were used for theoretical calculations of spectral albedos and HDRFs.

The spectral albedo and HDRF were observed using a grating spectrometer FieldSpecFR (ASD Inc., USA). The scanning spectral range of the spectrometer was 0.35–2.5 μm with a spectral resolution of 3 nm at $\lambda = 0.35$–$1.0\,\mu$m and 10 nm at $\lambda = 1.0$–$2.5\,\mu$m. Measurements of downward and upward fluxes were necessary to obtain the albedo. However, accurate measurement of the downward flux was very difficult under clear conditions (Warren et al., 1986). Thus, we developed an albedo observation system (Fig. 5 in Aoki et al., 2000) using the white reference standard (WRS) SRT-99 (Labsphere Inc., USA). The WRS was attached horizontally to the tip of a pipe stretching from a mount set on the top of a tripod. Downward and upward flux densities were observed by directing the optical fiber tip of the spectrometer downward to the upper surface of the WRS and upward to the bottom surface of the WRS, respectively.

To calculate spectral albedo and HDRF, a multiple scattering radiative transfer model for the atmosphere–snow system (Aoki et al., 1999) was used. In the model, snow grains were assumed to be mutually independent ice particles; and the radiative transfer calculations were based on Mie theory for single scattering, and doubling and adding method for multiple scattering omitting polarization. The other details of the observational conditions, instrumentation, and a radiative transfer model were described in Aoki et al. (2000).

5.5.2 Spectral albedo

The spectral albedo observed on February 23, 1998, was compared with the theoretical values in Fig. 5.9. At the wavelengths where a large standard deviation was observed, the energy of the downward solar flux was weaker than the sensitivity of the detector of our spectrometer. This is due mainly to the low sensitivity of the detector itself ($\lambda \sim 0.95\,\mu$m and $\lambda > 1.8\,\mu$m) and the energy of the downward solar flux was low because of the atmospheric gaseous absorption at $\lambda \sim 0.4$, 1.4, 1.9, and 2.5 μm. Except for these wavelengths, low values of standard deviation (less than 0.005) were obtained. The theoretical spectral albedos were calculated for four snow models (Fig. 5.10), in which layer thickness, effective snow grain radius (r_{eff}), and mass concentration of impurities (c) were varied.

Model-1 (Fig. 5.10) consists of a pure snow layer with $r_{\text{eff}} = 55\,\mu$m and a semi-infinite snow depth. Grain size was determined by comparing the observed spectral albedo and the theoretical values at $\lambda > 1.4\,\mu$m. This is based on the theoretical calculation by Warren and Wiscombe (1980), in which there are no significant effects of impurities on spectral albedo in this wavelength region because of the high value of $m_{\text{im}}(\lambda)$ for the ice and its much larger volume fraction than with impurities. The theoretical albedo agrees well with the observation in this wavelength region. The value of $r_{\text{eff}} = 55\,\mu$m agrees with r_2 better than r_1, using the definitions of r_1 and r_2 presented in section 5.5.1. In the region of $\lambda < 1.4\,\mu$m, the observed albedo is lower than the theoretical value. Uncertainty of the parameters (e.g., total snow depth, vertical profiles of snow grain size, and impurities) could be the reasons for this discrepancy between observation and calculation. To clarify this, the observed albedo was compared with the theoretical one calculated using Model-2 without impurities, in which snow grain size, depth, and density were determined from the snow pit work data. The measured snow grain size (particularly r_2) varied

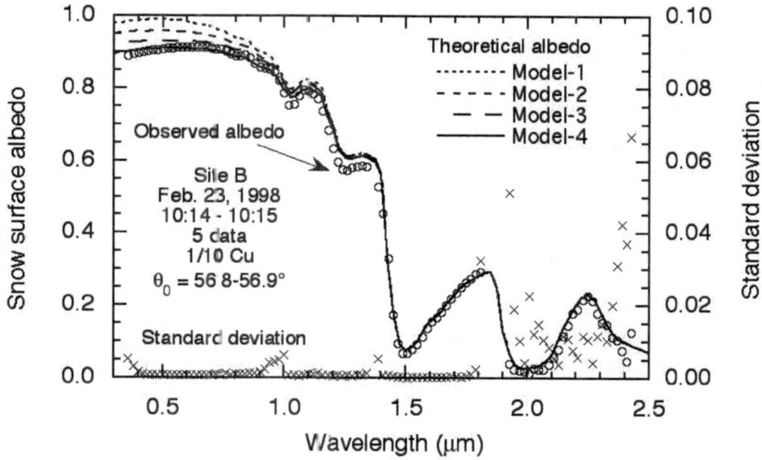

Fig. 5.9. Spectral snow surface albedo (left ordinate) and standard deviation (right ordinate) observed on February 23, 1998, in eastern Hokkaido, Japan, and theoretically calculated spectral albedos for the four types of snow models presented in Fig. 5.10. The plotted observed albedo is an averaged value of five spectral measurements, and the standard deviation is calculated from these five spectra. The observed spectral data with a standard deviation of less than 0.1 are presented (Aoki et al., 2000).

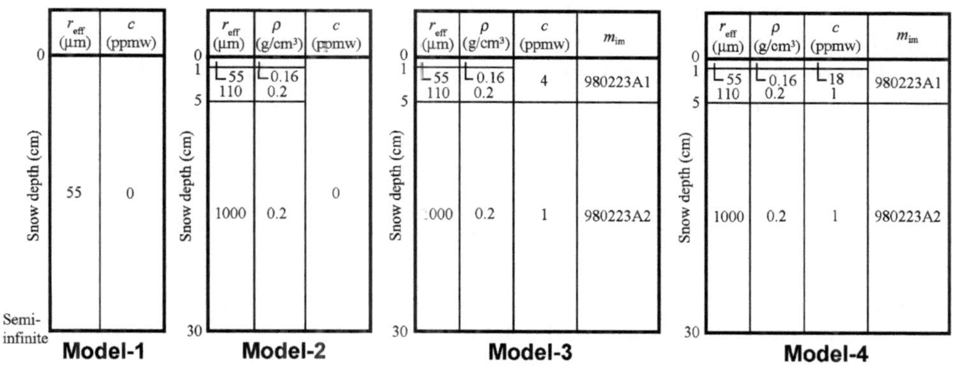

Fig. 5.10. Snow models for which the theoretical spectral albedos were compared with the measurement on February 23, 1998, where r_{eff} is the effective snow grain radius, ρ is the snow density, and c is the concentration of snow impurities. Column m_{im} indicates the snow sample from which the imaginary part of the refractive index of impurities was derived (Aoki et al., 2000).

drastically near the surface. To simulate this condition, in Model-2 we assumed three pure layers with $r_{\mathrm{eff}} = 55\,\mu\mathrm{m}$ (0–1 cm), $110\,\mu\mathrm{m}$ (1–5 cm), and $1000\,\mu\mathrm{m}$ (5–30 cm). In the second layer, the value of $r_{\mathrm{eff}} = 110\,\mu\mathrm{m}$ was assumed to be twice that of $r_{\mathrm{eff}} = 55\,\mu\mathrm{m}$ in the top layer, according to the measurements of r_2. For snow density, the observed value $\rho = 0.16\,\mathrm{g/cm}^3$ was assumed for the topmost layer and a constant value $\rho = 0.2\,\mathrm{g/cm}^3$ was assumed for the two lower layers. The visible albedo is reduced in Model-2, but is not sufficiently low due to the snow impurities. The measured concentration of impurities was 4.0 parts per million by weight (ppmw) in the 0–5 cm layer, and 1.2 ppmw in the 5–10 cm layer. We made Model-3 the same as Model-2, except for impurities whose concentration was assumed to be $c = 4$ ppmw for the top two layers and $c = 1$ ppmw for the bottom layer, based on measured snow impurity concentrations. The theoretical albedo for Model-3 is further reduced in the region of $\lambda < 1.0\,\mu$m, but does not yet agree with the observed values. Next, we considered that the snow impurities might have been concentrated at the surface, and Model-4 was created, with the concentration of impurities assumed to be $c = 18$ ppmw for the top (0–1 cm) layer and $c = 1$ ppmw for the two lower layers. The concentration in the top layer was set so as to keep the total column amount of impurities in the layers of 0–5 cm, as determined from the measured value of 3.6 μg/cm^2, where it was 3.7 μg/cm^2 for Model-4. The best agreement between the observed albedo and the theoretical value is obtained with Model-4. These results suggest that only a thin top layer, rather than the whole depth of 0–5 cm of snow, was highly contaminated due to dry fallout of atmospheric aerosols. A small discrepancy in the albedo between observed and calculated values is noted in only the region of $1.2 < \lambda < 1.4\,\mu$m. Similar results were found in the data of the other days, although the reason is not clear at this stage.

Theoretical calculations of spectral albedo using snow models similar to Model-4 were performed for February 22, 24, and 25, and compared with the observations presented in Figs. 5.11(a), (b), and (c). The snow models assumed for these three days are depicted in Figs. 5.12(a), (b), and (c), respectively. The effective grain radius and snow density were determined by the same method as for February 23 (Figs. 5.9 and 5.10.) On February 22, the best agreement was obtained using a value of c close to the measurement. However, on February 24 the value of c for which the theoretical albedo agrees with the observation was 50 ppmw in the topmost layer of 0–0.5 cm. The total column amount of impurities in the layer of 0–5 cm was 5.7 μg/cm^2, which is somewhat greater than the measured value (4.1 μg/cm^2). This may be due to an error in the thickness of snow sampling because it is not easy to obtain a precise snow sample in the layer of 0–5 cm. For February 25, good agreement is obtained for $c = 20$ ppmw with a total column amount of impurities of 5.8 μg/cm^2, in the layer of 0–5 cm (measured value of 6.0 μg/cm2). From these analyses, we can delineate the daily change in snow contamination as follows. On February 22, just after the snowfalls (February 20 and February 21), the snow surface was comparatively clean and had a high visible albedo. The snow impurities gradually became concentrated at the surface due to dry fallout of the atmospheric aerosols and snow densification. The measured concentration of snow impurities in the layers below 5 cm depth was higher on February 22 than on the other days, possibly due to the difference in observation sites or an error in the thickness of snow sampling. However, even if the value of $c = 1$ ppmw is assumed for 5–30 cm on February 22, the spectral albedo increases, at maximum, by only 0.001 at $\lambda = 0.5\,\mu$m. In the above analyses, small discrepancies between observed and calculated albedos are still found at around $\lambda = 1.3\,\mu$m for all cases in Fig. 5.11, as was found on February 23. These results suggest a systematic error in the calculation and/or observation in this work, but it has not yet been clarified. According to Sergent et al. (1998), the optically equivalent radius determined from the hemispherical-directional reflectance measured at $\lambda = 0.99\,\mu$m makes the theoretical value of

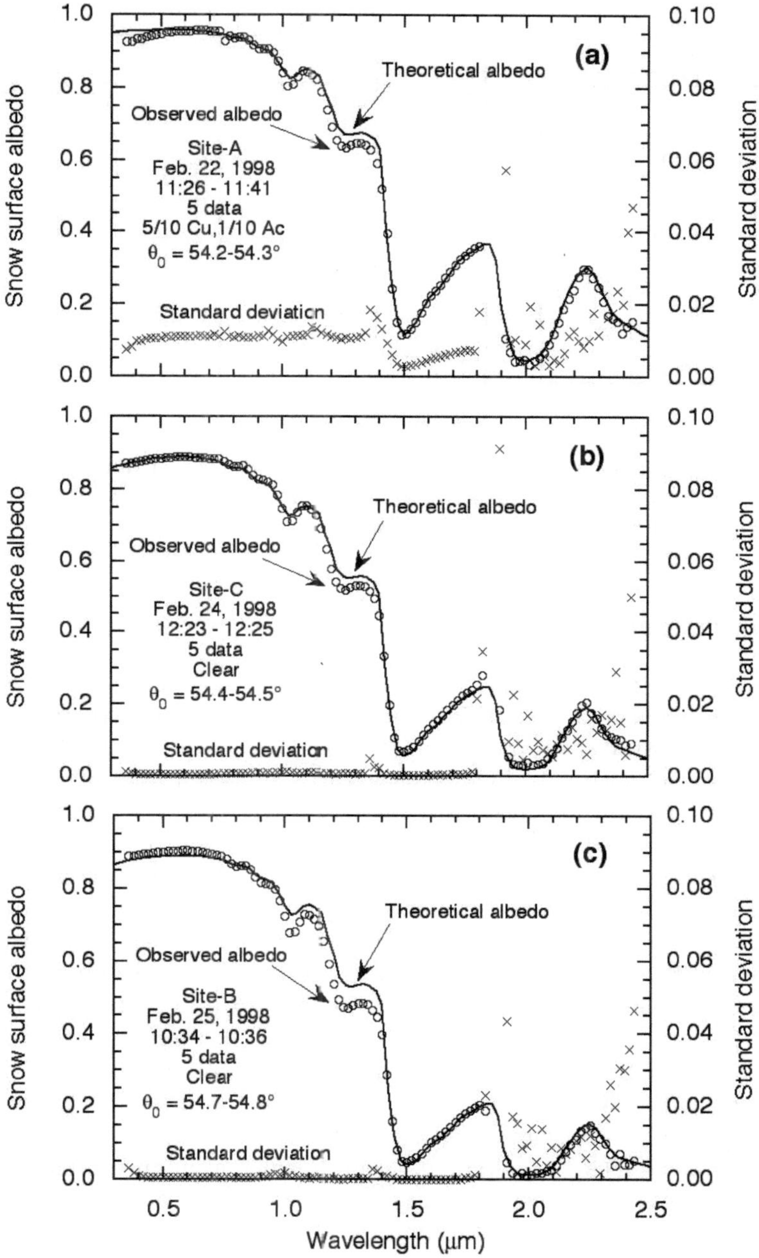

Fig. 5.11. Same as Fig. 5.9 for Model-4, but for (a) February 22, (b) February 24, and (c) February 25, 1998 (Aoki et al., 2000).

the spectral hemispherical-directional reflectance agree with the measurement in a wide region of $\lambda = 0.9\text{–}1.45\,\mu\text{m}$. The extension of such measurement to longer wavelengths would clarify the uncertainty in this study.

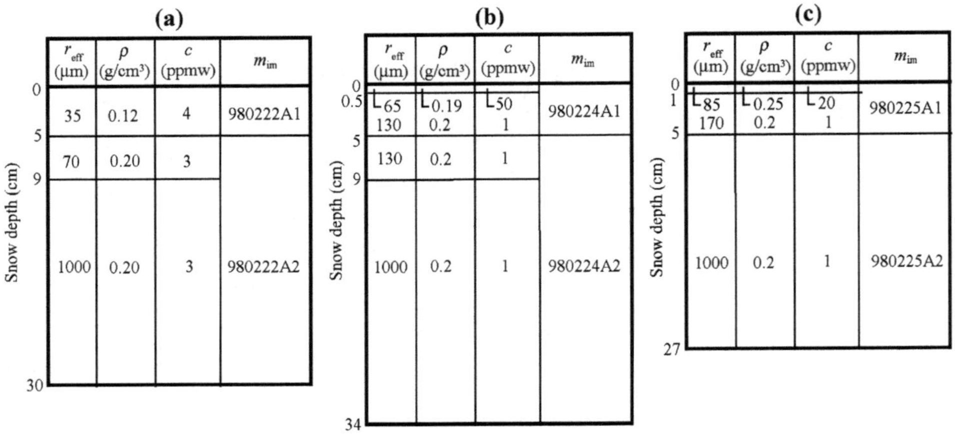

Fig. 5.12. Snow models on (a) February 22, (b) February 24, and (c) February 25, 1998, for which the theoretical spectral albedos were calculated and compared with the measurements in Fig. 5.11 (Aoki et al., 2000).

5.5.3 Observation of HDRF

The HDRF observation was carried out on the same snow surface as for the albedo measurement on February 25. This observation was made using two types of optical heads: a bare optical fiber whose field of view (FOV) was 25° for full angle, and a foreoptics head with a 1° FOV attached to the optical fiber. The bare optical fiber was set up as for the albedo observation system where the azimuth and zenith angles were controlled by turning the azimuth direction of the mount and the viewing direction of the optical fiber arm, respectively. The foreoptics head was attached to a goniostage (angle-setting device) on a tripod on the snow surface. Since the distance from the snow surface to the foreoptics was 1.0 m, even a slight undulation of the snow surface could affect the HDRF pattern observed with the foreoptics, especially at a low viewing zenith angle (θ'). In contrast, at high θ' the FOV of the optical fiber was too broad to measure the target accurately. We thus made a composite HDRF pattern from the measurements obtained with the optical fiber for $\theta' \leq 70°$ and from those obtained with the foreoptics for $\theta' \geq 80°$.

It is necessary to measure $F_\lambda^\downarrow(\theta_0, \phi_0)$ (Eqs. (5.3) and (5.4)) to calculate HDRF. However, it was difficult to obtain an accurate value using measurement with the foreoptics and WRS because the foreoptics head (60 mm $\phi \times$ 120 mm) was larger than the WRS (150 mm × 150 mm), and it obstructed the light coming from the zenith direction. Many studies involving HDRF observation used the anisotropic reflectance factor $f(\theta_0, \theta', \phi' - \phi_0)$, as defined by Eq. (5.5) (Taylor and Stowe, 1984a,b; Brandt et al., 1991; Grenfell et al., 1994; Warren et al., 1998). The merit of using $f(\theta_0, \theta', \phi' - \phi_0)$ is that it is not necessary to know the value of $F_\lambda^\downarrow(\theta_0, \phi_0)$ in Eq. (5.3). However, the complete angular measurement of $I_\lambda^\uparrow(\theta_0, \theta', \phi' - \phi_0)$ is required, and it was not obtained in our observations. We therefore calculated the normalized HDRF (NHDRF) by taking into account the nadir radiance. The composite NHDRF pattern was also calculated from each NHDRF obtained by means of the optical fiber and the foreoptics.

Fig. 5.13 presents the composite NBRDF observed for six wavelengths, where we used the display method employed previously by Taylor and Stowe (1984a,b) and Warren et al. (1998). Anisotropic reflection property is very significant at $\lambda = 1.64$ and $2.21\,\mu$m, while the NHDRF patterns are relatively flat in the visible region. The reason for this was explained by Leroux et al. (1999) as follows. Since the single scattering albedo is close to unity in the visible region due to the weak light absorption of ice, the BRDF is not influenced by the single scattering parameter, and vice versa in the near-infrared region. A similar result was obtained in the principal plane by Carlson and Arakelian (1993) in the measurement of anisotropic reflection in Antarctica. Our result was expected in view of the finding of Warren et al. (1998) that the BRDF pattern becomes more anisotropic with a relatively strong forward peak for $\lambda > 0.9\,\mu$m, where snow is more absorptive. In Fig. 5.13, the maximum NHDRF (2.2 at $\lambda = 0.52\,\mu$m and 16.2 at $\lambda = 1.64\,\mu$m) is observed at $\theta' = 85°$ in the forward scattered direction (the bottom of each map). In the side scattered directions (left and right directions on each map), the NBRDF decreases

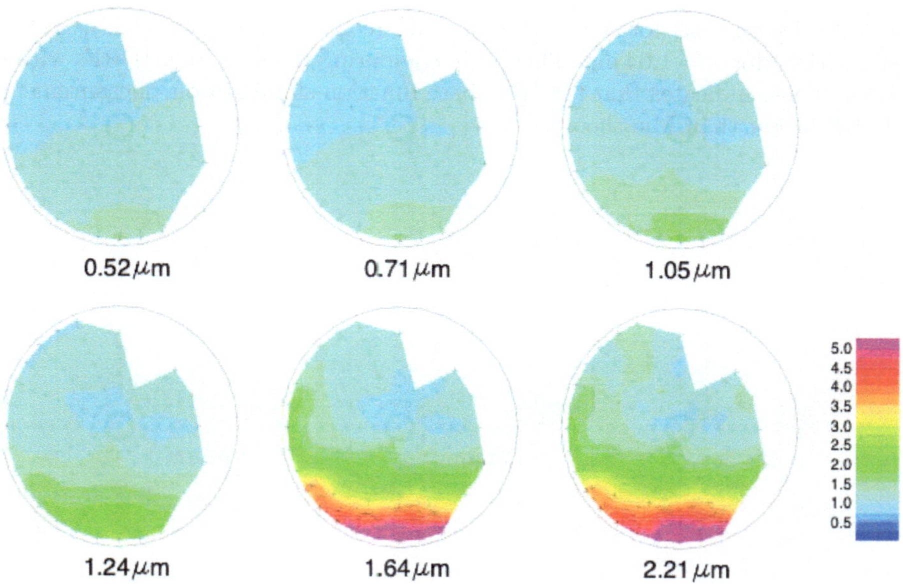

Fig. 5.13. Composite NHDRFs of snow for six wavelengths as obtained from measurements of anisotropic reflectance with an optical fiber of 25° field of view (FOV) for $\theta' \leq 70°$ and those obtained with foreoptics of a 1° FOV for $\theta' \geq 80°$. The measurements using the optical fiber were conducted from 1222 to 1258 LT ($\theta_0 = 54.0$–$56.0°$), and those using the foreoptics were conducted from 1320 to 1334 LT ($\theta_0 = 57.9$–$59.3°$) on February 25, 1998, in eastern Hokkaido. All reflectances are normalized by the value at the nadir. The plus signs on each NHDRF map indicate the observed points. The radial coordinate is proportional to the viewing angle θ', which is zero at the center of the circle (nadir) and is 90° on the circle. The sun comes from the upper direction of each map, and the bottom of each map is the forward scattering direction (Aoki et al., 2000).

to some degree with the viewing angle in the visible region and increases in the near-infrared region.

5.5.4 Theoretical calculations of HDRF and comparison with the measurements

Fig. 5.14 presents the theoretically calculated NHDRFs using two different phase functions under the same snow conditions as used in Fig. 5.11(c). The first NHDRF (left side) was calculated using the single scattering parameters calculated for spherical ice particles by Mie theory, and the second (right side) was calculated using the same parameters except for the phase function of Henyey–Greenstein (HG), which was calculated from the same asymmetry factor as in Mie theory. The most conspicuous difference between these two NHDRFs is the presence of a rainbow when using the Mie phase function, which can be seen clearly at $\lambda = 1.64\,\mu\text{m}$ and appears at any wavelength but is not clearly visible. On the other hand, no rainbow is seen at any wavelength in the measurements presented in Fig. 5.13. In addition, the maximum NHDRF, seen just below the horizon in the forward scattering direction, is higher in NHDRF using the Mie phase function than in that using the HG phase function. The values of NBRDF at $\theta' = 85°$ are 3.0 (Mie), 2.2 (HG), and 2.2 (observation) for $\lambda = 0.52\,\mu\text{m}$, and are 23.4 (Mie), 14.2 (HG), and 16.2 (observation) for $\lambda = 1.64\,\mu\text{m}$. Therefore, comparing theoretical NHDRFs with the measurements indicates that the HG phase function simulates our measurement of NHDRF better than Mie theory.

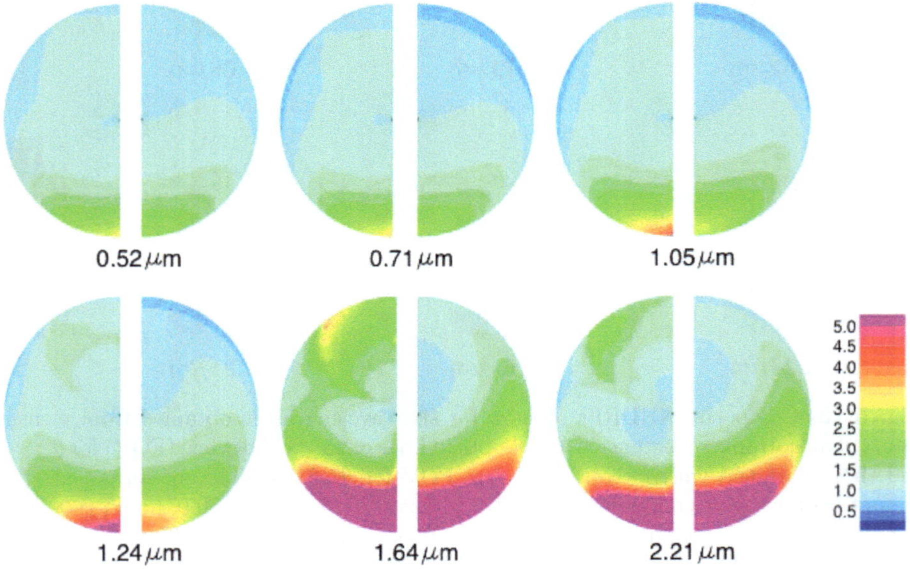

Fig. 5.14. Theoretical NHDRFs of snow for six wavelengths. The semicircular maps on the left side were calculated using the Mie phase function, and those on the right side using the Henyey–Greenstein phase function. The mesh points on each of the NHDRF maps indicate the calculation grid points (Aoki et al., 2000).

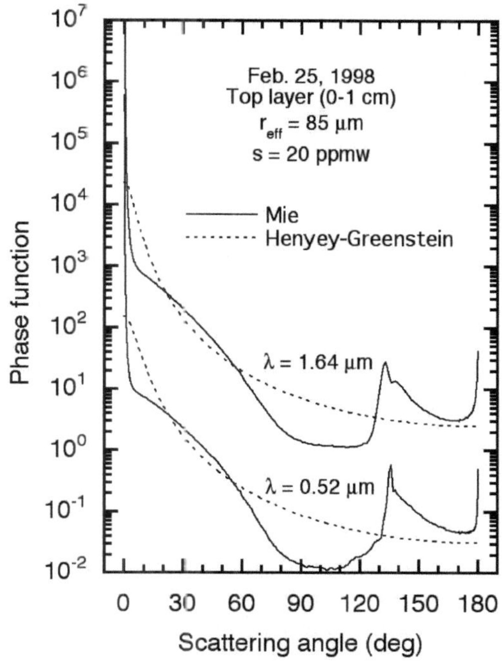

Fig. 5.15. Mie and Henyey–Greenstein phase functions of snow grains for $\lambda = 0.52$ and $1.64\,\mu$m in the top layer of the snow model (see Fig. 5.12(c)) of February 25, 1998. The curves for $\lambda = 1.64\,\mu$m are displaced upward by a factor of 10^2 (Aoki et al., 2000).

Fig. 5.15 depicts the Mie and HG phase functions of snow grains in the top layer of the snow model used in these NHDRF calculations. The peak at a scattering angle of 135° in the Mie phase function causes a rainbow of HDRF, and the forward scattering peaks in both phase functions are responsible for the maximum value of HDRF. These phenomena can easily be explained where single scattering by snow grains is dominant. Since multiple scattering by snow grains is dominant in the visible region due to the weak light absorption of ice, the pattern due to single scattering is reduced by the multiple scattering. However, in the near-infrared region, low-order scattering is dominant due to the strong light absorption of ice, resulting in the rainbow and maximum value in the forward scattering region in HDRF. Here, the shape of the phase function affects the HDRF of snow; thus, the snow grain shape also affects the HDRF. From these points of view, it seems that general hexagonal shapes are not suitable for HDRF calculation; the spherical shape is not suitable either because the halo is not usually seen on the snow surface itself, although it occasionally appears in the Antarctic Plateau (Warren et al., 1998). The shapes of snow grains differ from each other and change in the course of the aging of snow. A halo does not necessarily appear even in cirrus clouds. In measurements with a polar nephelometer by Gayet et al. (1998), examination of irregularly shaped cirrus clouds indicated a smooth phase function at forward scattering angles.

Studies on single scattering properties of nonspherical ice particles in clouds provide useful information for HDRF studies of the snow surface. One suggestive work was done by Macke et al. (1996a) who had calculated the phase function for randomized triadic Koch-fractals and found that it becomes smooth with increasing distortion. The complicated shape of the ice particles eliminates particular peaks, such as the halo or rainbow in the phase function. Smooth phase functions have been obtained in theoretical calculations for ice particles with inclusions of air bubbles or soot (Macke et al., 1996b; Mishchenko and Macke, 1997), imperfect hexagonal ice crystals (Hess et al., 1998), a randomly oriented oblate spheroid (Mishchenko and Travis, 1998), and a plate of hexagonal ice particles (Leroux et al., 1999). Comparison of the observations with these models is the next coming item of the present study. In addition, it will be necessary to study: (1) direct measurement of the phase function of snow particles at each stage of snow age and (2) calculation of single scattering for irregular ice particles.

We have seen in this section that the shape of the phase function is important for HDRF calculation. It is worth checking the effect of the phase function on albedo. We confirmed that there is almost no difference between spectral albedos theoretically calculated using the Mie and HG phase functions for the snow model of February 25, although the figure is not presented here (see Fig. 14 in Aoki et al., 2000). This result holds true except for large θ_0. Thus, the asymmetry factor is important for albedo, although the detailed shape of the phase function does not affect the albedo. We can safely say that the Mie phase function can be used for snow albedo calculations.

5.6 Effects of snow physical parameters on broadband albedos

5.6.1 Instrumentation, observational condition, and radiative transfer model

This section discusses the effects of snow physical parameters on broadband albedos by comparing the albedos continuously observed with those theoretically calculated with a radiative transfer model based on the results of Aoki et al. (2003). All field measurements were carried out in the 1999/2000 and 2000/2001 winters at the meteorological observation field (43° 49′ 21″ N, 143° 54′ 13″ E, 94 m a.s.l.) of the Kitami Institute of Technology in eastern Hokkaido, Japan. The surface condition for the snow-free period was flat with withered grass. In the radiation budget observation, the upward and downward components of radiant flux densities in shortwave ($\lambda = 0.305$–$2.8\,\mu$m), near-infrared ($\lambda = 0.695$–$2.8\,\mu$m), and longwave ($\lambda > 4\,\mu$m) spectral regions were measured using four pyranometers (MS-801, EKO Instruments, Japan) and two pyrgeometers (MS-200, EKO Instruments, Japan). To measure the near-infrared region, a cut-off filter dome at $\lambda = 0.695\,\mu$m was installed on the pyranometer. Each radiation component was sampled every ten seconds, and one-minute-averaged values were stored in a data logger. Visible radiation was determined by subtracting the near-infrared radiation from the shortwave radiation. Broadband albedos were calculated from 30-minute-averaged values of measured radiation components at every half-hour interval (01–30 and 31–00 in minutes). We analyzed only the data closest to the local solar noon (the data measured during

1131–1200 LT) to keep the observational condition uniform. Otherwise, when the sun is in an easterly or westerly position, the snow surface is shadowed by the two vertical frames supporting the instruments. The meteorological components measured with the radiation budget observation were air temperature and relative humidity at 1.0 m above the snow surface. The snow surface temperature was calculated from the observed longwave radiation data. These data were recorded every minute in a data logger. Snow depth was measured with a laser snow gauge, and precipitation with a rain gauge, every hour. Fig. 5.16 presents the daily variations of 30-minute-averaged albedos and representative snow and meteorological components at 1200 LT every day during all observation periods. The maximum snow depth in the 1999/2000 winter was 117 cm, and that in the 2000/2001 winter was 72 cm. The average snow depth in midwinter was around 60 cm in both winters. The air and snow temperatures were almost always below freezing point in December, January, and February. When the surface was not covered by snow, the near-infrared albedo was higher than the visible one. However, this relationship was reversed for the snow surface because of the difference in spectral variation of the albedo between withered grass and snow cover. Snow began to melt in March and disappeared in April.

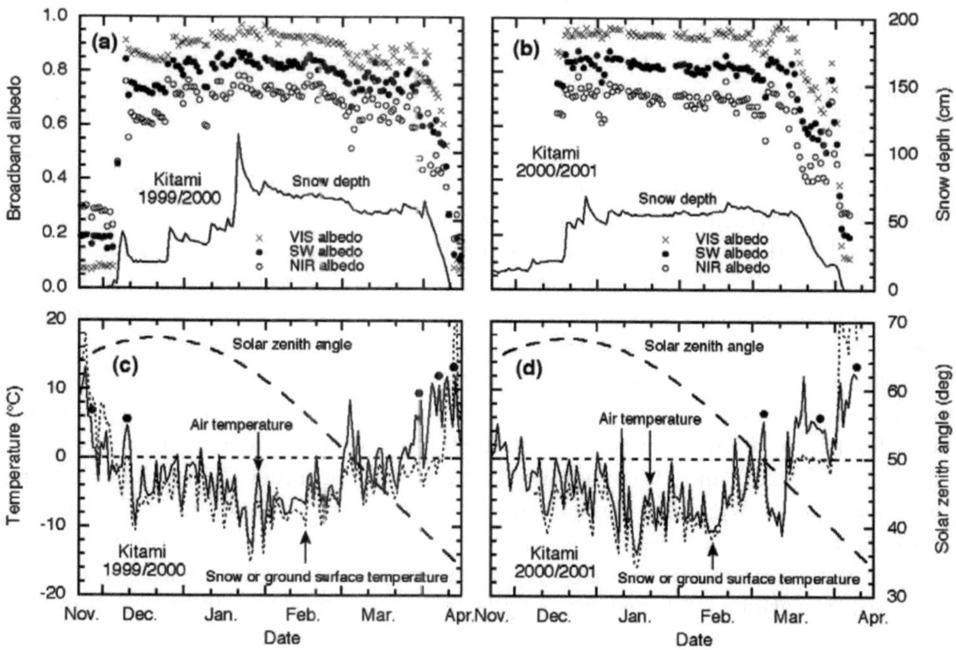

Fig. 5.16. (a, b) Broadband albedos in the visible (VIS), shortwave (SW), and near-infrared (NIR) regions measured in Kitami, and snow depth at 1200 LT. Albedos are the daily 30-min-averaged values from 1131 to 1200 LT. (c, d) Air temperature, snow or ground surface temperature, and solar zenith angle at 1200 LT. Dots above the curve of air temperature indicate rainfall. Observation periods are during two winters: (a, c) 1999/2000 and (b, d) 2000/2001 (Aoki et al., 2003).

Snow pit work was performed twice or three times a week for the components of snow type, temperature, density, and snow grain size in each snow layer, together with snow sampling for the measurement of mass concentration of snow impurities. In the measurements of snow grain size, two or three kinds of dimensions at the snow surface were measured with a handheld lens using the same method as in Aoki et al. (2000). These measurements were one-half length of the major axis of crystals or dendrites (r_1), one-half the branch width of dendrites or one-half the dimension of the narrower portion of broken crystals (r_2), and one-half the crystal thickness only for dendrites or plate-like crystals (r_3). For aggregate granular grains, one-half the dimension of the cluster and each grain's diameter were measured as r_1 and r_2. Aoki et al. (1998, 2000) concluded that the optically equivalent snow grain size was r_2 for new snow or faceted crystals from the spectrally detailed albedo measurements together with snow pit work. The snow impurities were filtered using a nuclepore filter, after melting the snow sample. We used a two-stage filtering system of nuclepore filters with different pore sizes of 0.2 and 5.0 μm for snow samples from the surface to three snow thicknesses (0–1, 0–5, and 0–10 cm). This procedure produced rough information of the impurity types and a vertical profile of the impurities in the snow. The majority of impurities were collected on the nuclepore filter with a pore size of 5.0 μm, and the main constituent was mineral dust particles. The concentrations of snow impurities were estimated by directly weighing the nuclepore filters before and after filtering, using a balance.

The effects of snow physical parameters on broadband albedos were investigated by comparing the measured albedos with those calculated using a multiple scattering model of radiative transfer for the atmosphere–snow system (Aoki et al., 1999, 2000). The appendix in Aoki et al. (2003) presents the calculated results of solar zenith angle dependence of albedos under the possible variation ranges of snow impurity concentrations (c), θ_0, and cloud cover when the albedos were measured. The model used the model atmosphere MW, a rural aerosol model with an optical thickness $\tau_a = 0.1$ at $\lambda = 0.5\,\mu$m, and the same model of water cloud as Aoki et al. (1999). The atmosphere was divided into 14 layers, and the snow into a single layer. Thus, the snow physical parameters are uniform from the surface to the bottom. The effects of θ_0 and cloud cover on albedos differ depending on the effective snow grain size (r_{eff}) and concentration of snow impurities. Albedos are therefore calculated for $r_{\mathrm{eff}} = 30\,\mu$m (new snow) and 1000 μm (granular snow), and $c = 1$ ppmw (background level) and 100 ppmw (very dirty case) of mineral dust. These values were determined from the snow pit measurements of r_2 for snow grain size and the concentration of impurities in the snow layer of 0–5 cm depth. For the size distribution of snow particles, we employed a log-normal distribution with a geometric standard deviation of 1.6 measured by Grenfell and Warren (1999) in Antarctica. For snow impurities, the coagulation mode of the mineral aerosol model (Hess et al., 1998) was used because the main constituent of snow impurities collected on the nuclepore filter from the snow sample was mineral dust.

5.6.2 Effects of the snow grain size on broadband albedos

The important snow physical parameters controlling the albedo are snow grain size and snow impurities (Fig. 5.2). We investigated the effects of these parameters measured from snow pit work on broadband albedos. Fig. 5.17 presents the broadband albedos as a function of snow grain size. The circles and crosses illustrate the differences in the dimensions of measured snow grain sizes r_2 and r_3. The curves denote the theoretically calculated broadband albedos for several snow impurity concentrations, solar zenith angles, and sky conditions (clear or cloudy). The curves represent the albedos under the possible variation ranges of c, θ_0, and cloud cover, when the snow albedo data we analyzed were measured. For example, for the visible region, the maximum albedo is theoretically expected when $c = 1\,\text{ppmw}$, $\theta_0 = 67°$, and the sky is clear; the minimum albedo is theoretically expected when $c = 100\,\text{ppmw}$, $\theta_0 = 40°$, and the sky is clear. Next, we consider the relationship between snow grain size and measured albedos (Fig. 5.17), where the measured albedos decrease with the snow grain size, especially for near-infrared albedos. This tendency agreed with those of theoretically calculated albedos for the data of r_2, while the albedos for r_3 are lower than the theoretical curves for both the visible and the near-infrared regions. This result confirms that the optically

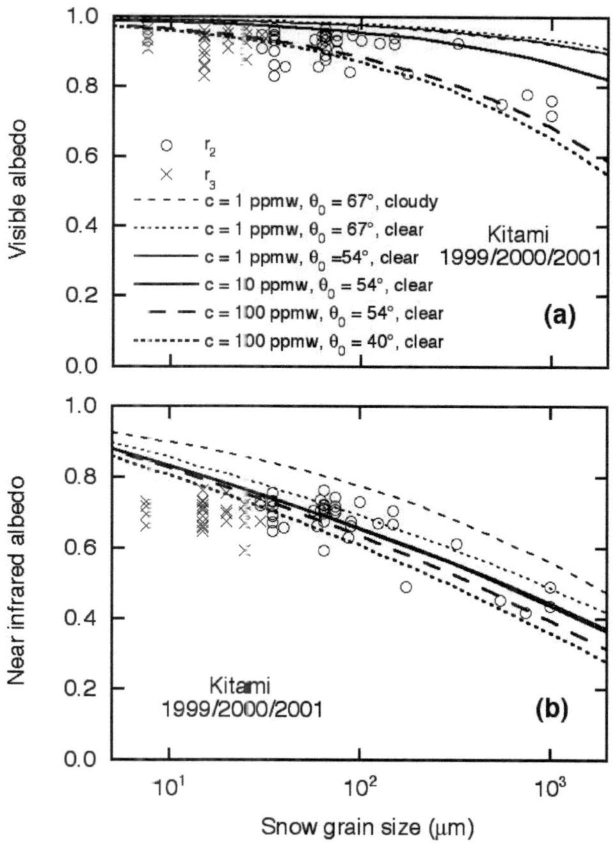

Fig. 5.17. (a) Visible albedo and (b) near-infrared albedo as a function of snow grain size. The circles and crosses indicate the differences in dimensions of measured snow grain size, r_2 and r_3, defined by Aoki et al. (2000, 2003). The curves denote the theoretically calculated broadband albedos for several snow impurity concentrations (c), solar zenith angles (θ_0), and sky conditions (clear or cloudy) (Aoki et al., 2003).

equivalent snow grain size is r_2, as indicated by Aoki et al. (1998, 2000). However, even for r_2, some data are out of the range of theoretically calculated albedos. One possible reason is that light absorption is weak in the snow impurity model (mineral dust) we employed, and the other possibility is discussed in the next section 5.6.3.

5.6.3 Effects of the snow impurities on broadband albedos

Fig. 5.18 depicts broadband albedos as a function of the concentration of snow impurities. The measured albedos decrease with the concentration of snow impurities. Warren and Wiscombe (1985) presented similar figures, in which the theoretical calculated shortwave albedo is reduced by 10% by soot of 0.1–1.0 ppmw contained in snow depending on snow grain size. However, the concentration of snow impurities that begins to reduce the albedo in our measurement is roughly two orders of magnitude higher due to the difference in absorption between soot and mineral

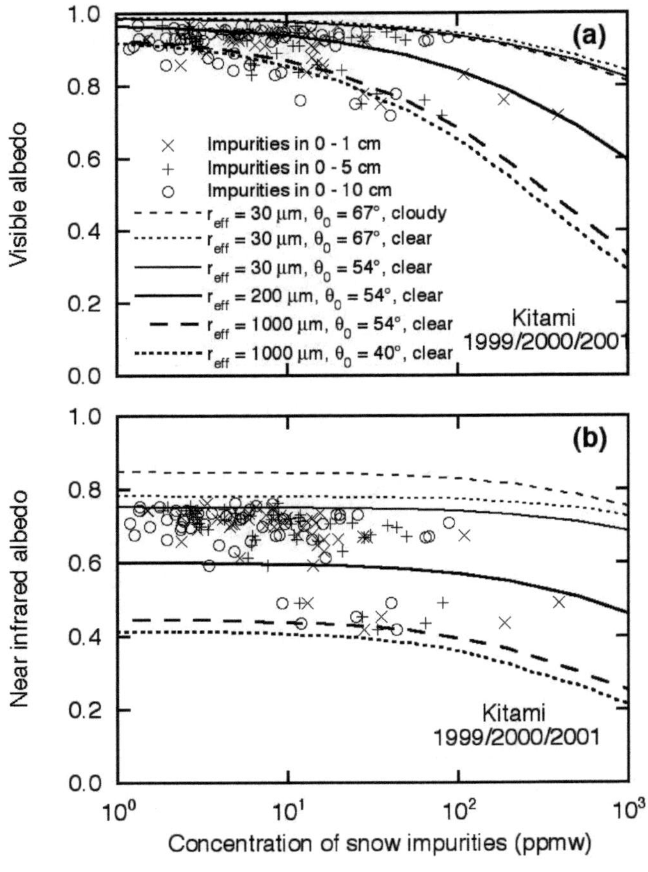

Fig. 5.18. (a) Visible albedo and (b) near-infrared albedo as a function of concentration of snow impurities (c). The crosses, plus signs, and circles indicate the differences in the snow sampling layer. The curves in the figure denote the theoretically calculated broadband albedos for several snow grain sizes (r_{eff}), solar zenith angles (θ_0), and sky conditions (clear or cloudy) (Aoki et al., 2003).

dust. In general, c is highest for the sampling snow layer of 0–1 cm depth, second for the layer of 0–5 cm depth, and lowest for the layer of 0–10 cm depth because of the contribution from the dry deposition of atmospheric aerosols (Aoki et al., 2000). Very high values of c (exceeding 100 ppmw) were recorded in the 0–1 cm layer (Fig. 5.18(a)), where the corresponding albedos fall in the theoretically calculated range. However, some visible albedos corresponding to c in the 0–10 cm layer were lower than any theoretical curve. Although this could also be explained by the weak absorption of the snow impurity model we employed, the inhomogeneity of impurities (high at the top layer) in snow layers near the surface could reduce the visible albedo (Aoki et al., 2000). On the other hand, the measured near-infrared albedos fall within the theoretically predicted range.

In both the snow grain size dependence and the impurity dependence of broadband albedos, the measured albedos fall roughly in the theoretically predicted range. In general, when the snow grain was large, the concentration of snow impurities was high (e.g., in March during the melting season). Moreover, there is an effect whereby a visible albedo reduction caused by snow impurities is enhanced by an increase of snow grain size (Warren and Wiscombe, 1980). To examine the relationships among the measured values of these parameters and albedos, we plotted the broadband albedos as a function of c for several ranges of snow grain size. In Fig. 5.19, the measured value of c is for the 0–5 cm layer, and the measured snow grain size is for r_2 at the surface. The visible albedos measured in each range of snow grain size except for $500\,\mu m \leq r_2$ do not correspond with the theoretical curves. On the contrary, the measured near-infrared albedos in each range of snow grain size clearly correspond with the theoretical curves. In the near-infrared region, light absorption by ice is stronger than in the visible region, so the near-infrared albedo contains information of the snow physical parameters near the surface. Snow grain size was measured at the snow surface, and the concentration of impurities was measured in the 0–5 cm layer, leading to the agreement of the measured near-infrared albedos with the theoretical values. In contrast, the visible albedo contains information on the snow physical parameters in the relatively deeper snow layers. If the snow physical parameters in the deeper layers were not the same as those in the surface layers, the visible albedo would deviate from the theoretical curves. For example, with a new snow cover of a few centimeters (small r_2) on old granular snow (large r_2), the near-infrared albedo would take a theoretically predicted value for small grain size. However, the visible albedo would be lower than the theoretically predicted value for all snow layers with small grain size due to the granular snow layer below. The fact that some measured visible albedos were lower than the theoretically calculated curves (Fig. 5.18(a)) mentioned in the previous paragraph might also be related to the inhomogeneity of the snow parameters.

5.7 Concluding remarks

The cryosphere is very sensitive to global warming because the surface albedos drastically change due to a phase change of the snow and ice surface. In the northern hemisphere, it has been confirmed that land snow cover and sea ice extent are decreasing and that the Greenland ice sheet is melting. Many climate models

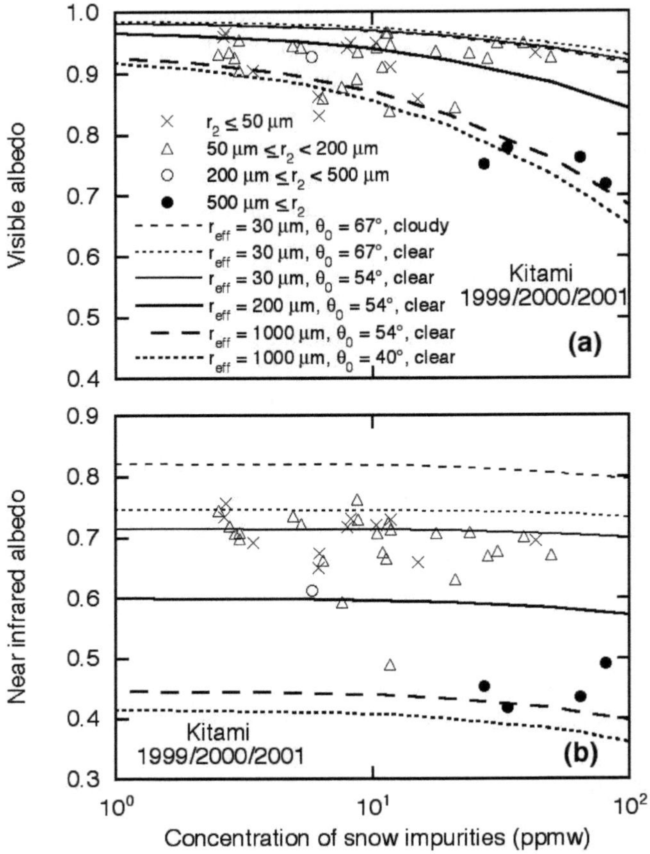

Fig. 5.19. (a) Visible albedo and (b) near-infrared albedo as a function of mass concentration of snow impurities in the top 5 cm layer. The curves denote the theoretically calculated broadband albedos for several kinds of snow grain sizes (r_{eff}), solar zenith angles (θ_0), and sky conditions (clear or cloudy). Measured albedo values are plotted by different characters for four ranges of snow grain sizes (r_2) (Aoki et al., 2003).

predict a large warming in the high latitudes of the northern hemisphere in this century, which is related to the reduced snow and sea ice cover. To accurately simulate the future climate in the cryosphere, a better understanding of snow optical properties, such as snow albedo and bidirectional reflectance, is necessary. The factors that affect snow albedo are essentially divided into two categories: (1) snow physical parameters and (2) external parameters (e.g., atmospheric conditions and solar zenith angle). The snow physical parameters affecting the albedo and reflectance are grain size, crystal shape, impurities (aerosols and algae), liquid water content, snow thickness, layer structure, and surface roughness. Snow grain size and impurities are especially important parameters controlling the albedo; thus, these snow parameters are important targets of satellite remote sensing. Bidirectional reflectance properties of snow surface are important for the satellite remote sensing of snow parameters because the satellite sensor observes the snow surface from a particular direction on the orbit at the time of measurement. An example of this application is the works by the ADEOS-II/GLI Cryosphere team, who have developed remote sensing algorithms to retrieve these snow physical parameters. The results were published as a series of three papers: Part I: Scientific basis (Stamnes et al., 2007), Part II: Validation results (Aoki et al., 2007b), and Part III: Retrieved results (Hori et al., 2007). Their results showed the snow surface temperature and grain size for the shallow layer agreed well with in situ measured values, while the mass fractions of snow impurities and grain size at the topmost layer were not accurate compared with the former two products. These snow parameters are expected to be used as indicators of climate change by long-period monitoring. However, further improvements of the algorithms are needed.

The Fourth Assessment Report of the Intergovernmental Panel on Climate Change (IPCC, 2007) estimated that the global annual mean radiative forcing due to BC on snow is $+0.1 \pm 0.1\,\mathrm{W\,m^{-2}}$, a value based mainly on estimates by Hansen and Nazarenko (2004) and Hansen et al. (2005), who used prescribed albedo reduction values from snow BC concentration data obtained in field experiments by Clarke and Noone (1985), along with more recent measurement data. Subsequently, some physically based snow albedo models were developed to calculate broadband albedos as functions of snow grain size and impurity concentration in GCM were developed (Jacobson, 2004 Flanner and Zender, 2005, 2006; Yasunari et al., 2011, Aoki et al., 2011). Their simulations indicated that the effect of snow impurities on albedo reduction and the heating effect of snow surface cannot be ignored locally and seasonally. On the other hand, in situ measurements of light-absorbing snow impurities have been performed since the 1980s from Greenland ice cores (e.g., Chýlek et al., 1995; McConnell et al., 2007) and from snow samples in Greenland (e.g., Hagler et al., 2007; Doherty et al., 2010), in the Arctic excluding Greenland (e.g., Perovich et al., 2009; Forsström et al., 2009; Doherty et al., 2010), in Antarctica (e.g., Warren and Clarke, 1990; Grenfell et al., 1994), and in midlatitudes (e.g., Fily et al., 1997; Kuchiki et al., 2009; Huang et al., 2010). These results indicate that the effect of BC concentration on albedo reduction is very limited in Greenland and ignorable in Antarctica, whereas in the other areas it cannot be ignored. Flanner et al. (2009) concluded that the effect on radiative forcing of the reduction of surface-incident solar energy (dimming) caused by atmospheric aerosols containing BC and organic matter is smaller than the effect of the reduction of snow albedo caused by deposition of such aerosols (darkening). The effects of snow impurities on the radiation budget at the snow surface have recently been investigated using improved snow albedo models and parameterizations. However, the albedo schemes and parameterizations used in many GCMs are still insufficiently validated. Thus, studies on the optical properties of snow from both theoretical and observational points of view are important.

Acknowledgments

We thank Dr Alexander Kokhanovsky of the University of Bremen for invaluable discussions and advices.

References

Anderson, G. P., S. A. Clough, F. X. Kneizys, J. H. Chetwynd, and E. P. Shettle, 1986: AFGL atmospheric constituent profiles (0–120 km), *AFGL-TR-86-0110*, Air Force Geophysics Laboratory, Hanscom, MA.

Aoki, Te., Ta. Aoki, and M. Fukabori, 1997: Approximations of phase function in calculating the spectral albedo of snow surface with multiple scattering, *Pap. Met. Geophys.*, **47**, 141–156.

Aoki, Te., Ta. Aoki, M. Fukabori, Y. Tachibana, Y. Zaizen, F. Nishio, and T. Oishi, 1998: Spectral albedo observation on the snow field at Barrow, Alaska, *Polar Meteorol. Glaciol.*, **12**, 1–9.

Aoki, Te., Ta. Aoki, M. Fukabori, and A. Uchiyama, 1999: Numerical simulation of the atmospheric effects on snow albedo with a multiple scattering radiative transfer model for the atmosphere-snow system, *J. Meteor. Soc. Japan*, **77**, 595–614.

Aoki, Te., Ta. Aoki, M. Fukabori, A. Hachikubo, Y. Tachibana, and F. Nishio, 2000: Effects of snow physical parameters on spectral albedos, and bidirectional reflectance of snow surface, *J. Geophys. Res.*, **105**, 10219–10236.

Aoki, Te., A. Hachikubo, and M. Hori, 2003: Effects of snow physical parameters on broadband albedos, *J. Geophys. Res.*, **108**, 4616, doi:10.1029/2003JD00 3506.

Aoki, Te., H. Motoyoshi, Y. Kodama, T. J. Yasunari, K. Sugiura, and H. Kobayashi, 2006: Atmospheric aerosol deposition on snow surfaces and its effect on albedos, *SOLA*, **2**, 013–016, doi:10.2151/sola.2006-004.

Aoki, Te., H. Motoyoshi, Y. Kodama, T. J. Yasunari, and K. Sugiura, 2007a: Variations of the snow physical parameters and their effects on albedo in Sapporo, Japan, *Ann. Glaciol.*, **46**, 375–381.

Aoki, Te., M. Hori, H. Motoyohi, T. Tanikawa, A. Hachikubo, K. Sugiura, T. J. Yasunari, R. Storvold, H. A. Eide, K. Stamnes, W. Li, J. Nieke, Y. Nakajima, and F. Takahashi, 2007b: ADEOS-II/GLI snow/ice products – Part II: Validation results using GLI and MODIS data, *Remote Sens. Environ.*, **111**, 274–290, doi:10.1016/j.rse.2007.02.035.

Aoki, T., K. Kuchiki, M. Niwano, Y. Kodama, M. Hosaka, and T. Tanaka, 2011: Physically based snow albedo model for calculating broadband albedos and the solar heating profile in snowpack for general circulation models, *J. Geophys. Res.*, **116**, D11114, doi:10.1029/2010JD015507.

Asano, S., and A. Uchiyama, 1987: Application of an extended ESFT method to calculation of solar heating rates by water vapor absorption, *J. Quant. Spectrosc. Radiat. Transfer*, **38**, 147–158.

Barkstrom, B. R., 1972: Some effects of multiple scattering on the distribution of solar radiation in snow and ice, *J. Glaciol.*, **11**, 357–368.

Barkstrom, B. R., and C. W. Querfeld, 1975: Concerning the effect of anisotropic scattering and finite depth on the distribution of solar radiation in snow, *J. Glaciol.*, **14**, 107–124.

Bergen, J. D., 1970: A possible relation between grain size, density, and light attenuation in natural snow cover, *J. Glaciol.*, **9**, 154–156.

Bergen, J. D., 1971: The relation of snow transparency to density and air permeability in a natural snow cover, *J. Geophys. Res.*, **76**, 7385–7388.

Berger, R. H., 1979: Snowpack optical properties in the infrared, CRREL Rep., 79–11, U.S. Army Cold Reg. Res. and Eng. Lab., Hanover, NH, 1–9.

Bindoff, N. L., J. Willebrand, V. Artale, A, Cazenave, J. Gregory, S. Gulev, K. Hanawa, C. Le Quéré, S. Levitus, Y. Nojiri, C. K. Shum, L. D. Talley, and A. Unnikrishnan, 2007: Observations: oceanic climate change and sea level. in Climate Change 2007: The

Physical Science Basis, Contribution of Working Group I to the Fourth Assessment Report of the Intergovernmental Panel on Climate Change [Solomon, S., D. Qin, M. Manning, Z. Chen, M. Marquis, K.B. Averyt, M. Tignor and H.L. Miller (eds.)]. Cambridge University Press, Cambridge, United Kingdom and New York, NY, USA.

Blanchet, J.-P., and R. List, 1987: On radiative effects of anthropogenic aerosol components in Arctic haze and snow, *Tellus.*, **39**B, 293–317.

Bohren, C. F., and B. R. Barkstrom, 1974: Theory of the optical properties of snow, *J. Geophys. Res.*, **79**, 4527–4535.

Brandt, R. E., T. C. Grenfell, and S. G. Warren, 1991: Optical properties of snow, *Antarct. J. U.S.*, **26**, 272–275.

Carlson, R. W., and T. Arakelian, 1993: Spectral bidirectional reflectance and energy absorption rates of antarctic snow, *Antarct. J. U.S.*, **28**, 256–258.

Carroll, J. J., 1982: The effect of surface striations on the absorption of shortwave radiation, *J. Geophys. Res.*, **87**, 9647–9652.

Choudhury, B. J., 1981: Radiative properties of snow for clear sky solar radiation, *Cold Reg. Sci. Technol.*, **4**, 103–120.

Choudhury, B. J., and A. T. C. Chang, 1979a: Two-stream theory of reflectance of snow, IEEE *Trans. Geosci. Electron.*, **GE-17**, 63–68.

Choudhury, B. J., and A. T. C. Chang, 1979b: The solar reflectance of a snow field, *Cold Reg. Sci. Technol.*, **1**, 121–128.

Choudhury, B. J., and A. T. C. Chang, 1981a: On the angular variation of solar reflectance of snow, *J. Geophys. Res.*, **86**, 465–472.

Choudhury, B. J., and A. T. C. Chang, 1981b: The albedo of snow for partially cloudy sky, *Boundary Layer Meteorol.*, **20**, 371–389.

Chýlek, P., V. Ramaswamy, and V. Srivastava, 1983: Albedo of soot-contaminated snow, *J. Geophys. Res.*, **88**, 10837–10843.

Chýlek, P., B. Johnson, P. A. Damiano, K. C. Taylor, and P. Clement, 1995: Biomass burning record and black carbon in the GISP2 Ice Core, *Geophys. Res. Lett.*, **22**, 89–92.

Comiso, J. C., and F. Nishio, 2008: Trends in the sea ice cover using enhanced and compatible AMSR-E, SSM/I, and SMMR data, *J. Geophys. Res.*, **113**, C02S07, doi:10.1029/2007JC004257.

Clarke, A. D., and K. J. Noone, 1985: Soot in the arctic snowpack: A cause for perturbations in radiative transfer *Atmos. Environ.*, **19**, 2045–2053.

Doherty, S. J., S. G. Warren, T. C. Grenfell, R. E. Brandt, A. D. Clarke, and R. E. Brandt, 2010: Light-absorbing impurities in Arctic snow, *Atmos. Chem. Phys.*, **10**, 11647–11680, doi:10.5194/acp-10-11647-2010.

Dunkle, R. V., and J. T. Bevans, 1956: An approximate analysis of the solar reflectance and transmittance of a snow cover, *J. Meteorol.*, **13**, 212–216.

Fily, M., B. Bourdelles, J. P. Dedieu, and C. Sergent, 1997: Comparison of in situ and Landsat Thematic Mapper derived snow grain characteristics in the Alps, *Remote Sens. Environ.*, **59**, 452–460.

Flanner, M. G., and C. S. Zender, 2005 Snowpack radiative heating: Influence on Tibetan Plateau climate, *Geophys. Res. Lett.*, **32**, L06501, doi:10.1029/2004GL022076.

Flanner, M. G., and C. S. Zender, 2006: Linking snowpack microphysics and albedo evolution, *J. Geophys. Res.*, **111**, D12208, doi:10.1029/2005JD006834.

Flanner, M. G., C. S. Zender, P. G. Hess, N. M. Mahowald, T. H. Painter, V. Ramanathan, and P. J. Rasch, 2009: Springtime warming and reduced snow cover from carbonaceous particles, *Atmos. Chem. Phys.*, **9**, 2481–2497.

Forsström, S., J. Ström, C. A. Pedersen, E. Isaksson, and S. Gerland, 2009: Elemental carbon distribution in Svalbard snow, *J. Geophys. Res.*, **114**, D19112, doi:10.1029/2008JD011480.

Gayet, J.-F., F. Auriol, S. Oshchepkov, F. Schröder, C. Duroure, G. Febvre, J.-F. Fournol, O. Crépel, P. Personne, and D. Daugereon, 1998: In situ measurements of the scattering phase function of stratocumulus, contrails, and cirrus, *Geophys. Res. Lett.*, **25**, 971–974.

Giddings, J. C., and E. LaChapelle, 1961: Diffusion theory applied to radiant energy distribution and albedo of snow, *J. Geophys. Res.*, **66**, 181–189.

Grenfell, T. C., and S. G. Warren, 1999: Representation of a nonspherical ice particle by a collection of independent spheres for scattering and absorption of radiation, *J. Geophys. Res.*, **104**, 31697–31709.

Grenfell, T. C., S. G. Warren, and P. C. Mullen, 1994: Reflection of solar radiation by the Antarctic snow surface at ultraviolet, visible, and near-infrared wavelengths, *J. Geophys. Res.*, **99**, 18,669–18,684.

Hagler, G. S. W., M. H. Bergin, E. A. Smith, J. E. Dibb, C. Anderson, and E. J. Steig, 2007: Particulate and water-soluble carbon measured in recent snow at Summit, Greenland, *Geophys. Res. Lett.*, **34**, L16505, doi:10.1029/2007GL030110.

Han, W., 1996: *Remote sensing of surface albedo and cloud properties in the Arctic from AVHRR measurements*, D. S. thesis, 129 pp, Univ. of Alaska, Fairbanks.

Hansen, J. E., 1969: Exact and approximate solutions for multiple scattering by cloud and hazy planetary atmosphere, *J. Atmos. Sci.*, **26**, 478–487.

Hansen, J. E., 1971: Multiple scattering of polarized light in planetary atmospheres. Part II. Sunlight reflected by terrestrial water clouds, *J. Atmos. Sci.*, **28**, 1400–1426.

Hansen, J., and L. Nazarenko, 2004: Soot climate forcing via snow and ice albedos, *Proc. Natl. Acad. Sci. U.S.A.*, **101**, 423–428.

Hansen, J., M. Sato, R. Ruedy, L. Nazarenko, A. Lacis, G. A. Schmidt, G. Russell, I. Aleinov, M. Bauer, S. Bauer, N. Bell, B. Cairns, V. Canuto, M. Chandler, Y. Cheng, A. Del Genio, G. Faluvegi, E. Fleming, A. Friend, T. Hall, C. Jackman, M. Kelley, N. Kiang, D. Koch, J. Lean, J. Lerner, K. Lo, S. Menon, R. Miller, P. Minnis, T. Novakov, V. Oinas, Ja. Perlwitz, Ju. Perlwitz, D. Rind, A. Romanou, D. Shindell, P. Stone, S. Sun, N. Tausnev, D. Thresher, B. Wielicki, T. Wong, M. Yao, and S. Zhang, 2005: Efficacy of climate forcings, *J. Geophys. Res.*, **110**, D18104, doi:10.1029/2005JD005776.

Hapke, B., 1993: *Theory of Reflectance and Emittance Spectroscopy*, Cambridge University Press, Cambridge, United Kingdom and New York, NY, USA, 455 pp.

Herber, A., L. W. Thomason, K. Dethloff, P. Viterbo, V. F. Radionov, and U. Leiterer, 1996: Volcanic perturbation of the atmosphere in both polar regions: 1991–1994, *J. Geophys. Res.*, **101**, 3921-3928.

Hess, M., P. Koepke, and I. Schult 1998: Optical properties of aerosols and clouds: The software package OPAC, *Bull. Am. Meteorol. Soc.*, **79**, 831–844.

Hori, M., Te. Aoki, K. Stamnes, and W. Li, 2007: ADEOS-II/GLI snow/ice products – Part III: Retrieved results, *Remote Sens. Environ.*, **111**, 291–336, doi:10.1016/j.rse.2007.01.025.

Huang, J., Q. Fu, W. Zhang, X. Wang, R. Zhang, H. Ye, and S. G. Warren, 2010: Dust and black carbon in seasonal snow across northern China, *Bull. Amer. Meteor. Soc.*, doi: 10.1175/2010BAMS3064.1 (in press).

IPCC, 2007: *Climate Change 2007: The Physical Science Basis*, Contribution of Working Group I to the Fourth Assessment Report of the Intergovernmental Panel on Climate

Change [Solomon, S., D. Qin, M. Manning, Z. Chen, M. Marquis, K. B. Averyt, M. Tignor and H. L. Miller (eds.)], Cambridge University Press, Cambridge, United Kingdom and New York, NY, USA.

Jacobson, M. Z., 2004: Climate response of fossil fuel and biofuel soot, accounting for soot's feedback to snow and sea ice albedo and emissivity, *J. Geophys. Res.*, **109**, D21201, doi:10.1029/2004JD004945.

Kokhanovsky, A. A., Te. Aoki, A. Hachikubo, M. Hori, and E. P. Zege, 2005: Reflective properties of natural snow: approximate asymptotic theory versus in situ measurements, *IEEE Trans. Geosci. Remote Sens.*, **43**, 1529–1535, doi: 10.1109/TGRS.2005.848414.

Kuchiki, K., Te. Aoki, T. Tanikawa, and Y. Kodama, 2009: Retrieval of snow physical parameters using a ground-based spectral radiometer, *Appl. Opt.*, **48**, 5567–5582.

Leroux, C., and M. Fily, 1998: Modeling the effect of sastrugi on snow reflectance, *J. Geophys. Res.*, **103**, 25,779–25,788.

Leroux, C., J.-L. Deuze, P. Goloub, C. Sergent, and M. Fily, 1998: Ground measurements of the polarized bi-directional reflectance of snow in the near-infrared spectral domain: Comparisons with model results, *J. Geophys. Res.*, **103**, 19721–19731.

Leroux, C., J. Lenoble, G. Brogniez, J. W. Hovenier, and J. F. De Haan, 1999: A model for the bidirectional polarized reflectance of snow, *J. Quant. Spectrosc. Radiat. Transfer*, **61**, 273–285.

Li, S., 1982: *A model for the anisotropic reflectance of pure snow*, M.S. thesis, 60 pp, Univ. of California, Santa Barbara.

Liljequist, G. H., 1956: Energy exchange of an Antarctic snow-field: Short-wave radiation, in Norwegian-British-Swedish Antarctic Expedition, 1949–52, Scientific Results, Vol. 2, Part 1A, Norsk Polarinstitutt, Oslo.

Macke, A., J. Mueller, and E. Raschke, 1996a: Single scattering properties of atmospheric ice crystals, *J. Atmos. Sci.*, **53**, 2813–2825.

Macke, A., M. I. Mishchenko, and B. Cairns, 1996b: The influence of inclusions on light scattering by large ice particles, *J. Geophys. Res.*, **101**, 23,311–23,316.

Marshall, S., and S. G. Warren, 1986: Parameterization of snow albedo for climate models, *Glaciological Data*, **18**, 215–223.

McConnell, J. R., R. Edwards, G. L. Kok, M. G. Flanner, C. S. Zender, E. S. Saltzman, J. R. Banta, D. R. Pasteris, M. M. Carter, and J. D. W. Kahl, 2007: 20th-century industrial black carbon emissions altered arctic climate forcing, *Science*, **317**, 1381, doi:10.1126/science.1144856.

Mellor, M., 1977: Engineering properties of snow, *J. Glaciol.*, **19**, 15–66.

Mishchenko, M. I., and A. Macke, 1997: Asymmetry parameters of the phase function for isolated and densely packed spherical particles with multiple internal inclusions in the geometric optics limit, *J. Quant. Spectrosc. Radiat. Transfer*, **57**, 767–794,.

Mishchenko, M. I., and L. D. Travis, 1998: Capabilities and limitations of a current FORTRAN implementation of the T-matrix method for randomly oriented, rotationally symmetric scatterers, *J. Quant. Spectrosc. Radiat. Transfer*, **60**, 309–324.

Motoyoshi, H., Te. Aoki, M. Hori, O. Abe, and S. Mochizuki, 2005: Possible effect of anthropogenic aerosol deposition on snow albedo reduction at Shinjo, Japan, *J. Meteorol. Soc. Japan*, **83A**, 137–148.

Painter, T. H., and Dozier J., 2004: Measurements of the hemispherical-directional reflectance of snow at fine spectral and angular resolution, *J. Geophys. Res.*, **109**, D18115, doi:10.1029/2003JD004458.

Perovich, D. K., T. C. Grenfell, B. Light, B. C. Elder, J. Harbeck, C. Polashenski, W. B. Tucker III, and C. Stelmach, 2009: Transpolar observations of the morphological properties of Arctic sea ice, *J. Geophys. Res.*, **114**, C00A04, doi:10.1029/2008JC004892.

Potter, J. F., 1970: The delta function approximation in radiative transfer theory, *J. Atmos. Sci.*, **27**, 943–949.

Sergent, C., C. Leroux, E. Pougatch, and F. Guirado, 1998: Hemispherical-directional reflectance measurements of natural snow in the 0.9–1.45 μm spectral range: Comparison with adding-doubling modeling, *Ann. Glaciol.*, **26**, 59–63.

Shaw, G. E., 1982; Atmospheric turbidity in the polar regions, *J. Appl. Meteorol.*, **21**, 1080–1088.

Stamnes, K., W. Li, H. Eide, Te Aoki, M. Hori, and R. Storvold, 2007: ADEOS-II/GLI snow/ice products – Part I: Scientific basis, *Remote Sens. Environ.*, **111**, 258–273, doi:10.1016/j.rse.2007.03.023.

Steffen, K., P. U. Clark, J. G. Cogley, D. Holland, S. Marshall, E. Rignot, and R. Thomas, 2008: Chapter 2. Rapid changes in glaciers and ice sheets and their impacts on sea level, in Abrupt Climate Change Science Program and the Subcommittee on Global Change Research, 60–142 (US Geological Survey).

Stroeve, J., M., M. M. Holland, J. Maslanik, W. Meier, T. Scambos, and M. Serreze, 2007: Arctic sea ice decline: faster than forecast, *Geophys. Res. Lett.*, **34**, L09501, doi: 10.1029/2007GL029703.

Tanikawa, T., Te. Aoki, M. Hori, A. Hachikubo, and M. Aniya, 2006: Snow bidirectional reflectance model using non-spherical snow particles and its validation with field measurements, *EARSeL eProc.*, **5**, 137–145.

Taylor, V. R., and L. L. Stowe, 1984a: Reflectance characteristics of uniform Earth and cloud surfaces derived from NIMBUS 7 ERB, *J. Geophys. Res.*, **89**, 4987–4996.

Taylor, V. R., and L. L. Stowe, 1984b: Atlas of reflectance patterns for uniform earth and cloud surfaces (NIMBUS-7 ERB–61 days), NOAA Tech. Rep. NESDIS 10, Natl. Oceanic and Atmos. Admin., Washington DC.

Uchiyama, A., 1992: Line-by-line computation of the atmospheric absorption spectrum using the decomposed Voigt line shape, *J. Quant. Spectrosc. Radiat. Transfer*, **47**, 521–532.

Warren, S. G., 1982: Optical properties of snow, Rev. Geophys. Space Phys., 20, 67–89. Warren, S. G., 1984: Optical constants of ice from the ultraviolet to the microwave, *Appl. Opt.*, **23**, 1206–1225.

Warren, S. G., and R. E. Brandt, 2008: Optical constants of ice from the ultraviolet to the microwave: A revised compilation, *J. Geophys. Res.*, **113**, doi:10.1029/2007JD009744.

Warren, S. G., and A. D. Clarke, 1986: Soot from Arctic haze: radiative effects on the Arctic snowpack, *Glaciological Data*, **18**, 73–77.

Warren, S. G., and A. D. Clarke, 1990: Soot in the atmosphere and snow surface of Antarctica, *J. Geophys. Res.*, **95**, 1811–1816.

Warren, S. G., and W. J. Wiscombe, 1980: A model for the spectral albedo of snow, II: Snow containing atmospheric aerosols, *J. Atmos. Sci.*, **37**, 2734–2745.

Warren, S. G., and W. J. Wiscombe, 1985: Dirty snow after nuclear war, *Nature*, **313**, 467–470.

Warren, S. G., T. C. Grenfell, and P. C. Mullen, 1986: Optical properties of Antarctic snow, *Antarct. J. U.S.*, **21**, 247–248.

Warren, S. G., W. Wiscombe, and J. F. Firestone, 1990: Spectral albedo and emissivity of CO_2-snow in Martian polar caps: model results, *J. Geophys. Res.*, **95**, 14717–14741.

Warren, S. G., R. E. Brandt, and P. O'Rawe Hinton, 1998: Effect of surface roughness on bidirectional reflectance of Antarctic snow, *J. Geophys. Res.*, **103**, 25,789–25,807.

Wendler, G., and J. Kelley, 1988: On the albedo of snow in Antarctica: Contribution to I. A. G. O., *J. Glaciol.*, **34**, 19–25.

Wiscombe, W. J., 1976: On initialization, error and flux conservation in the doubling method, *J. Quant. Spectrosc. Radiat. Transfer*, **16**, 637–658.

Wiscombe, W. J., 1977: The delta-M method: Rapid yet accurate radiative flux calculations for strongly asymmetric phase functions, *J. Atmos. Sci.*, **34**, 1408–1422.

Wiscombe, W. J., and S. G. Warren, 1980a: A model for the spectral albedo of snow, I: Pure snow, *J. Atmos. Sci.*, **37**, 2712–2733.

Wiscombe, W. J., and S. G. Warren, 1980b: Solar and infrared radiation calculations for the Antarctic Plateau using a spectrally-detailed snow reflectance model, in *International Radiation Symposium Volume of Extended Abstract*, 380–382, Colorado State University, Fort Collins.

Yamanouchi, T., 1983: Variation of incident solar flux and snow albedo on the solar zenith angle and cloud cover, at Mizuho Station, Antarctica, *J. Meteorol. Soc. Japan.*, **61**, 879–893.

Yamanouchi, T. and T. P. Charlock, 1995: Comparison of radiation budget at the TOA and surface in the Antarctic from ERBE and ground surface measurements. *J. Climate*, **8**, 3109–3120.

Yasunari, T. J., R. D. Koster, K. M. Lau, T. Aoki, Y. C. Sud, T. Yamazaki, H. Motoyoshi, and Y. Kodama, 2011: Influence of dust and black carbon on the snow albedo in the NASA Goddard Earth Observing System version 5 land surface model, *J. Geophys. Res.*, **116**, D02210, doi:10.1029/2010JD014861.

6 Measuring optical backscattering in water

James M. Sullivan, Michael S. Twardowski, J. Ronald, V. Zaneveld, and Casey C. Moore

6.1 Introduction

Knowledge of light scattering can provide important information on underwater radiative transfer and the nature and dynamics of suspended particulate matter within a water mass. As an inherent optical property (IOP), scattering is represented by the volume scattering function (VSF), $\beta(\theta)$, which describes the angular dependence (θ) of scattered light from an incident unpolarized beam. It is defined as the radiant intensity $dI(\theta)$, scattered from a volume element dV, in a unit solid angle centered in direction θ, per unit irradiance E, i.e. $\beta(\theta) = (1/E)dI(\theta)/dV$. The scattering coefficienti, b, is determined by integrating the VSF from 0 to π radians (0° to 180°) according to:

$$b = 2\pi \int_0^\pi \sin(\theta)\beta(\theta)\, d\theta\,,$$

while the backscattering coefficient, b_b is determined by integrating the VSF in the backward direction (over $\pi/2$ to π radians, or 90° to 180°) according to:

$$b_b = 2\pi \int_{\pi/2}^\pi \sin(\theta)\beta(\theta)\, d\theta\,.$$

The angular shape and magnitude of oceanic VSFs are dependent on the water and associated dissolved salts, density fluctuations associated with turbulent mixing, and the resident particle assemblage (including bubbles). The VSF of pure water with salts is known spectrally and as a function of angle, temperature, and pressure within about 2% (Twardowski et al., 2007; Zhang and Hu, 2009). Turbulence effects, manifested as refractive index discontinuities, are constrained to the very near forward angles of the VSF (typically <0.1°) and are most significant in density gradients undergoing significant mixing (Bogucki et al., 1998; Agrawal, 2005; Mikklesen et al., 2008). Remaining shape and magnitude variability in the VSF is due to suspended particles.

The VSF of a particle assemblage is dependent on both the size and shape of the particles, as well as the refractive index of the component structural material(s) of each particle and their internal distribution (e.g. Bohren and Huffman, 1983; Quinby-Hunt et al., 1989; Zaneveld and Kitchen, 1995; Mishchenko et al., 2000).

An assemblage with larger particles will generally enhance forward scattering relative to backscattering (e.g., Morel, 1973; Stramski and Kiefer, 1991; Twardowski et al., 2001; Dall'Olmo et al., 2009). Since reflection processes are more dominant to backscattering, as opposed to the diffraction processes more dominant to forward scattering, particle composition (i.e. refractive index differences) will have a greater effect on backscattering (Mobley, 1994; Twardowski et al., 2001; Jonasz and Fournier, 2007).

Accurate quantification of the backscattered light field has applications to numerous disciplines of oceanography (Stramski et al., 2004), including studies of particle dynamics (abundances, size and composition), biogeochemical cycling, and both passive and active remote sensing (Ulloa et al., 1994; Twardowski et al., 2001; Boss et al., 2004a, b; Sullivan et al., 2005; Loisel et al., 2007; Stramski et al., 2008; Dall'Olmo et al., 2009; Twardowski et al., 2012). For example, remote sensing reflectance is approximately proportional to $b_b/(a + b_b)$ (Gordon et al., 1975), where a is the absorption coefficient. The shape and spectral dependence of the VSF in the backward direction forms a critical component of the Bidirectional Reflectance Distribution Function (BRDF), i.e., the variation of the upward radiance field with respect to the viewing and incident illumination angles (Morel and Gentili, 1983; Gordon, 1989; Morel et al., 1995; Zaneveld, 1995; Voss et al., 2000; Voss and Morel, 2005). With respect to biogeochemical cycling, algorithms linking backscattering and particle characteristics can be used as proxies for parameters such as chlorophyll, particulate organic carbon (POC), and total suspended matter (TSM) concentrations, as well as biological productivity (e.g. Behrenfeld and Falkowski, 1997; Stramski et al., 1999; Stramska and Stramski, 2005; Sullivan et al., 2005; Stramski et al., 2008; Twardowski et al., 2001, 2012; Sun et al., 2009, Neukermans et al., 2012). When used in conjunction with satellite remote sensing, these algorithms can extract information from surface oceans synoptically on a global scale. Active remote sensing systems, such as light detection and ranging (lidar) can typically penetrate deeper into the ocean than passive sensors and provide additional information on the vertical structure of backscattering, attenuation, and characteristics of associated particle fields. Interpretation of lidar signal returns are dependent on knowledge of backscattering in the very near backward direction, i.e. $\beta(\pi)$ (Guenther, 1985; Churnside et al., 1998; Churnside and Donaghay, 2009).

The backscattering ratio (b_b/b), or the proportion of light scattered in the backward direction, provides additional information on both water mass and particle characteristics. For example, using the slope of the particle size distribution (PSD) and the particulate backscattering ratio (b_{bp}/b_p), Twardowski et al. (2001) developed a model to estimate the bulk refractive index of particles. Backscattering measurements in conjunction with other IOP measurements have been used to discriminate particle types and dynamics in oceanic thin layer studies (e.g. Sullivan et al., 2005, 2010, Churnside and Donaghay, 2009), while the sinking flux of particle aggregates from the spring bloom in the North Atlantic was quantified using backscattering measurements from autonomous underwater vehicles (AUVs) (Briggs et al., 2011).

With VSF measurements made over suitable angular resolution and range, computing b or b_b can be a straightforward integration; however, there are currently (and historically), a very limited number of sensors capable of VSF measurements over the required broad angular range (see review by Zhang et al., 2011). This is primarily due to the technical complexities of building such instruments, as well as a lack of focus perhaps on the VSF by the ocean optics community over the last several decades. Partly because of this complexity, interest evolved in the possibility of making a single VSF measurement weighted broadly as a function of angle in the backward direction to accurately estimate b_b (Oishi, 1990; Maffione and Dana, 1997; Boss and Pegau, 2001; Sullivan and Twardowski, 2009). These studies observed exceedingly low variability in the shape of backward scattering phase functions, lending credibility to using a single broadly weighted VSF measurement to approximate b_b. These findings proved critical to backscattering sensor designs that have emerged since the late 1990s. Such sensors can be constructed in a very small form factor, at low power and low cost. Several commercial designs have now been in regular use in a wide variety of applications and deployment modes (e.g. moorings, AUVs, vertical profilers) and have proved valuable in improving our understanding of particle dynamics in the ocean (Twardowski et al., 2005).

Herein, we review the principles of operation, evaluate calibration and measurement protocols, and measurement uncertainties for backscattering devices making one or more measurements of the VSF in the backward direction. This review also includes an assessment of the robustness of the relationships between the b_b and VSF measurements in the backward direction, and *in situ* comparisons of VSF estimates.

6.2 Generic sensor description

In their essential form, sensors measuring the VSF consist of a light source projecting into the water and a detector that collects light scattered into an acceptance aperture (Fig. 6.1). The detector (D) and the light source (S) are separated by a distance SD. The angle formed by the center of the detector field-of-view (FOV) and the line SD is θ_d. The half-angle of the detector FOV cone is σ_d. The corresponding angles for the source light beam are θ_s and σ_s. The nominal scattering angle of the sensor is $\gamma = \theta_s + \theta_d$. The intersecting volume formed from the source and detector conical beams is the sample volume.

The basic optical-electrical components of a VSF sensor include a source, such as a light emitting diode (LED) or laser, that couples to a detector, such as a photodiode or photomultiplier tube. A matched interference filter is usually employed to block out-of-band ambient light and re-emitted source radiation from inelastic sources such as Raman scatter and fluorescence. The spectral response of the sensor is determined by convolving the spectral band output of the source with the spectral characteristics of the detector assembly including the bandpass of the detector interference filter, and is normally modeled well as a Gaussian shape with associated centroid wavelength and full-width-half-maximum (FWHM) response. Using lenses and apertures, source and detector optical assemblies can be constructed to establish the desired VSF measurement geometry.

Calibration of VSF measurements requires the determination of instrument specific coefficients, termed scaling factors, which relate the device's detector response to $\beta(\theta)$. The specific angular response or weighting function of the measurement is

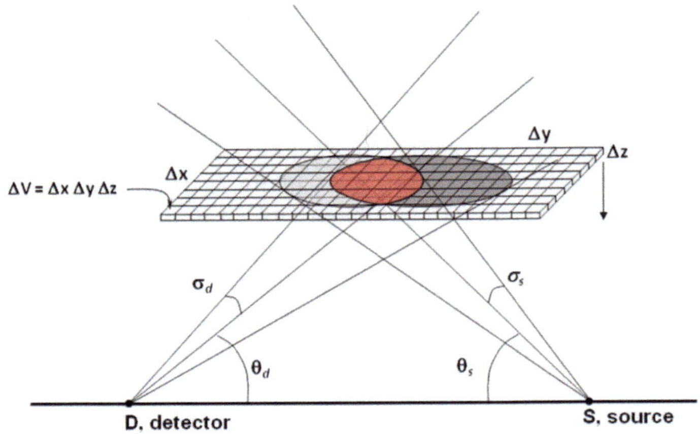

Fig. 6.1. Schematic drawing of backscattering sensor geometry. The intersection of the source and detector conical volumes and the $\Delta x \Delta y$ plane are ellipses, and the intersection of those ellipses (red shaded area) over Δz forms the effective sampling volume of the sensor.

dependent on the integrated relative signal response from each location within the sample volume. Three primary methods for calibrating VSF sensors are published in the literature. The first involves using a purified liquid standard such as benzene or toluene with a known VSF (Morel, 1966, 1974; Jackson et al., 1989). This is typically the preferred method of calibrating bench-top scattering devices, particularly those designed for quantifying colloids. The second, termed the 'plaque method' (Maffione and Dana, 1997; Dana and Maffione, 2002), measures the response of the instrument to a Lambertian target or plaque with a known reflectivity (e.g. Spectralon^{TM}), as a function of distance, z, perpendicular to the $x - y$ plane of the sensor face containing source and detector (e.g. Fig. 6.1). The angular response function of the sensor is equivalent to this signal measured over its entire scattering volume, i.e. as a function of z. A complete mathematical derivation of this method is presented in Maffione and Dana (1997) and will not be reviewed here. The third approach relies on microspherical beads with known scattering characteristics and requires derivation of the sensor's angular weighting function through numerical analysis of its optical geometry (Moore et al., 2000; Twardowski et al., 2007, 2012). The details of this method are presented below.

6.3 Bead method calibration

6.3.1 Overview

The scattering meter measurement may be defined as (ignoring spectral dependencies):

$$\bar{\beta}(\bar{\theta}, \Delta\theta) = \int_0^{\pi} \beta(\theta) W(\theta)\, d\theta \,,$$

where $\bar{\theta}$ is the centroid angle computed as:

$$\bar{\theta} = \int_0^{\pi} \theta W(\theta)\, d\theta \Big/ \int_0^{\pi} W(\theta)\, d\theta\,,$$

and $\Delta\theta$ is the FWHM bandwidth of the weighting function, $W(\theta)$. The weighting function describes the probability distribution function for collected light for a given scattering measurement based on the optical geometry of the sensor.

The objective of calibrating the scattering sensor is to obtain a scaling factor, *f*, that can be used to convert raw signal measured by a detector into $\beta(\bar{\theta})$, the VSF evaluated at a centroid angle, $\bar{\theta}$, for a specific weighting function. To achieve this, $W(\theta)$ must be known, and raw scattering signal (proportional to power) must be measured in a solution of known VSF. A dark offset, D, must also typically be determined to remove any background electronic bias signal. Here, raw digital signal, typically in volts or digital counts, with D subtracted is represented as $N(\bar{\theta})$.

6.3.2 Determination of the weighting function, $W(\theta)$

The source light beam and FOV of the detector are both conical volumes that intersect to form the sample volume. One method of determining $W(\theta)$ is to numerically partition this sample volume into many small elementary volumes $\Delta V = \Delta x \Delta y \Delta z$, and then assess the contribution to $W(\theta)$ for each ΔV (refer to Fig. 6.1). The simplest way to do this is to numerically step a Δz-plane parallel to the sensor face through the sample volume, assessing the contribution of the many ΔV's that occur at the intersecting source and detector ellipses for each step. This method relies on a precise knowledge of the source and detector beam geometries. Note that the drawing in Fig. 6.1 shows the source and detector beams originating as points rather than discrete area cross-sections, as is the case in most practical implementations of the sensor. In the numerical analysis, beams emerging from the sensor face can be traced back to points (S and D, respectively) and an imaginary SD line is drawn, representing the 'effective' sensor face and z-plane (refer to Figs. 6.1 and 6.2). Note that for some geometries with large scattering angles and very wide beam spreads, the sampling volume may not be discrete and theoretically can be infinite. Letting Z be the distance from the SD line to the intersection of the middle of the source and detector beams, contributions to the returned signal (and the weighting function) become negligible at a z-plane of about $15Z$ because of the $1/r^2$ effect in irradiance propagation for the source beam and scattered light. In the numerical analysis, a Z_{max} must be set and this can be varied accordingly to assess where the contributions from farther z-planes become negligible.

The first step in computing $W(\theta)$ is determining if a specific $\Delta V = \Delta x \Delta y \Delta z$ above the sensor face is in the sample volume. With z in the same plane as the sensor face and SD line, from simple geometry we see that:

$$r_s(x, y, z) = \sqrt{x^2 + y^2 + z^2}\,,$$
$$r_d(x, y, z) = \sqrt{x^2 + (SD - y)^2 + z^2}\,,$$

and

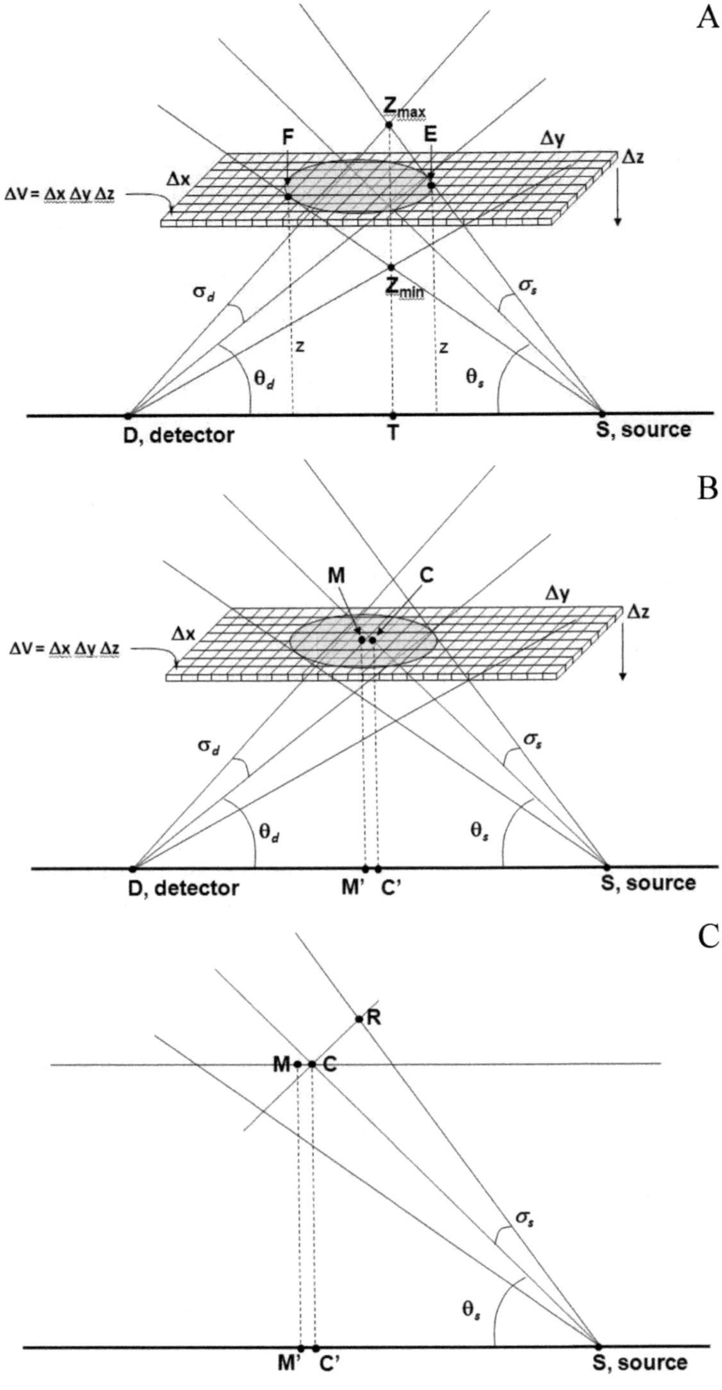

Fig. 6.2. Schematic drawings for derivation of sensor geometries

$$\cos(\pi - \gamma) = \frac{r_d^2 + r_s^2 - SD^2}{2r_d\,r_s}$$

where r_s and r_d are the distances from the source (S) to ΔV and the detector (D) to ΔV, respectively, and γ is the scattering angle formed between the light beam from S to ΔV and from ΔV to D. Additionally,

$$\tan\alpha_s = \frac{z}{\sqrt{x^2 + y^2}},$$

and

$$\tan\alpha_d = \frac{z}{\sqrt{(SD - y)^2 + x^2}}.$$

The solid angle of the source beam is,

$$\Omega_s = \pi(\tan\sigma_s)^2,$$

and similarly for the detector FOV,

$$\Omega_d = \pi(\tan\sigma_d)^2.$$

The intersection of the source beam cone and a plane defined by z is an ellipse (if the illuminated area of the plane is finite). If the semi-major and semi-minor axes of the ellipse are given by a_{majs} and a_{mins}, respectively, where the additional 's' subscript indicates the source parameters, the equation of the source beam ellipse at level z is derived as follows (refer to Figs. 6.2(A) and 6.2(B)):

$$SF = z\cot(\theta_s - \sigma_s),$$

$$SE = z\cot(\theta_s + \sigma_s),$$

and

$$2a_{majs} = z\big[\cot(\theta_s - \sigma_s) - \cot(\theta_s + \sigma_s)\big].$$

The minor axis does not occur at the center of the cone, C, but rather at the middle of the ellipse, M. The projection of C on the SD plane is C' and the projection of M is M'. Hence,

$$C'S = z\cot\theta_s,$$

and

$$SM' = SE + a_{majs}.$$

Thus,

$$C'M' = CM = SM' - C'S = SE + a_{majs} - z\cot\theta_s.$$

From a side view of the source beam, we can see that (refer to Fig. 6.2(C)):

$$\frac{CR}{CS} = \tan\sigma_s,$$

$$CR = CS\,\tan\sigma_s,$$

and

$$CS = \frac{z}{\sin\theta_s},$$

$$CR = \frac{z\ \tan\sigma_s}{\sin\theta_s}.$$

The equation of our ellipse is:

$$\frac{x^2}{a^2_{mins}} + \frac{y^2}{a^2_{majs}} = 1,$$

with $(x,\ y) = 0$ at M.

The chosen coordinate system sets $z = 0$ at the SD plane, $y = 0$ at the plane passing through S orthogonally to the SD line, and $x = 0$ at the plane passing through the SD line orthogonally to the SD plane (i.e., z-plane).

We already know a_{majs}. We need to determine a_{mins} for $x = CR$, and $y = -MC$ at the ellipse. Substituting these into the ellipse equation yields:

$$a^2_{mins} = \frac{CR^2}{1 - [CM/a_{majs}]^2}.$$

The center of the ellipse is at $x = 0$, $y = SM'$. The equation of the ellipse is thus:

$$\frac{x^2}{a^2_{mins}} + \frac{(y - SM')^2}{a^2_{majs}} = 1.$$

All relevant parameters have been derived. Any volume ΔV at (x, y, z) is illuminated by the source if the value of (x, y) in the left-hand side of the equation returns a value ≤ 1.

Derivation of the detector FOV ellipse is similar, where the terms G and H are analogous to the terms E and F, respectively, in the source derivation above. We will use the 'd' subscript to indicate detector FOV parameters:

$$\begin{aligned}
2a_{majd} &= z[\cot(\theta_d - \sigma_d) - \cot(\theta_d + \sigma_d)],\\
DG &= z\cot(\theta_d + \sigma_d),\\
DH &= z\cot(\theta_d - \sigma_d),\\
DM' &= DG + a_{majd},\\
CR &= \frac{z\tan\sigma_d}{\cos\theta_d},\\
CM &= DM' - z\cot\theta_d,
\end{aligned}$$

and

$$a^2_{mind} = \frac{CR^2}{1 - [CM/a_{majd}]^2}.$$

The equation of the detector ellipse at level z is:

$$\frac{x^2}{a_{mind}^2} + \frac{[y-(SD-DM')]^2}{{a_{majd}}^2} = 1\,.$$

All points (x, y, z) are within the detector FOV if their value returns ≤ 1 for the left-hand side of the equation.

It is of use in the programming to determine the limits of the illuminated and detected volumes (refer to Fig. 6.2):

$$\frac{Z_{max}}{DT} = \tan(\theta_d + \sigma_d)\,,$$

and

$$\frac{Z_{min}}{TS} = \tan(\theta_s - \sigma_s)\,.$$

DT plus TS are equal to SD, which is known. Thus,

$$Z_{min} = SD/\left[\frac{1}{\tan(\theta_d - \sigma_d)} + \frac{1}{\tan(\theta_s - \sigma_s)}\right]\,,$$

and

$$Z_{max} = SD/\left[\frac{1}{\tan(\theta_d + \sigma_d)} + \frac{1}{\tan(\theta_s + \sigma_s)}\right]\,.$$

The source maximum, Y_{maxs}, is equal to $SF = SE + 2a_{majs}$; the source minimum, Y_{mins}, is equal to SE; the detector maximum, Y_{maxd}, is equal to $SG = SD - DG = SH + 2a_{majd}$; the detector minimum, Y_{mind}, is equal to $SH = SD - DG - 2a_{majd}$. The maximum of x (X_{max}) is the smaller of a_{mind} or a_{mins}. The volume of interest (effective sampling volume) is thus defined by,

$$\left\{\begin{array}{l} x: 0 \text{ to } X_{max} \text{ (multiply by 2 to get the complete volume)} \\ y: \text{the larger of } Y_{mins} \text{ and } Y_{mind} \text{ to the smaller of } Y_{maxs} \text{ and } Y_{maxd} \\ z: Z_{min} \text{ to } Z_{max}\,. \end{array}\right\}$$

To obtain the weighting function, $W(\theta)$, the returned power is determined for each elementary volume found to reside in the sample volume, and integrated over the illuminated area (detector footprint) so that the power detected, P_d, is related to the source beam power, P_s, according to:

$$P_d(x, y, z) = P_s \beta(\gamma(x, y, z)) W(x, y, z)\,,$$

Powers P_d and P_s need to be solved at each ΔV to solve for $W(x, y, z)$. The solid angle for P_s is defined as,

$$\Omega_s = \pi \tan^2 \sigma_s\,.$$

The irradiance at ΔV is determined as follows: an area $\Delta A = \Delta x \Delta y$ at an angle, $(90 - \alpha_s)$, to the center of the source beam has a solid angle of:

$$\Omega(\Delta V) = \frac{\Delta x \Delta y \cos(90 - \alpha_s(x, y, z))}{r_s^2(x, y, z)}\,.$$

The power received by $\Delta V(x,y,z)$ is

$$P(\Delta V) = P_s \frac{\Omega(\Delta V)}{\Omega_s} .$$

This power of the beam traveling from the source to (x,y,z) is attenuated by

$$e^{-cr_s(x,y,z)} .$$

Thus, the irradiance, or power per unit area, at (x,y,z) is

$$E(x,y,z) = \frac{P(\Delta V)}{\Delta x \Delta y} = \frac{P_s \cos(90 - \alpha_s(x,y,z))}{r_s^2(x,y,z)\Omega_s} e^{-cr_s(x,y,z)} .$$

The scattering angle between the light ray from the source to $\Delta V(x,y,z)$ and the light ray from ΔV to the detector is defined by $\gamma(x,y,z)$.

Since, by definition,

$$\beta(\gamma) = \frac{dI(\gamma)}{E\, dV} ,$$

by rearrangement,

$$dI(\gamma) = \beta(\gamma)\, E\, dV .$$

Hence,

$$dI\big(\gamma(x,y,z)\big) = \beta\big(\gamma(x,y,z)\big)\, E(x,y,z)\Delta x \Delta y \Delta z, \;\; \text{(units of place Watts sr}^{-1}\text{)} ,$$

and

$$dI = P/\Omega_d(x,y,z) .$$

The power received at the detector due to scattering at ΔV is then:

$$P_d(x,y,z) = dI\big(\gamma(x,y,z)\big)\Omega_d(x,y,z)\, e^{-cr_d(x,y,z)} .$$

Ω_d is the detector solid angle for ΔV:

$$\Omega_d\,(x,y,z) = \frac{\Delta x \Delta y \cos\big(90 - \alpha_d(x,y,z)\big)}{r_d^2(x,y,z)} .$$

Substituting, we obtain:

$$\begin{aligned} P_d(x,y,z) =& \beta\big(\gamma(x,y,z)\big)\, E(x,y,z)\, \Delta x \Delta y \Delta z \\ & \times \frac{\Delta x \Delta y \cos\big(90 - \alpha_d(x,y,z)\big)}{r_d^2(x,y,z)} e^{-cr_d(x,y,z)} , \end{aligned}$$

The weighting function $W(x,y,z)$ can now be solved, setting P_s at unity for simplicity:

$$W(x,y,z) = E(x,y,z)\Delta x \Delta y \Delta z\, \Omega_d\, e^{-cr_d}$$

where $W(x,y,z)$ is the weighting function for $\Delta V(x,y,z)$ with the associated scattering angle γ. The sum of all $W(x,y,z)$ for angles γ yields the total weighting function and is normalized to the total sum of all weights (so that the value of P_s is irrelevant):

$$\sum_x \sum_y \sum_z W(x, y, z) = 1 .$$

Weighting functions can be determined as described for each unique source–detector pair dependent on instrument design.

The above assumes that the incident power, P_s, is equivalent throughout its beam cross-section. With many sources such as LEDs and lasers, the source image approximates a Gaussian intensity distribution. Such an intensity distribution can be applied as an additional weighting in the numerical analysis.

6.3.3 Determining theoretical phase functions

With the weighting function known, a hydrosol with a known $\beta(\theta)$ must be used to calibrate the sensor. This hydrosol can be created in the laboratory with a solution of purified water and polystyrene microspherical beads with National Institute of Standards and Technology (NIST) traceable size distributions and known complex refractive index. Using the bead manufacturer's reported value for the real part of the refractive index and Ma et al., (2003) for the imaginary part is recommended. Microspheres have been used in calibrating many different scattering devices (e.g., Volten et al., 1998; Lee and Lewis, 2003; Sullivan and Twardowski, 2009; Twardowski et al., 2011). Particle size distribution functions for the beads are modeled according to a Gaussian shape using the mean size and standard deviations reported from the bead manufacturer. Particulate phase functions, $\tilde{\beta}(\theta) = [\beta_p(\theta)/b_p]$, for these bead particles can be determined with Lorenz–Mie theory (Bohren and Huffman, 1983). Phase functions should be computed for at least 300 evenly spaced particle sizes spanning ± 3 standard deviations from the mean to ensure an accuracy better than 1%. The phase function for each particle size is then weighted according to the size distribution function, and summed to give the phase function for the entire population. By convolving the computed phase function with the sensor weighting functions (section 6.3.2), the appropriately weighted phase function, $\tilde{\beta}(\bar{\theta}_i, \Delta\theta_i)$ can be determined. The effect of spectral response associated with each unique source and detector can be additionally considered by imposing a spectral weighting in the phase function computation, although for sufficiently narrow spectral weightings (10–20 nm FWHM), associated errors are typically <2% (Twardowski et al., 2007).

6.3.4 Experimental calibration and application

For the laboratory part of the calibrations, a clean sensor is mounted in a covered test tank filled with 0.2 μm filtered fresh water from a polishing water purification system with resin cartridges to remove organic substances. The inside walls of the tank should have a minimally reflective surface (flat black) and the tank should accommodate an open cylindrical volume in front of the sensor face large enough to ensure any reflections from the tank wall will not contaminate the signal. Measurements with the sensor can then be taken over a concentration series of serially added microspherical beads. Beads should be sonicated prior to use with no shaking before addition to the tank, as recommended by the manufacturer. In addition to the volume scattering measurements, the total scattering coefficienti b can be measured concurrently using a beam attenuation meter. Polystyrene bead absorption is negligible in comparison to bead scattering, so that the c measurement can represent b with an accuracy better than 1%. The $\beta(\theta)$ may then be obtained for any concentration of beads by multiplying the theoretical phase function by measured b, with the provision that the b component of the theoretical phase function is computed using an acceptance angle matching the attenuation measurement. In practice, to maximize accuracy, experimental data should be collected at several bead concentrations and used to obtain a least-squares linear slope between b and the raw signal N from the VSF measurement. Since we are only concerned with the response of the sensors to the bead additions (i.e., the slope), no background subtractions are necessary, i.e., the observed responses are only a function of added beads. The slope can be considered an experimental phase function, $\tilde{\beta}_e(\bar{\theta}_i)$, in terms of counts $\mathrm{m^{-1}}$ or $\mathrm{V\,m^{-1}}$. Calibration scaling factors, $f(\bar{\theta}_i)$ in units of $\mathrm{m^{-1}\,sr^{-1}}$ counts^{-1} or $\mathrm{m^{-1}\,sr^{-1}\,V^{-1}}$ are then obtained by dividing the theoretically derived phase function, $\tilde{\beta}(\bar{\theta}_i)$, (see section 6.3.3) by the experimental phase function, $\tilde{\beta}_e(\bar{\theta}_i)$. At this point, basic calibration of the sensor is complete. Note that the concentration of beads does not need to be known with this calibration method.

To process field measurements or lab measurements of unknown solutions, raw signal from the sensor is converted to $\beta(\bar{\theta}_i)$ ($\mathrm{m^{-1}\,sr^{-1}}$), by first subtracting the background dark offset, D, and then multiplying the resulting value by the calibration scaling factor, $f(\bar{\theta}_i)$. Note that corrected measurements of $\beta(\bar{\theta}_i)$ include the VSF of pure water and any associated dissolved salts. The full relationship between the raw signal (after dark offset subtraction) and $\beta(\bar{\theta}_i)$ is:

$$N(\bar{\theta}_i)\, f(\bar{\theta}_i) = N(\bar{\theta}_i)\, \tilde{\beta}(\bar{\theta}_i)/\tilde{\beta}_e(\bar{\theta}_i) = N(\bar{\theta}_i)\, \tilde{\beta}(\bar{\theta}_i) \,/\, [N(\bar{\theta}_i) \,/\, b] = b\, \tilde{\beta}(\bar{\theta}_i) = \beta(\bar{\theta}_i)\,.$$

6.3.5 Dependence of the scattering signal on attenuation

There is attenuation of a sensor's incident and scattered beams over its pathlength, i.e. the distance from the light source to the sample volume to the detector. Attenuation losses result in scattering measurements being underestimates of the true scattering of the solution. Expanding on the formula above to include the attenuation effect (Twardowski et al., 2012):

$$\beta(\bar{\theta}_i) = N(\bar{\theta}_i) f(\bar{\theta}_i)\, e^{L[b_p \varepsilon + a_{pg} + a_w]}\,,$$

where $N(\bar{\theta}_i)$ is the raw signal after dark offset subtraction, $f(\bar{\theta}_i)$ is the scaling factor with units $\mathrm{m^{-1}\,sr^{-1}}$ counts^{-1} or $\mathrm{m^{-1}\,sr^{-1}\,V^{-1}}$, L is the total pathlength from source window to sample volume to detector window, b_p is the particulate scattering coefficient, ε is the fraction of scattering by particles along the optical path (other than the primary scattering event in the sample volume) that is not ultimately measured by the detector, a_{pg} is the absorption by particulate and dissolved material, and a_w is the absorption by pure water with any dissolved salts. Each elementary scattering volume has a specific pathlength, which can be numerically determined from geometry in the weighting function calculations (see section 6.3.3). Integrating over all elementary volumes in the sample volume gives the total dependence on attenuation and an effective pathlength, L_e, for the attenuation effect. The error due to attenuation along the path is minimized when the effective pathlength L_e is as small as possible. The attenuation dependence can be viewed as two separate components: light lost due to scattering and light lost due to absorption. The pure water contribution is constant and the a_{pg} contribution can be measured independently if necessary. The scattering contribution is minimized when ε is as small as possible. The scattering fraction ε can be represented as

$$\frac{\int_0^\pi W_\varepsilon(\theta)\beta(\theta)\, d\theta}{\int_0^\pi \beta(\theta)\, d\theta},$$

where W_ε is the angular weighting function for ε. To minimize ε, W_ε should be small, particularly at the near-forward angles where $\beta(\theta)$ is strongly peaked in natural waters. This can be achieved through sensor geometry by using widely dispersed source and detector beams to provide a relatively large sample volume target for light scattered along the path from the source, and a large footprint detector to provide a large target for light scattered along the path to the detector. Such large targets collect more light at forward scattering angles, thus reducing the attenuation effect. Such a geometry, however, produces broad angular weighting functions for the scattering measurement. These can be desirable for trying to resolve integrated backscattering, where, for a sufficiently compact sensor with broad source and detector beams, the attenuation effect can be negligible and thus be virtually ignored. However, when resolving $\beta(\theta)$ at fine angular increments, narrow weighting functions are necessary, which require relatively narrow source and detector beams, which will increase ε. For such sensors, the attenuation effect requires full correction (Twardowski et al., 2012).

6.4 Derivation of b_b from VSF measurements at single or multiple angles

For a single measurement of the VSF in the backward direction, b_b can be estimated using a conversion coefficient, termed a χ factor, for that measurement geometry. The χ factors are based on both modeled and/or measured VSF shape analysis in the backward direction (Oishi, 1990; Maffione and Dana, 1997; Boss and Pegau, 2001; Chami et al., 2006; Berthon et al., 2007; Sullivan and Twardowski, 2009) and generally must assume a constant shape in the particulate phase function in all water types unless the weighting function response of the VSF measurement approximates the $\sin(\theta)$ dependence in the integration of $\beta(\theta)$ to obtain b_b $(= 2\pi \int_{\pi/2}^{\pi} \sin(\theta)\beta(\theta)\, d\theta)$ (Haubrich et al., 2011). Note that as the weighting function becomes broader in the backward direction, the calculation of b_b becomes less susceptible to changes in the shape of the phase function. Sullivan and Twardowski (2009) found a remarkable consistency (<5% RMSE deviation) in shapes of particulate phase functions in the backward direction using a dataset consisting of over three million VSF measurements collected throughout a wide variety of both coastal and oceanic environments (Fig. 6.3(A), Table 6.1). Specifically, they found

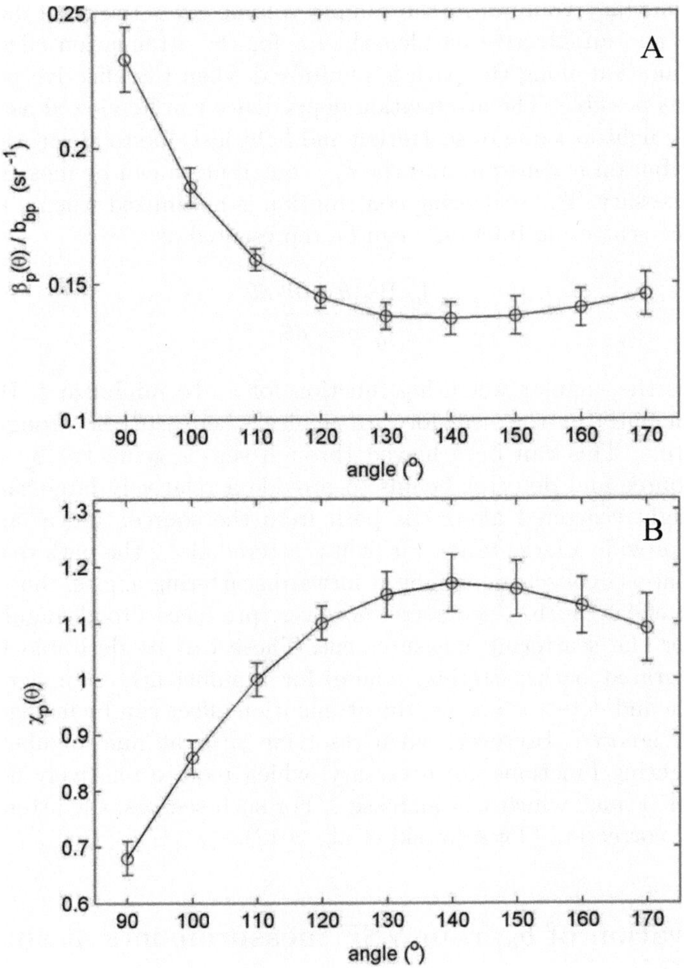

Fig. 6.3. The average (circles) and standard deviations (error bars through circles) of particulate phase functions (Fig. 6.3a) and χ_p factors (Fig. 6.3b) in the backward direction (adapted from Sullivan and Twardowski, 2009). The phase function and χ_p values represent the average of some three million VSFs measurements from a number of both coastal and oceanic environments

the minimum in the angular variability of the particulate VSF in the backward direction was between 110° and 120° (~2% RMSE deviation), while the maximum variability at other angles was 5% or less. These *in situ* measurements were consistent with the original modeled results of Oishi (1990). Comparisons of the Sullivan and Twardowski (2009) average particulate phase function ($\lambda \sim 658$ nm) to a number of other particulate phase functions in the backward direction are shown in Table 6.1. These phase functions are from *in situ* measurements (Petzold, 1972; Boss and Pegau, 2001; Chami et al., 2006; Berthon et al., 2007), analytical models (Fournier and Forand, 1994) and laboratory-based phytoplankton culture studies (Whitmire et al., 2010). All values were normalized to the backscattering coefficient

Table 6.1. The average (avg.) and standard deviation as percent (σ %) of particulate phase functions in the backward direction, $\beta_p(\theta)$ / b_{bp}, from Sullivan and Twardowski, 2009 (ST), Fournier and Forand, 1994 (FF), Petzold, 1972 (P), Boss and Pegau, 2001 (BP), Berthon et al., 2007 (B), Whitmire et al., 2010 (W) and Chami et al., 2006 (C). All measurements were normalized to the backscattering coefficient with the contribution of the backscattering by water removed

Source	Angle: (°)	90	100	110	120	130	140	150	160	170
ST	avg.	0.233	0.186	0.159	0.145	0.138	0.137	0.138	0.141	0.146
	σ%	5.2	3.8	2.5	2.8	3.6	4.4	5.1	5.0	5.5
FF	avg.	0.235	0.185	0.157	0.143	0.138	0.139	0.143	0.148	0.152
	σ%	8.0	4.8	1.7	1.1	3.4	5.1	6.3	7.0	7.3
P	avg.	0.232	0.185	0.159	0.148	0.137	0.131	0.133	0.150	0.169
	σ%	4.4	1.9	2.5	4.0	5.1	3.3	2.9	3.7	13.1
BP	avg.	0.224	0.177	0.155	0.142	0.136	0.135	0.141	0.159	0.257
	σ%	4.3	2.5	3.1	4.2	3.3	3.5	4.2	6.4	34.8
B	avg.	0.226	0.182	0.156	0.142	0.138	0.140	0.144	0.152	0.174
	σ%	3.7	2.6	1.0	2.6	2.1	1.8	4.2	3.0	6.4
W	avg.	0.238	0.192	0.152	0.140	0.128	0.129	0.136	0.164	0.155
	σ%	7.9	14.9	7.8	9.4	10.6	11.1	12.9	14.8	32.4
C	avg.	0.189	0.155	0.135	0.123	0.123	0.132	0.147	0.169	0.227
	σ%	5.2	4.5	5.3	5.5	5.4	5.1	5.3	5.8	11.5

with the contribution of backscattering by water removed (i.e. $\beta_p(\theta)/b_{bp}$). The Petzold (1972) values are the average of his three turbid (San Diego Harbor, USA) measurements ($\lambda \sim 520$ nm). This average Petzold phase function is commonly used in radiative transfer modeling studies (e.g. Gordon, 1993; Mobley, 1994). The Boss and Pegau (2001) values are average data from both field measurements off coastal New Jersey and Mie modeling ($\lambda \sim 555$ nm). The Chami et al. (2006) values are average measurements from the Black Sea ($\lambda \sim 555$ nm), while the Berthon et al. (2007) values are the average of measurements from three different seasons in the northern Adriatic Sea ($\lambda \sim 510$ nm). The range of values of the two inputs of size distribution slope (assuming power law) μ and refractive index n_p for the Fournier–Forand modeled phase functions were between 3 and 4 (at 0.1 increments) for μ, and between 1.02 and 1.18 (at 0.02 increments) for n_p, and the value here is the average of this array output. This input array range (i.e. PSD slopes between 3 and 4, and refractive indices between 1.02 and 1.18) is a good representation of the ranges found in most natural waters (Twardowski et al., 2001; Sullivan et al., 2005; Jonasz and Fournier, 2007). Finally, the Whitmire et al. (2010) values are the average of phase functions from twelve different phytoplankton cultures, representing a wide diversity of size and shape ($\lambda \sim 555$ nm). Except for the values of Chami et al. (2006), the remaining particulate phase functions are within 5% of each other, except at 170°. We should note here that making accurate VSF measurements in the near back angles (170° to 180°) can be difficult owing to internal reflection issues in some instrument designs (Berthon et al., 2007; Sullivan and Twardowski, 2009). However, even with a small increase in the natural variability in the particulate phase functions in the near backward direction, given the sin(θ) weighting in the integration to obtain b_b, this uncertainty will have a minimal effect on the accuracy of the b_b estimate. Similar to the Sullivan and Twardowski (2009) results, the percent variability in the average of these particulate phase functions has a minimum between 110° and 120°. These data strongly suggest that the assumption of constant shape and spectral independence for natural oceanic particulate phase functions (and the use of representative χ factors) is reasonable for most oceanic environments (see section 6.5.6 for further discussion).

As suggested by Boss and Pegau (2001), when using a χ conversion factor, the contribution to the total VSF measurement (β_t) by pure seawater (β_{sw}) should first be removed to yield particulate VSF measurements, β_p, (i.e. $\beta_p = \beta_t - \beta_{sw}$). Backscattering by pure seawater can be assumed constant with a very different phase function shape relative to natural particle populations. The values of Zhang et al. (2009) for $\beta_{sw}(\theta)$ and the backscattering coefficient of pure seawater (b_{bw}), with dependencies on temperature and salinity and a depolarization ratio of 0.039, are expected to be the most accurate at this time. Particulate conversion factors (i.e. χ_p, Fig. 6.3(B) and Table 6.2a) can then be used to estimate the particulate backscattering coefficient (b_{bp}) according to:

$$b_{bp} = 2\pi\, \chi_p(\theta)\, \beta_p(\theta)\,.$$

Averaged values of χ_p in the range $\pi/2$ to π must equal 1. The χ_p function must be convolved with a sensor's angular weighting function to derive a χ_p for that specific VSF measurement (see Table 6.2b for example). The total backscattering coefficient, b_b, can be determined by adding back in the contribution from pure seawater as a backscattering coefficient (i.e. b_{bw}).

Table 6.2a. The average (avg.) and standard deviation (σ) of χ_p factors from Sullivan and Twardowski (2009)

Angle: (°)	90	100	110	120	130	140	150	160	170
avg.	0.684	0.858	1.000	1.097	1.153	1.167	1.156	1.131	1.093
σ%	0.034	0.032	0.026	0.032	0.044	0.049	0.054	0.054	0.057

Table 6.2b. The Sullivan and Twardowski (2009) χ_p factors convolved with the weighting functions at the centroid angles for the ECO-VSF (104°, 130° and 151°) and ECO-BB (124°) sensors

ECO centroid angle θ(°)	104	124	130	151
$\chi_p(\bar{\theta})$	0.89	1.076	1.104	1.138

For multiple VSF measurements in the backward direction where the shape of the VSF is adequately resolved, the VSF can be simply integrated to provide b_b. Where only a few measurements are made over different angular ranges, b_b can be obtained by integrating over a polynomial fit to the available data. For example, using a backscattering sensor measuring the VSF at centroid angles 100°, 125°, and 150°, Sullivan et al. (2005) used the following steps to derive b_{bp}. First, the three $\beta_p(\bar{\theta}_i)$ measurements were multiplied by $2\pi\sin(\bar{\theta}_i)$ to add the weighting of the spherical integration. Since $2\pi\sin(\pi) = 0$, four values in the backward hemisphere could be used in the fitting procedure with a regression anchor of zero at 180°. The resulting values were then fit with a third order polynomial and integrated from $\pi/2$ to π (90° to 180°) to yield b_{bp}.

We should note here a recently described method to measure b_b that is not dependent on single (or multiple) angle VSF measurements (i.e. Haubrich et al., 2011). These authors have designed a relatively simple instrument that produces a $\sin(\theta)$ weighting function by using a circular detector aperture surrounding the optical axis of the source beam exit window. Theoretical analysis has shown that b_b may be resolved with a maximum uncertainty of only a few percent and potentially <1 percent with this technique. To date, this instrument is not yet commercially available.

6.5 Analysis of measurement uncertainties

Uncertainties in backscattering measurements can arise from multiple sources, including reproducibility in sensor machining and components, calibrations, instrument wear and age, and environmental dependencies due to ambient temperature, light, and fluctuations in electromagnetic interference. These factors can affect measurements in two ways: by introducing uncertainties in the sensor's baseline response (i.e., a change in the detector background dark offset D), or by introducing uncertainties in the sensor's dynamic response (i.e., a change in the effective calibration scaling factor f). Since the background dark offset is nominally a constant, the significance of any long-term dark offset drift to measurement uncertainties will be dependent on the optical environment where the sensor is deployed; that is, it will have the greatest effect in clear, low scattering environments where the *in situ* signal will be small. On the other hand, changes in the dynamic response (scaling factor) will not be dependent on the optical environment. Scaling factors are multiplicative, so percent changes directly impact uncertainties defined in terms of percent.

6.5.1 Calibration

Uncertainty in calibration arises from assumptions made in the numerical computation of the angular weighting function, the input parameters for calculation of bead phase functions using Lorenz–Mie theory, and the methodological reproducibility for the laboratory measurements. With respect to the weighting function, accuracy is dependent on how close the geometric variables used in the numerical analysis match that of the actual sensor. For sensors with well-defined detector FOVs, with precisely set apertures and source beams that are well-behaved such as a laser, weighting function accuracy is expected to be excellent. The numerical analysis described here has been validated experimentally in such a system (Twardowski et al., 2012). For sources such as LEDs that can exhibit uneven intensity distributions and variability from LED to LED, weighting function accuracy is more difficult to evaluate. In such cases, accuracy can be assessed by replicating the calibration with multiple bead sizes to determine to what extent resultant scaling factors agree (they should be identical). Note this check evaluates the uncertainties of the complete calibration, not just the weighting function (see below). The uncertainty associated with weighting functions can be minimized by using bead sizes in calibration that have as flat a response in the backward direction as possible. Phase functions typically flatten as bead sizes decrease. For example, the phase function for 0.2 μm beads is nearly constant at 650 nm at angles greater than about 100°.

The Lorenz–Mie theory computation of phase functions requires the size distribution and complex refractive index of the bead dispersion. NIST-traceable polystyrene beads (ThermoScientific Inc.) are high quality and suitable for calibration. In terms of impacting the shape of the phase function, mean diameter is typically much more important than the standard deviation of the distribution. For beads with nominal mean diameter of $0.2 \pm 2\%$ μm (ThermoScientific catalog number 3200A), maximum possible uncertainties in phase function shape are observed in the near-backward at around 5%. For beads with nominal mean diameter of $2.0 \pm 1\%$ μm (ThermoScientific catalog number 4202A), maximum possible uncertainties in phase function shape are also observed in the near-backward at around 5%. As the angular weighting function becomes broader in the backward, this maximum possible uncertainty decreases substantially.

For the real refractive index of the polystyrene beads, several dispersion models have been published (Matheson and Saunderson, 1952; Nikolov and Ivanov, 2000; Ma et al., 2003). Values provided by Duke Scientific (now ThermoScientific) are considered the most accurate as these values provide theoretical attenuation spectra that closely match hyperspectral measurements. Using the Cauchy dispersion equation:

$$n_p = A + \frac{B}{\lambda^2} + \frac{C}{\lambda^4},$$

with λ in μm, the Duke Scientific values are $A = 1.5663$, $B = 7.85 \times 10^{-3}$, and $C = 3.34 \times 10^{-4}$. Uncertainties in the red portion of the visible are an estimated ± 0.001. The significance of refractive index uncertainty increases dramatically as bead size increases. For 0.2 μm beads, a 0.005 difference in refractive index only leads to a 0.2% maximum difference in phase function values in the backward direction. For 2 μm beads, the maximum phase function difference increases to 8%. For 25 μm beads the difference increases to >50%. This substantial uncertainty is strong justification for using small bead sizes in calibrations. Ma et al. (2003) is the only study to experimentally determine the imaginary portion of the refractive index n_i. From 370 to 700 nm, n_i ranged between 0.0003 and 0.0005, with a value of 0.0005 ± 0.0001 in the red. This uncertainty has a negligible impact on phase function shapes for both 0.2 μm and 2 μm beads. As stated previously, using the bead manufacturer's reported value for the real part of the refractive index and Ma et al., (2003) for the imaginary part is recommended. Further, for calibrations of sensors with blue or green wavelengths, 0.1 μm beads should be used in preference to 0.2 μm beads, as the phase function of 0.1 μm beads is flatter in these wavelength regions. The drawback to using 0.1 μm beads is higher concentrations of beads are required to get an adequate dynamic range in signal, and the beads are costly, thus 0.2 μm beads are generally preferred for red wavelength sensor calibrations.

ThermoScientific uses a variety of techniques to certify the reported diameter and distribution of their beads including optical microscopy with a NIST calibrated stage micrometer, electron microscopy, electrical resistance analysis, and photon correlation spectroscopy. Our laboratory has validated some of the reported size distributions with scanning electron microscopy. It is important to note that the primary application of these beads is to provide a size calibration for particle sizing devices. The critical parameter is the location of the distribution peak, or modal (also mean) diameter, and this is what is NIST certified. Such beads are not NIST certified as standards for the calibration of scattering sensors. While they still may serve this application effectively (and we think they do), a concern may be that, unlike many single particle size analyzers, all the particles potentially present in the bead solution influence bulk scattering in the calibration of volume scattering sensors. Contaminating particles larger than the bead size are probably not a significant contributor but small colloidal particles that are not detected by many of the validation methods above could be present. These particles have low scattering efficiencies, but their concentration is unknown and might be relatively high. Manufacturing pure water without any particulate contamination that may influence optical scattering is extremely difficult (Morel, 1974). The assumption is that the population of beads in the prepared solution would be in such high concentrations as to effectively render any contamination negligible. Because of their very small size, contamination is likely less a problem with bead distributions peaking in the submicrometer size range.

All laboratory procedures involved in the VSF and b measurements for the solutions of beads are subject to methodological errors in replication. In practice, if several calibration points are collected for a least-squares linear fit with careful attention to detail, the methodological uncertainty for the slope determination is ~1% (RMSE). Experimental calibration replication confirms these uncertainties.

6.5.2 Instrument resolution and electronic noise

Resolution and noise will be a function of optical components used and electronics. Commonly used optical components for backscattering measurements are LED sources and silicon photodiode detectors (Maffione and Dana, 1997; Moore et al., 2000; Twardowski et al., 2005). To assess uncertainties for a practical sensor embodiment, in this section and the following sections, we will consider the WET Labs (Philomath, OR, USA) ECO backscattering sensors. Sensor design geometry and associated angular weighting functions for ECOs are shown in Figs. 6.4 and 6.5(A), respectively.

WET Labs ECO sensors employ a 12-bit A/D converter, providing a 4096 count dynamic range. For a standard sensor, with electronic noise levels tuned to approximately 2 digital counts, resolutions of about 2×10^{-5} m^{-1} sr^{-1}, 8×10^{-6} m^{-1} sr^{-1}, and 4×10^{-6} m^{-1} sr^{-1}, are obtained with the ECO-BB sensor for blue, green, and red LED source measurements, respectively. Sensors can be tuned to higher or lower dynamic range. At all wavelengths, this resolution is 7–8% of the volume scattering of pure seawater, which is similarly strongly dependent on wavelength (Zhang et al., 2009). Improving resolution with increasing wavelength is a function of increasing LED power output and resulting scattered signal power.

The dynamic range of ECO sensors is also consequently wavelength dependent, where, with noise levels tuned to 2 digital counts, typical signal saturation for blue, green and red wavelength ECOs occurs in natural waters with b values of roughly $20 \pm 5\,\mathrm{m}^{-1}$, $10 \pm 5\,\mathrm{m}^{-1}$, and $5 \pm 1\,\mathrm{m}^{-1}$, respectively.

6.5.3 Long-term stability in background dark offsets (baseline noise)

Dark offsets are relatively straightforward to measure for WET Labs ECO sensors, so large amounts of data on offsets are available with which to assess long-term stability. Offsets are determined by completely covering the detector with opaque tape, leaving the light source fully exposed, and immersing the sensor in water. Measurements of dark offsets are taken over an appropriate time period (e.g., 1 min.) to determine both the average signal (D) and the variability about the average. Routine monitoring of dark offsets, especially during field deployments, is recommended, as it may reveal unanticipated variations due to specific instrumental/power setups and environmental conditions (Twardowski et al., 2007). For the highest accuracy, dark offsets should be measured *in situ*.

Sensor components affecting dark offsets are the silicon photodiode detector and associated electronics. Dark offsets have been found to be very stable over long time periods (years). For example, Dall'Olmo et al. (2009) recorded ECO dark offsets every other day during a one-month cruise and found they varied by ±1 count

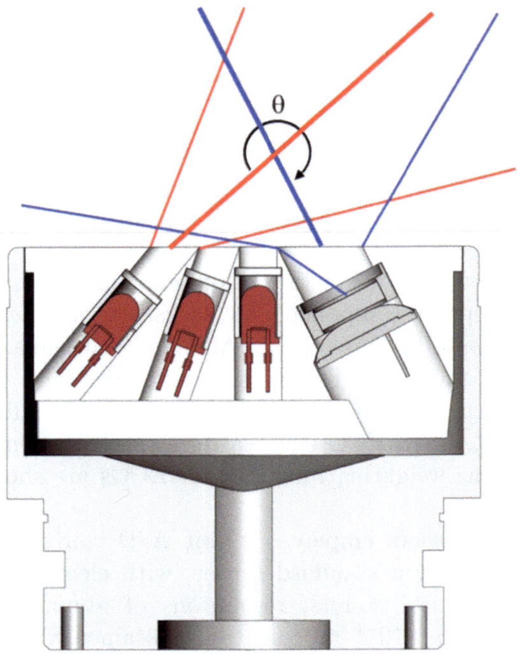

Fig. 6.4. Schematic drawing of an ECO-VSF sensor optical head measuring the VSF at three different geometries. Centroid angles of the measurements are at 104°, 130° and 151°. Three LED sources (shaded red) and a photodiode detector are shown, with an interference filter above the dome-shaped photodiode detector. Sketches of the beams for the 104° measurement with sources in red and detector FOV in blue are shown

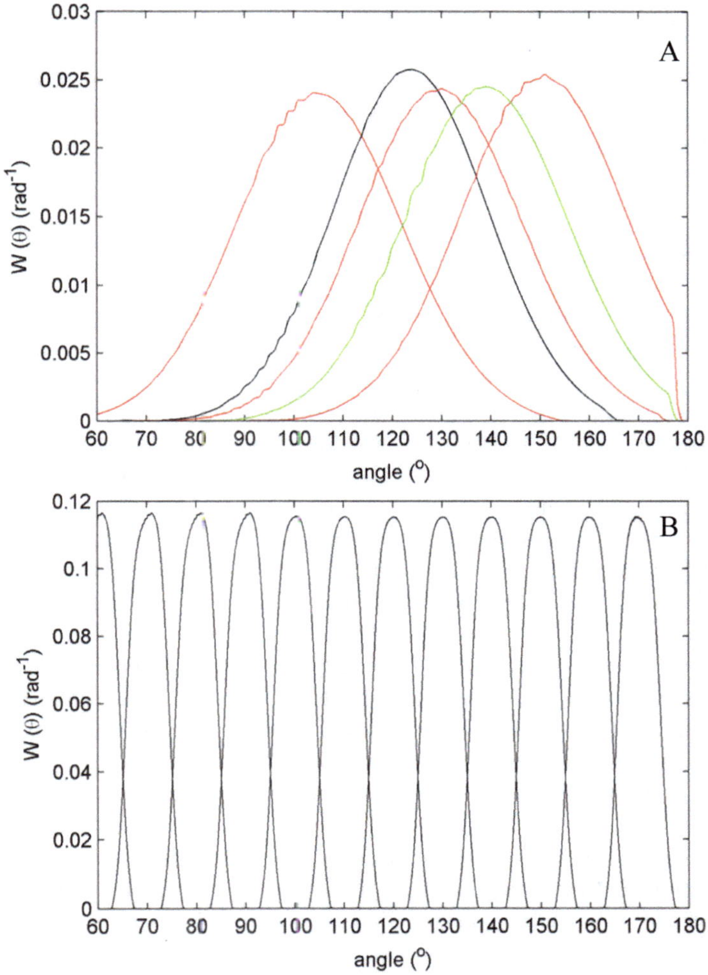

Fig. 6.5. (A) Numerically derived weighting functions, $W\ (\bar{\theta})$, for the WET Labs ECO-BB (black line), ECO-NTU (green line), and ECO-VSF (red lines) sensors. Centroid angles are 124° for the ECO-BB, 140° for the ECO-NTU, and 104°, 130° and 151° for the ECO-VSF. Centroid angles are simply the maximum of the weightings; other methods of computing centroid angles, such as finding the mid-angle in the weighting distribution, agree to within a degree. (B) Numerically derived $W\ (\bar{\theta})$ for each of the centroid angles of the WET Labs MASCOT VSF device over the same angular range

(less than the tuned resolution of the measurement). Additionally, dark offsets from two different ECO sensors (an ECO-VSF and ECO-BB3) were tracked over 6 and 8 years, respectively, by this research group during extensive field use and found to have rarely varied greater than ±2 counts from the original factory determinations. Most of these dark offsets were recorded *in situ* before ambient measurements, in waters ranging in temperature from 1°C to 30°C. Laboratory tests have confirmed the lack of any dependence in dark offsets with ambient temperature (see section 6.5.5.2). However, in rare instances (<0.1% of available data), dark offset determinations for both sensors were temporarily found to be 5 to 6 counts lower than normal. This drift may be due to a very rapidly changing thermal environment or a period of unique electromagnetic interference (such as a grounding or power system resistance problem), although the effect has not been reproducible in the laboratory.

6.5.4 Long-term stability in scaling factors

Scaling factors are a function of the source output and the dynamic response of the detector assembly. Any physical damage or fouling of the sensor face affects both components. Scratching of the source or detector windows may impact the transmittance of light through the sensor–water interface, especially if the difference in refractive index between the windows and water is large. Scratching may also alter the weighting function of the measurement. Depending on the sensor design, fouling of the sensor face can artificially increase or decrease the measured scattering signal. If the source and detector are far enough apart, the scattering signal will decrease. If the source and detector are very close, source light can scatter through the surface film directly into the detector, causing an artificial signal increase.

Sources of long-term drift due to changes in source intensity output, detector dynamic response and optical component degradation (e.g. interference filters) are difficult to quantify separately and are sensor and situation specific. However, assuming a clean and undamaged sensor, tracking the difference in scaling factors obtained through recalibrations over time should yield some indication of their cumulative magnitude. This type of analysis was conducted with ten different WET Labs ECO sensors of various designs and age. Although this was not a rigorous, long-term analysis (for example, drifts were not normalized to hours used), scaling factors in ECO sensors increased by about 8% y^{-1} for blue wavelength sensors, 1–2% y^{-1} for green wavelength sensors, and 3–4% y^{-1} for red wavelength sensors (Fig. 6.6(A)). For most of the ECO sensors studied, only two calibration points were available, and most calibrations were only 1 to 2 years apart, which is obviously not ideal for detailed characterization. However, a more extensive scaling factor calibration history, available from a red wavelength ECO-VSF, exhibited what is likely the shape function of long-term scaling factor drift. In this case, five calibrations were done approximately once per year over five years. The results reveal that the drift approximates a first-order exponential increase with time (Fig. 6.6(B)). This characteristic shape likely indicates that gradual dimming of LEDs with age is the causative mechanism, as LEDs are known to exhibit decreases in light output with age. The rates of drift for blue LED sensors appeared higher than those of green and red LED sensors. It is theorized that these higher drift rates may also include a 'yellowing' effect from the optical epoxy that is used in potting the optical components in the sensor face. Experiments are currently underway to test this theory.

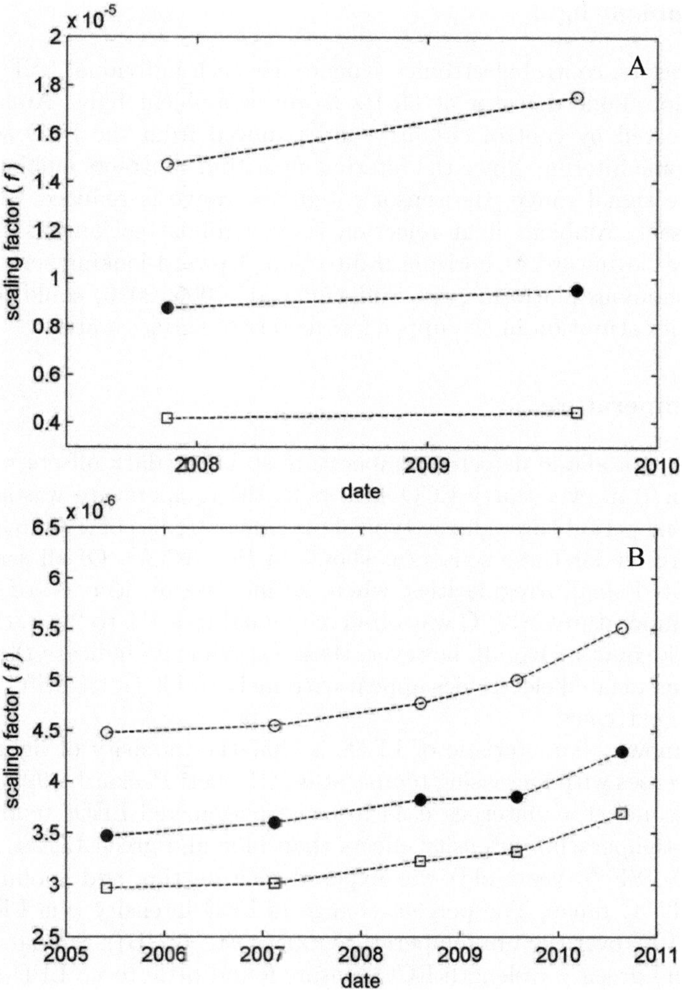

Fig. 6.6. (A) The change in calibration scaling factors (f) as a function of time for a representative ECO-BB3 sensor. The blue wavelength channel f (open circles) increased by about 10% y^{-1}, the green wavelength channel f (filled circles) by about 4% y-1. and the red wavelength channel f (open squares) by 3% y^{-1}. (B) The change in f as a function of time for a red wavelength ECO-VSF sensor. Overall, the three centroid angles, 104° (open circles), 131° (filled circles) and 151° (open squares) all increased by about 4% y^{-1}; however, the rate of change in f is increasing with time

6.5.5 Environmentally induced uncertainties

Along with the methodological uncertainties and long-term drift described above, evaluating sensor performance under different environmental conditions parameterizes short-term stability. Common environmental factors that could affect sensors include changes in ambient temperature and light, and electromagnetic interference (EMI). Each of these factors is addressed below for the WET Labs ECO sensors.

6.5.5.1 Ambient light

For ECO sensors, control electronics synchronize each individual LED light source with the photodiode detector at 60 Hz to reject ambient light. Ambient light is directly detected by control circuitry and removed from the LED source signal using electronic filtering. Since the filtering algorithm subtracts ambient light from the available signal range, the sensor's dynamic range is reduced when ambient light is present. Ambient light rejection is very robust as long as the sensor is oriented in a downward or horizontal direction. Upward-looking sensors (used on special autonomous platforms, e.g. Sullivan et al., 2005, 2010) could be subject to ambient light saturation in the upper few meters of surface waters.

6.5.5.2 Temperature

To examine photodiode detector temperature stability, dark offsets were continuously measured in over thirty ECO sensors as the temperature was slowly cycled over a 12-hour period through the typical oceanic extremes of 2°C to 35°C. A representative result for these sensors is shown in Fig. 6.7(A). Of all sensors tested, a few exhibited small irregularities, where an increase or decrease of several dark offsets at temperatures >30°C was observed, equal to a 1% to 2% variation in the baseline dark counts. Overall, however, these experiments indicate that sensor detectors and associated electronics appear extremely stable (<1% drift) over oceanic temperature extremes.

A well known characteristic of LEDs is that the intensity of light from these sources decreases with increasing temperature (Hewlett Packard, 1995). Depending on the semiconductor materials used in construction, red LEDs tend to be more sensitive to temperature-intensity effects than blue and green LEDs. A red wavelength ECO-VSF (6 years old) was exposed to a heating and cooling cycle over the 2°C to 35°C range. The percent change in LED intensity (via LED reference signal) was 10% over the full temperature range (Fig. 6.7(B)). Similar experiments with blue and green wavelength ECO sensors found little to no LED temperature dependency (2% or less over the entire temperature range). Other tests with instruments containing LED sources have confirmed these results. As the ECO sensors were calibrated around 20°C, blue and green wavelength ECO sensors would be expected to have little to no temperature uncertainty under normal or even extreme oceanic conditions, while red wavelength ECOs could have up to 5% uncertainty if operated at either temperature extreme. If an instrument is going to be routinely used in environments at the extremes of the oceanic temperature range, scaling factor calibrations should be conducted as close to the expected environmental temperature as possible for highest accuracy.

Note that temperature is also known to have an effect on LED peak spectral output, typically increasing (with increasing temperature) from 0.03 to 0.13 nm per °C, depending on LED semiconductor material (Reynolds et al., 1991). Thus, a 2 to 3 nm shift in LED peak spectral output could occur over the entire oceanic temperature range. Since ECO sensors typically employ bandpass interference filters in the 20 to 30 nm FWHM range, source output intensity should not be affected.

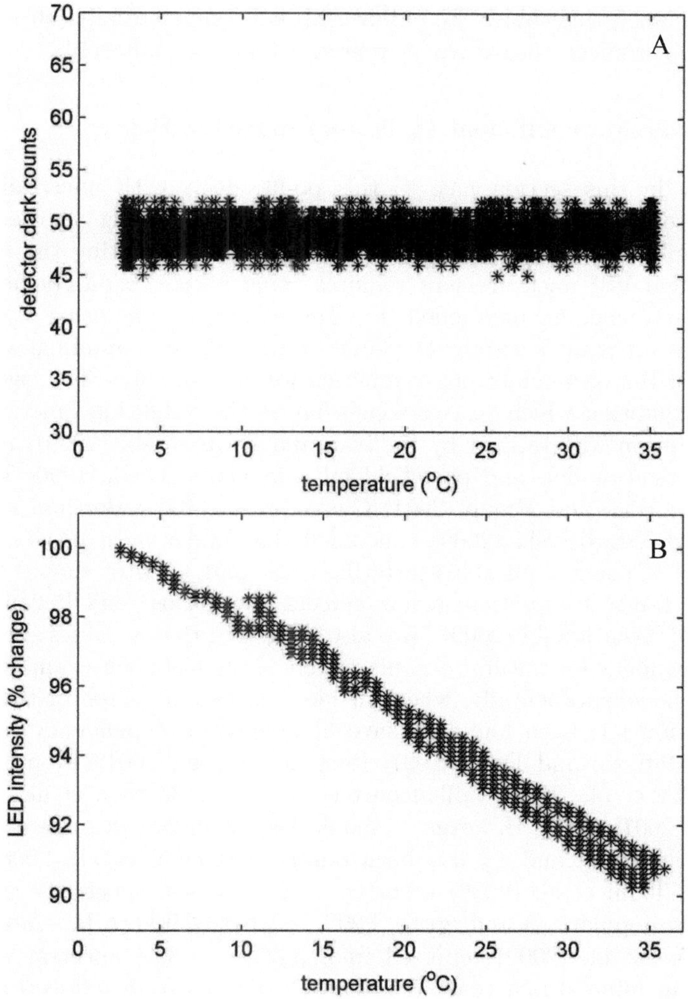

Fig. 6.7. (A) The typical response of detector dark offset as a function of temperature for an ECO sensor. (B) The percent change in light intensity as a function of temperature for a red wavelength LED in an ECO sensor

6.5.5.3 Electromagnetic interference (EMI)

Electromagnetic interference (EMI) is a common concern for sensors deployed in conductive seawater. ECO testing has shown vulnerability to external noise from other oceanographic instruments (e.g. pumps, other sensors, etc.) depending on deployment characteristics and system setup. Causes of the EMI have been linked to poor system grounding, stray voltage leaks and/or RS-232 frequency coupling. Typically, the interference may increase the standard deviation of detector signal up to a factor of 2 to 3, although the increased EMI does not change the averaged measurement value (i.e. the noise is normally distributed). Improved shielding in sensors and associated cables can reduce EMI noise substantially (50%). Since EMI is a white noise effect, measurement reproducibility is unaffected.

6.5.6 Conversion coefficient (χ factor) uncertainties

The review in this section has, to this point, dealt with uncertainties associated with sensor output in terms of the VSF (β), focusing on the WET Labs ECO example. However, as detailed in section 6.4, estimating the b_b using any sensor's single VSF measurement requires using a conversion coefficient (χ factor) for that angle. As mentioned, the key assumption in using a single set of χ factors to estimate accurate b_b values is that there is minimal variability in the shape of the particulate phase function for the chosen VSF measurement geometry, a condition which relaxes somewhat as the weighting function broadens. This assumption was verified by Sullivan and Twardowski (2009), in agreement with theoretical models and past field measurements (Oishi, 1990; Maffione and Dana, 1997; Boss and Pegau, 2001; Chami et al., 2006; Berthon et al., 2007). Sullivan and Twardowski (2009) concluded that under most oceanic conditions, estimates of b_b using a suitable scattering angle and χ factor should have a maximum uncertainty better than a few percent. Both Boss and Pegau (2001) and Sullivan and Twardowski (2009) have also suggested that χ values should be spectrally independent for natural oceanic waters. This has been examined both theoretically and experimentally, where in most cases, the shape of the particulate phase function has been found to have little spectral dependency (e.g. Ulloa et al., 1994; Maffione and Dana, 1997; Boss and Pegau, 2001; Twardowski et al., 2001; Mobley et al., 2002; Vaillancourt et al., 2004; Berthon et al., 2007; Whitmire et al., 2007, 2010). However, it should be noted that in some cases, spectral variability in b_{bp}/b_p and χ_p has been observed (Chami et al., 2006; McKee et al., 2009). Chami et al. (2006) reported that χ_p varied spectrally $\pm 6\%$ for natural particle populations and up to $\pm 20\%$ in phytoplankton laboratory cultures, while McKee et al. (2009) reported finding 25% to 30% spectral variability in b_{bp}/b_p within mineral-rich coastal waters. Deviations from a spectrally independent phase function may occur, for example, in waters with steeply peaked size distributions. A majority of studies have confirmed the spectral independence, however. In addition to the Sullivan and Twardowski (2009) study, Berthon et al. (2007), using a later generation of the VSF instrument employed by Chami et al. (2006), found very little (1% to 2%) spectral dependency in natural particle populations in the Mediterranean Sea. Similarly, Whitmire et al. (2007) found spectrally independent b_{bp}/b_p in a diverse oceanic dataset containing measurements collected from five different regions around the world, while Whitmire et al. (2010) found no significant spectral variability in χ_p for twelve different phytoplankton cultures with a wide variety of shapes and sizes. Although more work is needed to better define those cases where the shape of the phase function may deviate from spectral independency, the majority of theoretical and experimental studies to date support the use of a single set of χ_p values for natural oceanic waters.

6.5.7 Measurement uncertainty summary

As a general rule, uncertainties in calibrating sensors measuring volume scattering in the backward direction decrease as bead size decreases. For 0.1-μm beads with a nearly flat phase function response in the visible spectrum in the backward direction (so that measurement geometry uncertainty is minimized), the only significant error in calibration is the 2% uncertainty in mean bead diameter, which results in a maximum possible uncertainty of 5% in the phase function shape near 180°. For broad weighting functions, this maximal uncertainty decreases, so, in practice, an uncertainty of about 2% would appear reasonable for sensors such as the WET Labs ECO. Propagated with an estimated uncertainty of about 1% in laboratory determinations of linear slope between raw sensor signal and total scattering b, resulting scaling factors have an expected uncertainty of 2–3%.

Using 2.0-μm beads in calibration will increase uncertainty because the phase function shape is not as flat in the backward and the uncertainty in bead refractive index may be as high as 8%. However, if we assume the Duke Scientific uncertainty in the red of 0.001 is reasonable, then the maximal uncertainty due to the real part of refractive index is only 3%. Propagating error, we thus may expect uncertainties in calibration with 2.0 μm beads to be 1–2% higher than that for 0.1 μm beads, but that does not include weighting function uncertainty that potentially becomes an issue because of the introduced variability in phase function shape with the 2.0 μm beads.

Because of the multiple sources of possible uncertainty in calibrations, and the difficulty in characterizing some uncertainties, independent validation provides a valuable check for the overall uncertainty in calibrations. This can be accomplished by (1) calibrating the same device multiple times with different bead size distributions to check for consistency in derived scaling factors and (2) intercomparison with other sensors that have similar and dissimilar measurement geometries. For various sensor geometries measuring volume scattering throughout the backward hemisphere, we have observed worst-case differences of 5% when the same sensor was calibrated with both 0.1-μm and 2-μm NIST traceable beads. For different sensors with the same measurement geometry calibrated with the same beads, worst-case agreement was also 5%. Intercomparison of sensors with different measurement geometries is addressed in sections 6.6 and 6.7.

The above discussion all relates to calibration with high-quality NIST traceable bead distributions. ThermoScientific and other manufacturers also sell non-NIST traceable bead products for size analyzer calibrations with mean diameter and standard deviation of the distribution reported on the bottle. However, the uncertainty in these distribution parameters is not known. Solutions of non-NIST traceable beads also are less purified than their NIST traceable counterparts, with up to an order of magnitude more background solids (Duke Scientific, now ThermoScientific, personal communication, 2006). Concentrations of beads in solution are also much higher, however. Nevertheless, comparison of scaling factors derived from calibrations with both NIST and non-NIST traceable 2.0 μm diameter beads have shown reasonable agreement with a worst-case error of 8% for WET Labs ECO sensors.

Long-term drift for ECO sensors appears to be driven by dimming of the LED source, with observed annual rates of drift approximately 8–10%, 1–2%, and 3–4%

for blue, green, and red LED sensor measurements, respectively, although these rates will be dependent on sensor usage. Annual sensor recalibration ensures green and red scattering measurements drift no more than a few percent, whereas blue scattering measurements would need to be calibrated more frequently to obtain a similar uncertainty. The only significant environmental effect for the ECO sensor was determined to be the temperature dependency of the red LED source, with a 10% change in response over the full temperature range expected for typical ocean conditions. Overall, the most stable ECO scattering measurements are made with green LED sources.

6.6 Sensor comparisons in the field

Comparing measurements of volume scattering from sensors with different optical components, beam geometries, resolution, and optical pathlengths can be a valuable way to evaluate assumptions in calibration. During a number of field efforts, the ECO-VSF and ECO-BB3 sensors discussed in section 6.5.2 were co-deployed with the WET Labs Multi-Angle SCattering Optical Tool (MASCOT) measuring the VSF at 658 nm over a $\bar{\theta}$ angular range extending from 10° to 170° in 10° increments (Fig. 6.5(B)). Details of the MASCOT, its use and calibration can be found in Sullivan and Twardowski (2009) and Twardowski et al. (2012). The pathlength of the MASCOT is 20 cm for each channel, much longer than that of the ECO sensors, and requires a more exact attenuation correction (Twardowski et al., 2012). The MASCOT uses a 30-mW laser diode source, has a 14-bit dynamic range, and samples at 20 Hz. Estimated accuracy in VSF measurements with MASCOT are 2% to 3%. The spectral centroid of the ECO-VSF and the red LED channel of the ECO-BB3 were within 1 nm of the MASCOT's. All sensors were calibrated with 0.2-μm NIST traceable beads.

Measurements from all three sensors were collected in a wide variety of water types from five different coastal and oceanic environments: off the coast of Oahu, Hawaii, during March of 2007; the New Jersey bight during May of 2007 and July of 2008; the Santa Barbara Channel off the coast of Santa Barbara, California, during September 2008; and the Ligurian Sea off Liguria, Italy, during October of 2008. Collectively, these measurements represented over 5 000 one-meter binned VSF samples from vertical profiles taken at each of the locations. Comparative measurements of β_p from the three sensors were very similar over a large dynamic range of VSFs (Fig. 6.8(A)).

The mean β_p values from all field measurements show that the ECO-VSF and ECO-BB3 were 5% to 10% higher than those of the MASCOT interpolated at their specific centroid angles (Fig. 6.8(B)). However, while ECOs resolve angular scattering in the backward direction with relatively broad source beams and detector FOVs (large $\Delta\theta$ for W) to minimize uncertainties in deriving backscattering coefficients from a single measurement (i.e. Fig. 6.5(A)), the MASCOT angular weighting functions are more discrete (small $\Delta\theta$ for W) (i.e. Fig. 6.5(B)). Results are thus not directly comparable unless the MASCOT VSF is appropriately weighted to the ECO geometry:

$$\beta(\bar{\theta}_{ECO}) = \int_0^{\pi} \beta_M(\theta) W_{ECO}(\theta)\, d\theta\,,$$

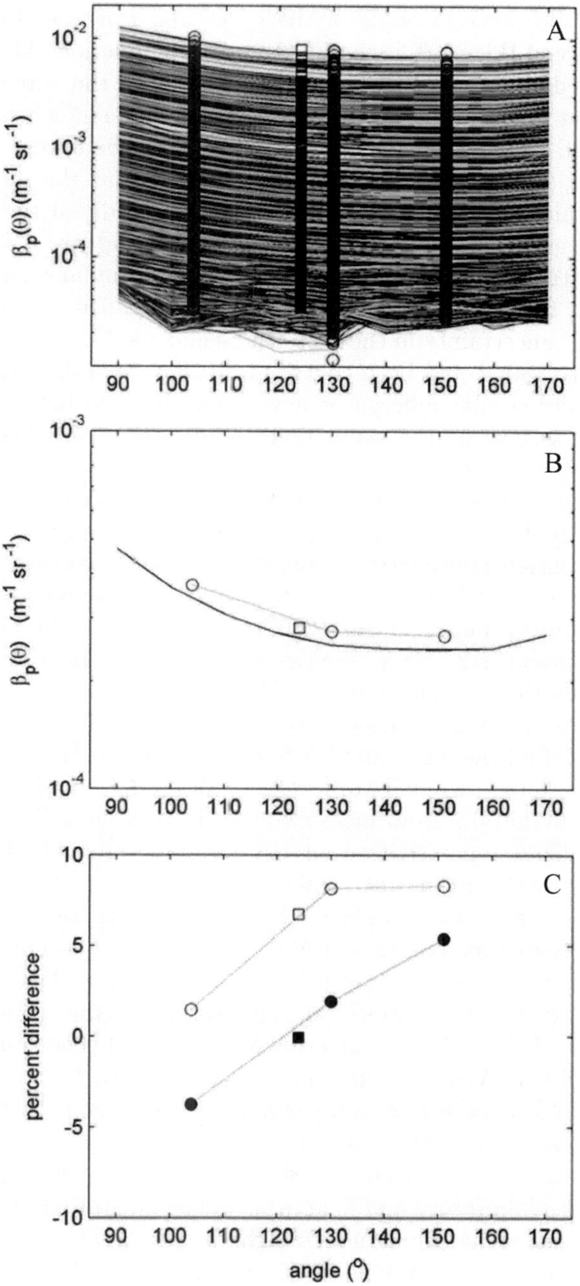

Fig. 6.8. (A) The VSFs in the backward angles from all field sites (see text) for the MASCOT (solid lines), ECO-VSF (open circles) and ECO-BB3 (open squares). (B) The average values of all field data for the MASCOT (solid line), ECO-VSF (open circles) and ECO-BB3 (open square). (C) The percent difference for the ECO-VSF (circles) and ECO-BB3 (square) relative to the MASCOT measurements before convolving the averaged MASCOT VSF with the ECO weighting functions (open symbols) and after (filled symbols)

where $\bar{\theta}_{ECO}$ is the centroid angle of the ECO weighting function, $\beta_M(\theta)$ is the MASCOT VSF, and $W_{ECO}(\theta)$ is the ECO weighting function. After convolving the MASCOT VSF data to the ECO weighting functions, the three independent instruments agreed to within ±5% (Fig. 6.8(C)). The difference between the sensor's measurements at centroid angles 124° and 130° were the lowest (<2% difference) and were higher (5%) at 104° and 151°, suggesting that the most accurate calibrations are attainable at centroid angles near the middle of the backward. Since the ECO measurement centered at 104° includes some forward scattering where the VSF shape in natural waters starts to increase rapidly, any uncertainty in the shape of the weighting function in this region becomes amplified. For the measurement centered at 151°, uncertainty in the VSF shape and the ECO weighting function in the range 170° to 180° could both contribute to the small disagreement. It should be noted that very similar differences have been observed between the MASCOT and other ECO sensors at the same angles, indicating that these differences are non-random.

Direct *in situ* comparisons of backscattering estimates from ECO sensors and other commercial devices that have a similar design concept, but use different centroid angles, weighting functions, calibration procedures (e.g. 'plaque method'; Maffione and Dana, 1997) and methods to estimate b_b, have agreed within 10% or better. For example, Loisel et al. (2007) found that ECO and Hydroscat^{TM} (HOBI Labs, Tucson, AZ, USA) sensors agreed to within a few percent during co-deployments in the English Channel and North Sea. These comparison measurements were taken over a large range in b_b from ~0.003 to 0.06 m^{-1}. Similarly, Boss et al. (2004a,b) found the instruments agreed to within ~2% in co-deployments off the coast of New Jersey, while Twardowski et al. (2007) found agreement between the two sensors within 3% from measurements in very clear South Pacific waters. McKee et al. (2009) reported that in UK coastal waters, ECO and Hydroscat measurements agreed within 10%, while Chami et al. (2005) reported a similar agreement between the sensors in Black Sea deployments. Boss et al. (2007) found that ECO and Hydroscat sensors agreed to within 3% in co-deployments in Crater Lake, OR, an optically clear and deep (~600 m) sub-alpine lake. During measurements off coastal New Jersey, Boss et al. (2004a,b) found that comparisons between an ECO sensor and the Volume Scattering Meter (VSM) (Lee and Lewis, 2003), an instrument capable of VSF measurements from ~0.6° to 177° in 0.3° increments, agreed within 10% or better. A comparison between VSM, ECO and Hydroscat measurements taken during the summer of 2000 at the LEO-15 coastal observatory site reported that sensors agreed within 8% (Mobley et al., 2002). Balch et al. (2009) found that in both laboratory and field measurements, an ECO-VSF agreed within a few percent of the values from an EOS light scattering photometer (Wyatt Technologies Corp., Santa Barbara, CA, USA), a bench-top VSF instrument capable of measurements from 10° to 170°. While these sensors compare very well in side-by-side measurements, it is not surprising that there are small differences. Notably, even if all aspects of calibration, instrument age, measurement uncertainties, etc. are known and corrected in these data, differences in weighting functions alone, if not corrected for in the comparison, could result in several percent difference between sensor's $\beta(\theta)$ values.

6.7 Conclusions

Determining the backscattering coefficient requires measurement of the VSF in the backward direction at one or more angular weighting distributions. If the VSF is resolved at a sufficient number of angles, it may be simply integrated to provide b_b. With only a single VSF measurement in the backward, an approximation must be made, although this is expected to have excellent accuracy because of the remarkable consistency in the shape of the VSF observed for natural waters. Additionally, as the weighting function of the VSF measurement broadens in the backward, it will experience less dependence on changes in the shape of the particulate phase function. Theoretical analysis has shown virtually no expected dependence (<1%) on phase function shape in deriving b_b using a recently developed single measurement volume scattering geometry employing a $\sin(\theta)$ weighting (Haubrich et al., 2011).

With careful attention to detail in sensor design and construction, weighting function derivation, calibration bead choice, theoretical computation of phase functions for calibration beads, and laboratory measurements, uncertainties of 2% or better should be attainable. This is the uncertainty expected for a sensor such as the MASCOT discussed in the previous section (Twardowski et al., 2012). For sensors such as the ECO that use broad weighting functions with LED sources that may exhibit some variability in the distribution of LED output intensity, uncertainties may be as high as 5%, assuming recent calibration. These estimated uncertainties were verified in sensor calibration replication and inter-comparisons of sensors using different measurement geometries and calibration methods. The lowest rates of temporal drift were observed for ECO measurements employing green LED sources. In addition to inter-comparisons of the MASCOT and ECO sensors (section 6.6), ECO sensor b_b values have been compared with values from other sensors with agreement within 10% and often better than 3% (e.g. Boss et al., 2004a,b, 2007; Chami et al., 2005; Loisel et al., 2007; Twardowski et al., 2007). Further validation of ECO backscattering measurements have been demonstrated through optical closure between *in situ* backscattering and passive solar reflectance measurements using radiative transfer computations (Mobley et al., 2002; Tzortziou et al., 2006; Twardowski et al., 2007; Gordon et al., 2009). Considering the small differences between the MASCOT and ECO sensors (Fig. 6.8(C)), a hopeful aspect is that these differences are non-random, indicating that further improvement is possible in the future by accounting for small bias error(s) that remain embedded in some aspect of calibration.

Acknowledgements

The authors would like to thank Scott Freeman, Heather Groundwater, Alex Derr, Cale Wetzel, Marvin Johnson, Wes Strubhar and Dan Morrissette for their assistance in collecting data for this review. We also thank David McKee, Emmanuel Boss and Alexander Kokhanovsky for helpful comments on the manuscript. We are grateful to acknowledge support for this work from ONR contracts N00014-06-C-0027, N000140510064, N0001405C0418, N00014-02-C-0173, ONR MURI grant N000140911054, NASA contract NNX09AV55G, and WET Labs.

References

Agrawal, Y. C., 2005: The optical volume scattering function: Temporal and vertical variability in the water column off the StateplaceNew Jersey coast, *Limnol. Oceanogr.* **50**, 1787–1794.

Balch, W. M., A. J. Plueddeman, B. C. Bowler, and D. T. Drapeau, 2009: Chalk-Ex– Fate of $CaCO_3$ particles in the mixed layer: Evolution of patch optical properties, *J. Geophys. Res.*, **114**, C07020.

Beardsley, G. F., Jr., H. Pak, K. Carder, and B. Lundgren, 1970: Light scattering and suspended particles in the eastern equatorial Pacific Ocean, *J. Geophys. Res.*, **75**, 2837–2845.

Behrenfeld M. J., and P. G. Falkowski, 1997: Photosynthetic rates derived from satellite-cased chlorophyll concentration, *Limnol. Oceanogr.*, **42**, 1–20.

Berthon, J.-F., E. Shybanov, M. E.-G. Lee, and G. Zibordi, 2007: Measurements and modeling of the volume scattering function in the coastal northern placeAdriatic Sea, *Appl. Opt.*, **46**, 5189–5203.

Beutell, R. G., and A. W. Brewer, 1949: Instruments for the Measurement of the Visual Range, *J. Sci. Instrum.*, **26**, 357–359.

Bogucki, D. J., J. A. Domaradzki, D. Stramski, and J. R. Zaneveld, 1998: Comparison of near-forward light scattering on oceanic turbulence and particles, *Appl. Opt.*, **37**, 4669–4677.

Bohren, C. F., and D. R. Huffman, 1983: *Absorption and Scattering of Light by Small Particles*, John Wiley, New York.

Boss, E., and W. S. Pegau, 2001: Relationship of light scattering at an angle in the backward direction to the backscattering coefficient, *Appl. Opt.*, **40**, 5503–5507.

Boss, E., M. S. Twardowski, and S. Herring, 2001: Shape of the particulate beam spectrum and its inversion to obtain the shape of the particle size distribution, *Appl. Opt.*, **40**, 4885–4893.

Boss, E., W. S. Pegau, M. Lee, M. S. Twardowski, E. Shybanov, G. Korotaev, and F. Baratange, 2004a: The particulate backscattering ratio at LEO-15 and its use to study particle composition and distribution, *J. Geophys. Res.*, **109**, C0101410.1029/ 2002JC001514.

Boss, E., D. Stramski, T. Bergmann, W. S. Pegau, and M. Lewis, 2004 b: Why should we measure the optical backscattering coefficient?, *Oceanography*, **17**, 44–49.

Boss, E., R. Collier, G. Larson, K. Fennel, and W. S. Pegau, 2007: Measurements of spectral optical properties and their relation to biogeochemical variables and processes in Crater Lake, Crater Lake National Park, OR, *Hydrobiologia*, **574**, 149–159.

Briggs, N., M. J. Perry, I. Cetinic, C. Lee, E. D'Asaro, A. Gray, and E. Rehm, 2011: Underwater gliders observe an aggregate 1 flux event in high vertical and temporal resolution during the North Atlantic spring bloom, *Deep Sea Res. I*, **58**, 1031–1039.

Chami, M., E. B. Shybanov, T. Y. Churilova, G. A. Khomenko, M. Lee, O. V. Martynov, G. A. Berseneva and G. K. Korotaev, 2005: Optical properties of the particles in the Crimea coastal waters (Black Sea), *J. Geophys. Res.*, **110**, C11020, doi:10.1029/2005JC003008.

Chami, M., E. Marken, J. J. Stamnes, G. Khomenko, and G. Korotaev, 2006: Variability of the relationship between the particulate backscattering coefficient and the volume scattering function measured at fixed angles, *J. Geophys. Res.*, **111**, C05013.

Churnside, J. H., and P. L. Donaghay, 2009: Thin scattering layers observed by airborne lidar, *ICES J. Mar. Sci.*, **66**, 778–789.

Churnside, J. H., V. V. Tatarskii, and J. J. Wilson, 1998: Oceanographic lidar attenuation coefficients and signal fluctuations measured from a ship in the Southern California Bight, *Appl. Opt.*, **37**, 3105–3112.

Dall'Olmo, G., T. K. Westberry, M. J. Behrenfeld, E. Boss, and W. H. Slade, 2009: Significant contribution of large particles to optical backscattering in the open ocean, *Biogeosciences*, **6**, 947–967.

Dana, D. R., and R. A. Maffione, 2002: Determining the backward scattering coefficienti with fixed-angle backscattering sensors – Revisited, *Ocean Optics XVI*, CityplaceSanta Fe, StateNM.

Fournier, G. R., and J. L. Forand, 1994: Analytic phase function for ocean water. *Proc. SPIE*, **2258**, 194–201.

Gordon, H. R., 1989: Dependence of the diffuse reflectance of natural waters on the sun angle, *Limnol. Oceangr.*, **34**(8), 1484–1489.

Gordon, H. R., 1993: Sensitivity of radiative transfer to small-angle scattering in the ocean: Quantitative assessment, *Appl. Opt.*, **32**, 7505–7511.

Gordon, H. R., O.B. Brown, and M. M. Jacobs, 1975: Computed relationships between the inherent and apparent optical properties, *Appl. Opt.*, **14**, 417– 427.

Gordon, H. R., M. R. Lewis, S. D. McLean, M. S. Twardowski, S. A. Freeman, K. J. Voss, and G. C. Boynton, 2009: Spectra of particulate backscattering in natural waters, *Opt. Express*, **17**, 16192–16208.

Guenther, G. C., 1985: Airborne laser hydrography: system design and performance factors, United States National Ocean Service, Office of Charting and Geodetic Services. National Technical Information Service, Rockville, MD. Springfield, VA.

Haubrich, J. M., Musser, J., and E. S. Fry, 2011: Instrumentation to measure the backscattering coefficient b_b for arbitrary phase functions, *Appl. Opt.*, **50**(21), 4134–4147.

Hewlett Packard, 1995: Application Brief I-012, Temperature compensation circuit for constant LED intensity, 2 pp.

Jackson, C ., L. M. Nilsson, and P. J. Wyatt, 1989: Characterization of biopolymers using a multi-angle light scattering detector with size-exclusion chromatography, *J. Appl. Polym. Sci., Appl. Polym. Symp.*, **43**, 99–114.

Jerlov, N. G., 1961: Optical measurements in the eastern North Atlantic. *Medd. Oceanogr. Inst. Göteborg, Ser. B*, 8.

Jerlov, N. G., 1968: *Optical Oceanography*. Elsevier, New York.

Jerlov, N. G., 1976: *Marine Optics*. Elsevier, New York.

Jonasz, M., and G. R. Fournier, 2007: *Light Scattering by Particles in Water: Theoretical and Experimental Foundations*, Academic Press.

Kopelevich, O. V., and V. I. Burenkov, 1971: The nephelometric method for determining the total scattering coefficienti of light in sea water, *Izv. Atmos. Oceanic Phys.*, **7**, 1280–1289.

Kullenberg, G., 1968: Scattering of light by Sargasso seawater. *Deep Sea Res.*, **15**, 423–432.

Kullenberg, G., 1972: A comparison between observed and computed light scattering functions, 2, *Rep. 19*, University of Copenhagen, Copenhagen.

Lee, M. E., and M. R. Lewis, 2003: A new method for the measurement of the optical volume scattering function in the upper ocean, *J. Atmos. Ocean. Technol.*, **20**, 563–571.

Loisel, H., X. Meriaux, J.-F. Berthon, and A. Poteau, 2007: Investigation of the optical backscattering to scattering ratio of marine particles in relation to their biogeochemical composition in the eastern English Channel and southern North Sea, *Limnol. Oceanogr.*, **52**(2), 739–752.

Ma, X., J. Q. Lu, R. S. Brock, K. M. Jacobs, P. Yang, and X. Hu, 2003: Determination of complex refractive index of polystyrene microspheres from 370 to 1610 nm, *Phys. Med. Biol.*, **48**, 4165–4172.

Maffione, R. A., and D. R. Dana, 1997: Instruments and methods for measuring the backward scattering coefficienti of ocean waters, *Appl. Opt.*, **36**, 6057–6067.

Matheson, L. A., and J. L. Saunderson, 1952: *Optical and Electrical Properties of Polystyrene, Styrene: Its Polymers, Copolymers and Derivatives*, R. H. Boundy and R. F. Boyer, eds., Reinhold Publishing Corporation, New York, pp. 517–546.

McKee, D., M. Chami, I. Brown, V. Sanjuan Calzado, D. Doxaran, and A. Cunningham, 2009: Role of measurement uncertainties in observed variability in the spectral backscattering ratio: a case study in mineral-rich coastal waters, *Appl. Opt.*, **48**, 4663–4675.

Mikkelsen, O.A., T. G. Milligan, P. S. Hill, R. J. Chant, C. F. Jago, S. E. Jones, V. Krivtsov, and G. Mitchelson-Jacob, 2008: The influence of schlieren on in situ optical measurements used for particle characterization, *Limnol. Oceanogr.: Methods*, **6**, 133–143.

Mishchenko, M. I., J. W. Hovenier, and L. D. Travis, 2000: *Light Scattering by Nonspherical Particles*, Academic Press, San Diego, CA.

Mobley, C. D., 1994: *Light and Water: Radiative Transfer in Natural Waters*, Academic Press, San Diego, CA.

Mobley, C. D., L. K. Sundman, and E. Boss, 2002: Phase function effects on oceanic light fields, *Appl. Opt.*, **41**, 1035–1050.

Moore, C., M. S. Twardowski, and J. R. V. Zaneveld, 2000: The ECO VSF: a multiangle scattering sensor for determination of the volume scattering function in the backward direction, *Ocean Optics XV*, Monaco, 16–20 October.

Morel, A., 1966: Etude expérimentale de la diffusion de la lumière par l'eau, le solutions de chlorine de sodium et l'eau de mer optiquement pure. *J. de Chimie Physique*, **10**, 1359–1366.

Morel, A., 1973: Diffusion de la lumiere par les eaux de mer. Resultats experimentaux et approche theorique, in *Optics of the Sea*, AGARD Lecture Series, No. 61, NATO, pp. 3.1.1–3.1.76.

Morel, A., 1974: Optical properties of pure water and pure seawater, in *Optical Aspects of Oceanography*, N. G. Jerlov and E. Steeman, eds. (Academic Press, London), pp. 1–24.

Morel, A., and B. Gentili, 1983: Diffuse reflectance of oceanic waters. II. Bidirectional aspects, *Appl. Opt.*, **32**, 6864–6879.

Morel, A., K. Voss, and B. Gentilli, 1995: Bidirectional reflectance of oceanic waters: A comparison of modeled and measured upward radiance fields, *J. Geophys. Res.*, **100**, 13,143–13,150.

Neukermans, G., H. Loisel, X. Mériaux, R. Astoreca, and D. McKee, 2012: In situ variability of mass-specific beam attenuation and backscattering of marine particles with respect to particle size, density, and composition, *Limnol. Oceanogr.*, **57**(1), 124–144.

Nikolov, I. D., and C. D. Ivanov, 2000: Optical plastic refractive measurements in the visible and the near-infrared regions, *Appl. Opt.*, **39**, 2067–2070.

Oishi, T., 1990: Significant relationship between the backward scattering coefficienti of sea water and the scatterance at 120°, *Appl. Opt.*, **29**, 4658–4665.

Petzold, T. J., 1972: Volume scattering functions for selected ocean waters, Tech. Rep., Scripps Institution of Oceanography, 72–78.

Prentice, J. E., A. D. Weidemann, W. S. Pegau, K. J. Voss, M. E. Lee, E. Shybanov, O. Martynov, A.E. Laux, A. Briggs, and G. Chang, 2002: Laboratory comparisons of optical scattering instrumentation, *Ocean Optics XVI*, Santa Fe, New Mexico.

Quinby-Hunt, M. S., A. J. Hunt, K. Lofftus, and D. Shapiro, 1989: Polarized-light scattering studies of marine chlorella, *Limnol. Oceanogr.*, **34**, 1587–1600.

Reynolds, K. J., J. P. De Kock, L. Tarassenko and J. T. B. Moyle, 1991: Temperature dependence of LED and its theoretical effect on pulse oximetry, *Br. J. Anaesth.*, **67**(5), 638–643.

Stramska, M., and D. Stramski, 2005: Variability of particulate organic carbon concentration in the north polar Atlantic based on ocean color observations with Sea-viewing Wide Field-of-view Sensor (SeaWiFS), *J. Geophys. Res.*, **110**, C10018, doi:10.1029/2004JC002762.

Stramski, D., and D. A. Kiefer, 1991: Light scattering by microorganisms in the open ocean, *Prog. Oceanogr.*, **28**, 343–383.

Stramski, D., R. A. Reynolds, M. Kahru, and B. G. Mitchell, 1999: Estimation of particulate organic carbon in the ocean from satellite remote sensing, *Science*, **5425**, 239–241.

Stramski, D., E. Boss, D. Bogucki, and K. J. Voss, 2004: The role of seawater constituents in light backscattering in the ocean, *Progr. Oceanogr.*, **61**(1), 27–56.

Stramski, D., R. A. Reynolds, M. Babin, S. Kaczmarek, M. R. Lewis, R. Röttgers, A. Sciandra, M. Stramska, M. S. Twardowski, and H. Claustre, 2008: Relationships between the surface concentration of particulate organic carbon and optical properties in the eastern South Pacific and eastern Atlantic Oceans, *Biogeosciences*, **5**, 171–183.

Sullivan, J. M., and M. S. Twardowski, 2009: Angular shape of the oceanic particulate volume scattering function in the backward direction, *Appl. Opt.*, **48**(35), 6811–6819.

Sullivan, J. M., M. S. Twardowski, P.L. Donaghay, and S. Freeman, 2005: Using optical scattering to discriminate particle types in coastal waters, *Appl. Opt.*, **44**, 1667–1680.

Sullivan, J. M., P. L. Donaghay, and J. E. B. Rines, 2010: Coastal thin layer dynamics: consequences to biology and optics, *Cont. Shelf Res.*, **30**(1): 50–65. doi:10.1016/j.csr.2009.07.009.

Sun, D. Y. Li, Q. Wang, J. Gao, H. Lv, C. Le, and C. Huang, 2009: Light scattering properties and their relation to the biogeochemical composition of turbid productive waters: a case study of Lake Taihu, *Appl. Opt.*, **48**, 1979–1989.

Twardowski, M. S., E. Boss, J. B. Macdonald, W. S. Pegau, A. H. Barnard, and J. R. V. Zaneveld, 2001: A model for estimating bulk refractive index from the optical backscattering ratio and the implications for understanding particle composition in case I and case II waters, *J. Geophys. Res.*, **106**(C7), 14129–14142.

Twardowski, M. S., M. R. Lewis, A. H. Barnard, and J. R. V. Zaneveld, 2005: Water instrumentation and platforms for ocean color remote sensing applications, in *Remote Sensing and Digital Image Processing, 1*, Vol. 7: Remote Sensing of Coastal Aquatic Environments, pp. 69–100.

Twardowski, M. S. H. Claustre, S. A. Freeman, D. Stramski, and Y. Huot, 2007: Optical backscattering properties of the 'clearest' natural waters, *Biogeosciences*, **4**, 1041–1058.

Twardowski, M. S., X. Zhang, S. Vagle, J. Sullivan, S. Freeman, H. Czerski, Y. You, L. Bi, and G. Kattawar, 2012: The optical volume scattering function in a surf zone inverted to derivesediment and bubble particle subpopulations, *J. Geophys. Res.*, **117**, doi:10.1029/2011JC007347.

Tyler, J. E. and W. H. Richardson, 1958: Nephelometer for the measurement of volume scattering function in situ, *J. Opt. Soc. Am.*, **48**, 354–357.

Tzortziou, M., J. R. Herman, C. L. Gallegos, P. J. Neale, A. Subramaniam, L. W. Harding, Jr., and Z. Ahmad. 2006: Bio-optics of the Chesapeake Bay from measurements and radiative transfer closure. *Est. Coast. Shelf Sci.*, **68**, 348–362.

Ulloa, O., S. Sathyendranath, and T. Platt. 1994: Effect of the particle-size distribution on the backscattering ratio in seawater, *Appl. Opt.*, **30**, 7070–7077.

Vaillancourt, R. D., C. W. Brown, R. R. L. Guillard, and W. M. Balch, 2004: Light backscattering properties of marine phytoplankton: relationships to cell size, chemical composition, and taxonomy, *J. Plankton Res.*, **26**, 191–212.

Volten, H., J. F. de Haan, J. W. Hovenier, R. Schreurs, and W. Vassen, 1998: Laboratory measurements of angular distributions of light scattered by phytoplankton and silt, *Limnol. Oceangr.*, **43**, 1180–1197.

Voss, K., and A. Morel, 2005: Bidirectional reflectance function for oceanic waters with varying chlorophyll concentrations: Measurements versus predictions. *Limnol. Oceanogr.*, **50**(2), 698–705.

Voss, K. J., A. Chapin, M. Monti, and H. Zhang, 2000: Instrument to measure the bidirectional reflectance distribution function of surfaces, *Appl. Opt.*, **39**, 6197–6206.

Whitmire, A. L., E. Boss, T. J. Cowles, and W. S. Pegau, 2007: Spectral variability of the particulate backscattering ratio, *Opt. Express*, **15**, 7019–7031.

Whitmire, A. L., W. S. Pegau, L. Karp-Boss, E. Boss, and T. J. Cowles, 2010: Spectral backscattering properties of marine phytoplankton cultures, *Opt. Express*, **18**, 15073–15093.

Zaneveld, J. R. V., 1995: A theoretical derivation of the dependence of the remotely sensed reflectance on the inherent optical properties, *J.Geophys Res.*, **100**(C7), 13, 135–13, 142.

Zaneveld, J. R. V., and J. C. Kitchen, 1995: The variation in the inherent optical properties of phytoplankton near an absorption peak as determined by various models of cell structure, *J. Geophys. Res.*, **100**, 13, 309–13, 320.

Zhang, X., and L. Hu, 2009: Scattering by pure seawater at high salinity, *Opt. Express*, **17**(15), 12685–12691.

Zhang, X., L. Hu, and M.-X. He, 2009: Scattering by pure seawater: effect of salinity, *Opt. Express*, **17**, 5698–5710.

Zhang, X., M. S. Twardowski, and M. Lewis, 2011: Retrieving composition and sizes of oceanic particle subpopulations from the volume scattering function, *Appl. Opt.*, **50**, 1240–1259.

7 Molecular light scattering by pure seawater

Xiaodong Zhang

The general theory of the molecular scattering, and the scattering by water and seawater in particular, will be reviewed. For a pure liquid, free from any foreign particles, the statistical thermal motion of the molecules gives rise to the scattering of light. The resultant fluctuation of the density of the water and fluctuation in the orientations of the water molecules bring about fluctuation of the optical dielectric constant, which in turn causes scattering. For solutions such as seawater, additional scattering is due to the fluctuation of the concentration of sea salts. The latest measurements of key thermodynamic parameters of water and the development of a recent theoretical model allow the scattering by seawater to be investigated in detail as a function of temperature and salinity. The model results agree with laboratory observations within experimental error.

7.1 Introduction

Water is the only substance on Earth that can exist naturally in three different states as a solid (ice), liquid (water) or gas (vapor). Yet, the scattering of light by the same water molecules, but in different states is different. For example, if the intensity of light scattered at an angle of θ_s is 100 units for a given amount of water vapor, then when the vapor condenses into liquid, the scattering intensity would be about 2 units, and when the liquid further freezes into ice, the scattering intensity would be nearly zero, as long as θ_s is not in the direction of refraction. Generally speaking, the reduced scattering intensity results from destructive interference. For a vapor (or gases in general), each molecule acts as a scattering particle whose position is far from, and random to the others. Therefore, scattering by gases is directly additive with no interference. For pure ice (or any crystal) il luminated with light at a wavelength much greater than the separation of the lattice planes, no light is scattered, because it is always possible to pair two scattering planes so that destructive interference occurs. There are two exceptions to this general description. At the refraction direction, the scattered light is always constructive. When the wavelength (λ) of the incident radiation is roughly equal to the distance (d) separating the scattering planes, scattering can be observed at Bragg angles, defined as $\sin(\theta_s/2) = n\lambda/2d$, where n is an integer.

For scattering by liquid water, most of scattered light goes through destructive interference. However, Brownian motion produces transient optical inhomogeneity in the liquid, the presence of which allows a small fraction of the scattered radiation to escape destructive interference and to be observed. Fundamentally, all scattering originates from the interaction of photons with molecules; however, the theoretical approach in quantifying the scattering by gases, liquids and solids is different, as summarized in Table 7.1.

Table 7.1. Summary of typical treatment of scattering by water in different states

	Scattering center	Interference	Angular distribution
Gases (vapor)	Molecules in random motion	Not considered	Approx. isotropic
Liquid (water)	Small volumes in random motion[1]	Not considered	Approx. isotropic
	Molecules[2]	Considered	
Solid (ice)	Lattice planes	Considered	Refraction and reflection

[1] Einstein-Smoluchowski theory considering the liquid as a continuous medium with localized fluctuation in density, which can be calculated from thermodynamics (Smoluchowski, 1908; Einstein, 1910).
[2] Zimm (1945) considered explicitly the scattering by individual molecules and addressed the interference using the results from statistical mechanics describing the correlations between molecular separations. Zimm's molecular theory of the scattering reproduces the results of Einstein and Smoluchowski but has the advantage that it does not yield infinite scattering at the critical point.

7.2 General theory of scattering

7.2.1 Isotropic particles

Consider a plane wave polarized in the YZ plane incident upon an isotropic particle of spherical polarizability α situated at the origin of the coordinate system (Fig. 7.1) in a medium of refractive index n. Only the electric field of the incident light is of interest here and its amplitude is E_{0Z}. The instantaneous electric field of incident light: $E = E_{0Z}\cos(k_0 y - \omega_0 t)$, where $k_0 = 2\pi n/\lambda_0$, λ_0 is the wavelength in vacuum, and $\omega_0 = 2\pi\nu_0$ is the angular frequency. The electric field E interacts with electrons in an atom or molecules, inducing an electric dipole moment, which oscillates at the angular frequency ω_0. For an isotropic scatterer, for which α is independent of orientation, the induced dipole moment is:

$$P_Z = \alpha E = \alpha E_{0Z}\cos(k_0 y - \omega_0 t)\,. \tag{7.1}$$

An oscillating dipole produces radiation, which at large distances, i.e., where $R \gg \lambda_0$, the electric field E_{SZ} of the scattered light is proportional to $d^2P/dt^2 = -\omega_0^2 P$

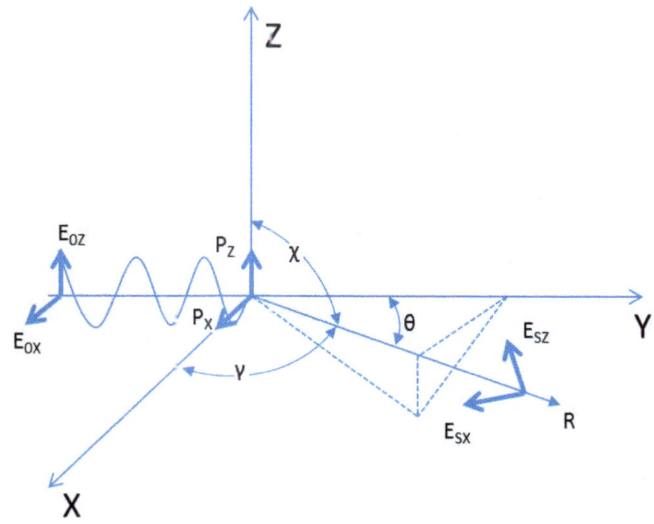

Fig. 7.1. Scattering geometry.

and inversely proportional to distance R,

$$E_{SZ} = \frac{-\omega_0^2 P_Z \sin(\chi)}{4\pi\varepsilon_0 c^2 R}, \tag{7.2}$$

where χ is the angle between the induced moment P_Z and R, ε_0 is the permittivity of free space, and c is the speed of light. The instantaneous irradiance, $E_{irr} = c\varepsilon_0 E^2$. Since all measurements require times that are much longer than the oscillation period of the radiation field, the cycle averaged E_{irr} is of interest. Therefore,

$$\langle E_{irr}\rangle = c\varepsilon_0 \left\langle E^2 \right\rangle = \frac{c\varepsilon_0}{2} E^2 . \tag{7.3}$$

The factor 0.5 in Eq. (7.3) results from averaging the term $\cos^2(k_0 y - \omega_0 t)$ in Eq. (7.1) over sampling period that is much longer than the oscillation period of the electric field.

Combining Eqs. (7.1)–(7.3), the scattered irradiance, E_{irr-s},

$$E_{irr-s} = \frac{c\varepsilon_0}{2} E_{0Z}^2 \left[\frac{\omega_0^4 \alpha^2 \sin^2(\chi)}{(4\pi\varepsilon_0)^2 c^4 R^2} \right] . \tag{7.4}$$

Note that the incident irradiance,

$$E_{irr-i} = \frac{c\varepsilon_0}{2} E_{0Z}^2 ,$$

and therefore,

$$\frac{E_{irr-s}}{E_{irr-i}} = \frac{\omega_0^4 \alpha^2 \sin^2(\chi)}{(4\pi\varepsilon_0)^2 c^4 R^2} = \frac{\pi^2 \alpha^2 \sin^2(\chi)}{\varepsilon_0{}^2 \lambda_0^4 R^2} . \tag{7.5}$$

Similarly, for an incident wave polarized in the XY plane with electric field amplitude of E_{0X}, the ratio of scattered irradiance to the incident irradiance is

$$\frac{E_{irr-s}}{E_{irr-i}} = \frac{\pi^2 \alpha^2 \sin^2(\gamma)}{\varepsilon_0{}^2\ \lambda_0^4\ R^2}, \tag{7.6}$$

where γ is the angle between the induced moment P_X and R. Therefore, for unpolarized incident light, the ratio of scattered irradiance to the incident irradiance would be

$$\frac{E_{irr-s}}{E_{irr-i}} = \frac{\pi^2 \alpha^2}{\varepsilon_0{}^2 \lambda_0^4 R^2} \frac{\sin^2(\chi) + \sin^2(\gamma)}{2} = \frac{\pi^2 \alpha^2}{\varepsilon_0{}^2 \lambda_0^4 R^2} \frac{1 + \cos^2(\theta)}{2}. \tag{7.7}$$

The second equality in Eq. (7.7) follows $\cos^2(\chi) + \cos^2(\gamma) + \cos^2(\theta) = 1$. Note, Eq. (7.7) can also be derived for polarized light if all the scattered light in the cone $\theta + d\theta$ is collected regardless of the angle χ (or γ).

Equation (7.7) describes scattering by one particle. For gases, each molecule can be considered as a particle, moving randomly relative to each other. For a volume V of gas with N molecules, the ratio of scattered irradiance at scattering angle θ to the incident irradiance is simply N multiply Eq. (7.7), i.e.

$$\frac{E_{irr-s}}{E_{irr-i}} = \frac{\pi^2 N \alpha^2}{\varepsilon_0{}^2 \lambda_0^4 R^2} \frac{1 + \cos^2(\theta)}{2}. \tag{7.8}$$

The polarizability is not a quantity that can be determined experimentally, but from Maxwell's equation it has been shown that it is related to the dielectric constant (also called relative permittivity), ε_r, and hence the square of the refractive index n,

$$\varepsilon_r = n^2 = 1 + \frac{N\alpha}{V\varepsilon_0}. \tag{7.9}$$

Substituting $N\alpha^2/\varepsilon_0{}^2$ in Eq. (7.8) with Eq. (7.9) and applying the definition that the volume scattering function,

$$\beta(\theta) = \frac{I(\theta)}{E_{irr-i} V} = \frac{1}{E_{irr-i} V} \frac{E_{irr-s} \Delta A}{\Delta \Omega} = \frac{E_{irr-s} R^2}{E_{irr-i} V},$$

where ΔA and $\Delta\Omega$ are the cross-sectional area and the solid angle of the scattered light,

$$\beta(\theta) = \frac{1}{2} \frac{\pi^2 \left(n^2 - 1\right)^2}{\lambda_0{}^4 N_0} \left(1 + \cos^2\theta\right), \tag{7.10}$$

where $N_0 = N/V$, the number of particles per unit volume.

In a pure liquid, the molecules are compacted densely enough that the movement of one molecule cannot be considered totally random to others, and therefore the interference of scattered light by each molecule has to be accounted for. To overcome this challenge, Smoluchowski (1908) and Einstein (1910) developed fluctuation theory, in which fluctuations in the number of particles in a given small volume element, which is small compared with the wavelength of light, but big enough to contain a large number of molecules, result in changes in density which will produce corresponding changes in the dielectric constant. Assuming in a sample of volume V, each small volume element is dV with an instantaneous polarizability $\alpha_V = \overline{\alpha} + \Delta\alpha_V$, where $\overline{\alpha}$ represents the time average of α_V, $\Delta\alpha_V$ the instantaneous fluctuation, the time average of which, $\overline{\Delta\alpha_V}$, is zero by definition. In Eq. (7.7), the scattered light is proportional to the square of polarizability; the contribution from a volume element is the time average of the α_V^2. Thus,

$$\overline{(\overline{\alpha} + \Delta\alpha_V)^2} = (\overline{\alpha})^2 + \overline{(\Delta\alpha_V)^2}\,. \tag{7.11}$$

The contribution from $(\overline{\alpha})^2$, whose value does not change for the sample, cancels exactly as in perfect crystals because we can always find another element volume whose distance to the volume element being considered is such that the scattered electric fields by the two volume elements are opposite in phase and cancel each other. Therefore the net scattering in Eq. (7.7) depends only on $\overline{(\Delta\alpha_V)^2}$, i.e, for liquid,

$$\frac{E_{irr-s}}{E_{irr-i}} = \frac{\pi^2\alpha^2}{\varepsilon_0{}^2\,\lambda_0^4\,R^2}\,\frac{\sin^2(\chi) + \sin^2(\gamma)}{2} = \frac{\pi^2\,\overline{(\Delta\alpha_V)^2}}{\varepsilon_0{}^2\,\lambda_0^4\,R^2}\,\frac{1+\cos^2(\theta)}{2}\,. \tag{7.12}$$

Following the same deduction from Eq. (7.7) to Eq. (7.10) for gases, but with $\overline{(\Delta\varepsilon_r)^2}$ or $\overline{(\Delta n^2)^2} = \overline{(\Delta\alpha_V)^2}$ and $N_0 = 1/dV$ as number of volume elements per unit volume, we have the volume scattering function for liquid,

$$\beta\,(\theta) = \frac{1}{2}\,\frac{\pi^2\overline{(\Delta n^2)^2}\,dV}{\lambda_0{}^4}\left(1+\cos^2\theta\right)\,. \tag{7.13}$$

The fluctuations of n^2 is a result of fluctuations of density ρ and temperature T,

$$\Delta n^2 = \left(\frac{\partial n^2}{\partial\rho}\right)_T \Delta\rho + \left(\frac{\partial n^2}{\partial T}\right)_\rho \Delta T\,. \tag{7.14}$$

But the second term is negligible compared with the first and may be neglected, so the fluctuations in dielectric constant are

$$\overline{(\Delta n^2)^2} = \left(\frac{\partial n^2}{\partial\rho}\right)_T^2 \overline{(\Delta\rho)^2}\,. \tag{7.15}$$

From thermodynamic statistics (e.g., Fabelinskii, 1968),

$$\overline{(\Delta\rho)^2} = \frac{\rho^2 k_B T \beta_T}{dV}\,, \tag{7.16}$$

where k_B is Boltzmann constant and β_T the isothermal compressibility. Inserting Eqs. (7.15) and (7.16) into Eq. (7.13), we have

$$\beta\,(\theta) = \frac{1}{2}\,\frac{\pi^2 k_B T \beta_T}{\lambda_0{}^4}\left(\rho\frac{\partial n^2}{\partial\rho}\right)^2\left(1+\cos^2\theta\right)\,. \tag{7.17}$$

7.2.2 Anisotropic particles

When $\theta = \pi/2$, scattered light is completely polarized vertically (this would be the case in Fig. 7.1, where R is in the XZ plane), because the angle $\gamma = 0$, and therefore the scattered light that is horizontally polarized vanishes. However, experimental observations indicate that the scattered light at $\theta = \pi/2$ is not completely polarized because a molecule is not an isotropic particle (with the exception of noble gases, probably) (e.g., Strutt, 1918; Cabannes, 1922). For an anisotropic particle, the induced dipole moment is not in the same direction as the electric field, i.e., P_Z in Fig. 7.1 would not align with the Z axis. Therefore, instead of Eq. (7.1), generally,

$$\begin{pmatrix} P_X \\ P_Y \\ P_Z \end{pmatrix} = \begin{pmatrix} \alpha_{xx} & \alpha_{xy} & \alpha_{xz} \\ \alpha_{yx} & \alpha_{yy} & \alpha_{yz} \\ \alpha_{zx} & \alpha_{zy} & \alpha_{zz} \end{pmatrix} \begin{pmatrix} E_{0X} \\ E_{0Y} \\ E_{0Z} \end{pmatrix}, \tag{7.18}$$

where α_{ij} $(i, j = x, y, z)$ are elements of the polarizability tensor, and normally, $\alpha_{ij} = \alpha_{ji}$. Depending on the coordinate system chosen for representation, the values of α_{ij} change. However, the mean polarizability α defined by

$$\alpha = \frac{1}{3}\left(\alpha_{xx} + \alpha_{yy} + \alpha_{zz}\right), \tag{7.19}$$

and the anisotropy β defined by

$$\beta^2 = \frac{1}{2}\left[\left(\alpha_{xx} - \alpha_{yy}\right)^2 + \left(\alpha_{yy} - \alpha_{zz}\right)^2 + \left(\alpha_{zz} - \alpha_{xx}\right)^2 + 6\left(\alpha_{xy}^2 + \alpha_{yz}^2 + \alpha_{zx}^2\right)\right]. \tag{7.20}$$

do not change. Also by definition, $\beta = 0$ for isotropic particles.

The orientations of molecules in gases or liquid, like their movements, are random too. The mean values for these anisotropic elements are:

$$\overline{\alpha_{xx}^2} = \overline{\alpha_{yy}^2} = \overline{\alpha_{zz}^2} = \alpha^2 + \frac{4}{45}\beta^2, \tag{7.21}$$

$$\overline{\alpha_{xy}^2} = \overline{\alpha_{yz}^2} = \overline{\alpha_{zx}^2} = \frac{\beta^2}{15}. \tag{7.22}$$

Refer again to Fig. 7.1 and assume that the scattered light is measured in the XY plane ($\chi = \pi/2$). This assumption will not lose generality because the mean states of polarizability (Eqs. (7.19) and (7.20)) do not depend on how the coordinate system is defined. The scattered light measured at the scattering angle θ:

$$\begin{aligned} E_{VV} &\propto \overline{\alpha_{zz}^2} \\ E_{VH} &\propto \overline{\alpha_{zy}^2} \\ E_{HV} &\propto \overline{\alpha_{xz}^2} \\ E_{HH} &\propto \overline{\alpha_{xx}^2}\cos^2\theta + \overline{\alpha_{xy}^2}\sin^2\theta, \end{aligned} \tag{7.23}$$

where the first and second subscripts indicate the polarization state of incident and scattered light, respectively. For vertically polarized incident light, the scattered light (E_{VV} or E_{VH}) does not depend on the scattering angle θ, for horizontally polarized incident light, the vertically polarized scattered light (E_{HV}) does not depend on θ, and only the horizontally polarized scattered light resulting from horizontally polarized incident light (E_{HH}) depends on θ through two parts: $\overline{\alpha_{xx}^2}\cos^2\theta$ for the contribution from OX component of the dipole moment and $\overline{\alpha_{xy}^2}\sin^2\theta$ for the contribution from OY component of the dipole moment. For natural unpolarized light, the total scattered light

$$\begin{aligned} E_{total} &= E_{VV} + E_{VH} + E_{HV} + E_{HH} \\ &\propto \frac{45\alpha^2 + 13\beta^2}{45}\left(1 + \frac{45\alpha^2 + \beta^2}{45\alpha^2 + 13\beta^2}\cos^2\theta\right) \end{aligned} \tag{7.24}$$

and the depolarization ratio is defined by ($\theta = \pi/2$)

$$\delta_n\left(\theta = \frac{\pi}{2}\right) = \frac{E_{VH} + E_{HH}}{E_{VV} + E_{HV}} = \frac{\overline{\alpha_{zy}^2} + \overline{\alpha_{xy}^2}}{\overline{\alpha_{zz}^2} + \overline{\alpha_{xz}^2}} = \frac{6\beta^2}{45\alpha^2 + 7\beta^2}\,. \tag{7.25}$$

It is apparent from Eq. (7.24), and as we have derived above, that for isotropic particles for which $\beta = 0$,

$$E_{total} \propto \alpha^2\left(1 + \cos^2\theta\right). \tag{7.26}$$

Taking the ratio of Eq. (7.24) to Eq. (7.26) for $\theta = \pi/2$,

$$\frac{E_{total}(\beta \neq 0, \theta = \pi/2)}{E_{total}(\beta = 0, \theta = \pi/2)} = \frac{45\alpha^2 + 13\beta^2}{45\alpha^2} = \frac{6 + 6\delta_n}{6 - 7\delta_n}\,. \tag{7.27}$$

This ratio is known as Cabannes factor (Cabannes, 1920), accounting for the increased scattering due to anisotropic nature of molecules. It should be noted that (1) both depolarization ratio and Cabannes factor are defined for $\theta = \pi/2$; (2) if the incident light is polarized, then values for both parameters will be different; and (3) the Cabannes factor was originally developed for gases, but it has been shown that it holds equally well for a liquid (Prins and Prins, 1956).

As shown in Eq. (7.24), the molecular anisotropy also affects the angular distribution of scattering (King, 1923; Martin, 1923), which becomes:

$$1 + \frac{45\alpha^2 + \beta^2}{45\alpha^2 + 13\beta^2}\cos^2\theta = 1 + \left(\frac{1 - \delta_n}{1 + \delta_n}\right)\cos^2\theta\,. \tag{7.28}$$

Combinations of Eqs. (7.10) and (7.17) with Eqs. (7.27) and (7.28) lead to the volume scattering function

$$\beta(\theta) = \frac{1}{2}\frac{\pi^2\left(n^2 - 1\right)^2}{{\lambda_0}^4 N_0}\frac{6 + 6\delta_n}{6 - 7\delta_n}\left(1 + \frac{1 - \delta_n}{1 + \delta_n}\cos^2(\theta)\right) \tag{7.29}$$

for gases, and

$$\beta(\theta) = \frac{1}{2}\frac{\pi^2 k_B T \beta_T}{{\lambda_0}^4}\left(\rho\frac{\partial n^2}{\partial\rho}\right)^2\frac{6 + 6\delta_n}{6 - 7\delta_n}\left(1 + \frac{1 - \delta_n}{1 + \delta_n}\cos^2(\theta)\right) \tag{7.30}$$

for a pure liquid, respectively.

7.2.3 Liquid solutions

For a liquid solution, the fluctuation in dielectric constant has an additional contribution from the fluctuation of solute concentration. Therefore, for a liquid solution, Eq. (7.14) should be modified

$$\Delta n^2 = \left(\frac{\partial n^2}{\partial \rho}\right)_T \Delta\rho + \left(\frac{\partial n^2}{\partial T}\right)_\rho \Delta T + \left(\frac{\partial n^2}{\partial C}\right)_\rho \Delta C\,, \tag{7.31}$$

where $C = m_s M_s / dV$ is the concentration of solute and m_s and M_s are the number of moles and the molecular weight of solute, respectively. The concentration fluctuation ΔC results in changes to the mixing ratio of solute and solvent within the volume element of dV, while the total mass and hence the density remains constant. This ensures that fluctuations due to density and concentration are independent of each other, such that (neglecting the fluctuation due to temperature)

$$\left(\overline{\Delta n^2}\right)^2 = \left(\frac{\partial n^2}{\partial \rho}\right)_T^2 \left(\overline{\Delta\rho}\right)^2 + \left(\frac{\partial n^2}{\partial C}\right)_T^2 \left(\overline{\Delta C}\right)^2 . \tag{7.32}$$

Scattering due to density fluctuations has been discussed above and for concentration fluctuation,

$$\overline{\Delta C^2} = \frac{k_B T}{\left(\frac{\partial^2 A}{\partial C^2}\right)_{T,V}}\,, \tag{7.33}$$

where A is the Helmholtz free energy. The change in Helmholtz free energy associated with the concentration change at constant temperature and volume is given by,

$$dA = \mu_0 dm_0 + \mu_s dm_s\,, \tag{7.34}$$

where μ denotes the chemical potential, and m the number of moles. The subscripts 0 and s denote properties defined for solvent and solute, respectively. Within the volume element dV, m_0 and m_s are related by,

$$dV = m_0 V_0^{'} + m_s V_s^{'}\,, \tag{7.35}$$

where V' denotes the partial molar volume. Since dV is held as a constant, differentiating Eq. (7.35) leads to,

$$dm_0 = -\frac{V_s^{'}}{V_0^{'}} dm_s\,. \tag{7.36}$$

The partial molar volume depends on the concentration too, but compared with the changes of concentration fluctuations of dm_0 and dm_s, their changes can be neglected. Therefore it is safe to assume that both $V_0^{'}$ and $V_s^{'}$ are constants during concentration fluctuations. Combining Eqs. (7.34) and (7.36) and the definition of C, we have

$$\left(\frac{\partial^2 A}{\partial C^2}\right)_{T,V} = \left(\frac{\partial \mu_s}{\partial C} - \frac{V_s^{'}}{V_0^{'}}\frac{\partial \mu_0}{\partial C}\right)\frac{dV}{M_s}\,. \tag{7.37}$$

Applying Gibbs–Duhem equation,

$$m_0 d\mu_0 + m_s d\mu_s = 0 \,. \tag{7.38}$$

Eq. (7.37) becomes

$$\left(\frac{\partial^2 A}{\partial C^2}\right)_{T,V} = -\frac{dV}{CV_0'}\left(\frac{\partial \mu_0}{\partial C}\right)_{T,V} . \tag{7.39}$$

By definition, chemical potential and activity of a species are related,

$$\mu = \mu^0 + RT \ln a \,, \tag{7.40}$$

where μ^0 is the chemical potential in a standard state and can be considered as a constant, a is the activity, and R is the gas constant. Differentiating Eq. (7.40) and inserting the result into Eq. (7.39),

$$\left(\frac{\partial^2 A}{\partial C^2}\right)_{T,V} = -RT\frac{dV}{CV_0'}\left(\frac{\partial \ln a_0}{\partial C}\right)_{T,V} . \tag{7.41}$$

Combining Eqs. (7.41), (7.33), (7.32), (7.13), we have the volume scattering function as a function of concentration fluctuations,

$$\beta(\theta) = \frac{1}{2}\frac{\pi^2}{N_A \lambda_0^4}\left(\frac{\partial n^2}{\partial C}\right)^2 \frac{CV_0'}{-\partial \ln a_0 \,/\partial C}\frac{6+6\delta_n}{6-7\delta_n}\left(1+\frac{1-\delta_n}{1+\delta_n}\cos^2\theta\right) , \tag{7.42}$$

where N_A is Avogadro's number.

7.2.4 Seawater

Seawater is a multi-component solution with dissolved components of sea salts either in their original form or as disassociated ions. Extending Eq. (7.42), which strictly speaking, only applies to a two-component solution, to a multi-component system involves the coupling of terms among any two solutes (Brinkman and Hermans, 1949; Kirkwood and Goldberg, 1950; Stockmayer, 1950), which is difficult, if not impossible, to evaluate or measure. However, if sea salts can be considered as one hypothetical compound thermodynamically, or equivalently, the molar ratios among dissolved components of sea salts in seawater remain constant during the fluctuations of concentration, Zhang et al. (2009) showed that Eq. (7.42) can still apply for seawater accounting for the scattering due to concentration fluctuations. Since the concentration of sea salts is typically measured as salinity, $S = C\rho$, Eq. (7.42) can be rewritten as

$$\beta\left(\theta\right) = \frac{1}{2}\frac{\pi^2}{N_A {\lambda_0}^4}\frac{M_0}{\rho}\frac{S\left(\partial n^2/\partial S\right)^2}{-\partial \ln a_0 \,/\partial S}\frac{6+6\delta_n}{6-7\delta_n}\left(1+\frac{1-\delta_n}{1+\delta_n}\cos^2\theta\right) , \tag{7.43}$$

where M_0 is the molecular weight of pure water. It should be noted that S in Eq. (7.43) represents the mass concentration of sea salts with a unit of g/kg and it differs from the Practical Salinity Unit $S(‰)$ in both definition and values, $S(\text{g/kg}) = 1.004 \times S(‰)$ (Millero et al., 2008).

Combining Eqs. (7.30) and (7.43), we have the total volume scattering function for pure seawater,

$$\beta(\theta) = \beta_d(\theta) + \beta_c(\theta) , \tag{7.44}$$

where the scattering due to density fluctuations is

$$\beta_d(\theta) = k_B T \beta_T \left(\rho \frac{\partial n^2}{\partial \rho}\right)^2 f(\lambda_0, \theta, \delta_n) , \tag{7.45}$$

and the scattering due to concentration fluctuations

$$\beta_c(\theta) = \frac{M_0}{N_A \rho} \frac{S(\partial n^2/\partial S)^2}{-\partial \ln a_0 \,/\partial S} f(\lambda_0, \theta, \delta_n) . \tag{7.46}$$

where the function f summarizes the dependence of scattering on the wavelength of incident light in a vacuum (λ_0), the angle of scattering (θ) and the depolarization ratio for unpolarized light (δ_n),

$$f(\lambda_0, \theta, \delta_n) = \frac{\pi^2}{2{\lambda_0}^4} \frac{6 + 6\delta_n}{6 - 7\delta_n} \left(1 + \frac{1 - \delta_n}{1 + \delta_n} \cos^2 \theta\right) . \tag{7.47}$$

For pure water with $S = 0$, $\beta_c = 0$. It should be noted that for seawater, the density term β_d is different from that for pure water because β_T, ρ, and n are all functions of salinity. The total scattering coefficienti for seawater, b, is

$$b = \int_0^{\pi} 2\pi\beta(\theta) \sin(\theta)\, d\theta = \frac{8\pi}{3} \frac{2 + \delta_n}{1 + \delta_n} \beta(90) , \tag{7.48}$$

and the total backscattering coefficient, b_b, is

$$b_b = \int_{\pi/2}^{\pi} 2\pi\beta(\theta) \sin(\theta)\, d\theta = \frac{b}{2} . \tag{7.49}$$

7.3 Brief review and discussion

Morel (1966, 1968) measured the scattering by pure water and pure seawater of salinity 38.4 ‰ at five wavelengths of 366, 405, 436, 546 and 578 nm at 20°C (Fig. 7.2). Compared with the few other experimental studies of pure water (Kraut and Dandliker, 1955; Mysels and Princen, 1959; Cohen and Eisenberg, 1965; Kratohvil et al., 1965; Pethica and Smart, 1966; Parfitt and Wood, 1968), Morel's determinations gave the lowest values, likely an indication of the quality of the experiment as particle-free pure water is difficult to prepare. There have been no other published measurements investigating the effect of sea salts on scattering. The experimental error of these measurements was 2%.

Theoretical modeling of pure water scattering had been challenged by a lack of high-precision characterizations of the relevant thermodynamic quantities (Zhang and Hu, 2009), in particular, the density fluctuation term, $(\partial n^2)/\partial \rho$, and the value of depolarization ratio, δ_n. For seawater, an additional challenge is a lack of a theoretical model explicitly formulating the effects of sea salts (Zhang et al., 2009). In the following sections, these challenges, along with some other relevant issues, are briefly reviewed and discussed.

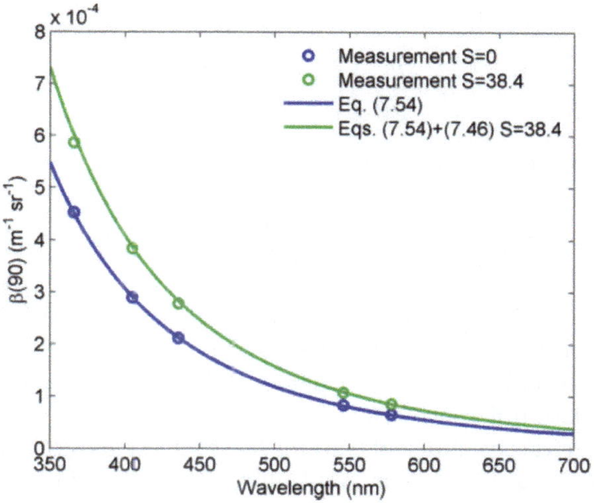

Fig. 7.2. Measured and modeled spectral volume scattering function at 90° ($\beta(90)$) at 20°C for pure water and pure seawater of $S = 38.4$ ‰. For modeling, $\delta_n = 0.039$. The agreements between the measurements and predictions are −1% and 1% for water and seawater, respectively, all within the experimental error of 2%.

7.3.1 Density derivative

From the classic Lorentz–Lorenz equation or Clausius–Mossotti relation,

$$\frac{n^2-1}{n^2+2} = \text{const.} \times \rho\,, \tag{7.50}$$

it can be easily derived that

$$\rho\frac{\partial n^2}{\partial \rho} = \frac{(n^2-1)(n^2+2)}{3}\,. \tag{7.51}$$

Inserting Eq. (7.51) into Eq. (7.30), the scattering for pure water (or due to density fluctuations alone) can be written as

$$\beta_d(\theta) = k_B T \beta_T \frac{\left(n^2-1\right)^2\left(n^2+2\right)^2}{9} f\left(\lambda_0, \theta, \delta_n\right)\,. \tag{7.52}$$

While Eq. (7.52) is attractive in that it unifies the scattering by gas and liquid because for gases, $\beta_T = 1/P$, where P is pressure, and $n^2+2 \approx 3$, it predicts poorly for water, overestimating measurements by more than 30% (Zhang and Hu, 2009). Actually, the use of the Lorentz–Lorenz equation or its variations (such as Laplace equation) failed to give a satisfactory agreement for a variety of liquids (Kerker, 1969), because the constant assumed in the Lorentz-Lorenz equation (Eq. (7.50)) is actually a function of temperature (Eisenberg, 1965; Beysens and Calmettes, 1977). Upon reviewing the earlier studies, Morel (1974) suggested the density derivative be replaced with the pressure derivative, i.e.,

$$\left(\rho\frac{\partial n^2}{\partial \rho}\right)_T = \frac{2n}{\beta_T}\left(\frac{\partial n}{\partial P}\right)_T ,$$

which can be measured relatively more easily. With this,

$$\beta_d(\theta) = \frac{4k_B T n^2}{\beta_T}\left(\frac{\partial n}{\partial P}\right)_T^2 f(\lambda_0, \theta, \delta_n) . \tag{7.53}$$

Equation (7.53) has also been used by Shifrin (1988) and Buiteveld et al. (1994) to estimate scattering by pure water. Using the most recent experimental results, estimates by Buiteveld et al. (1994) agree with Morel's measurements for pure water with a relative difference of 6%.

However, the refractive index of water is least sensitive to the change in pressure as compared to T, λ, and S (Austin and Halikas, 1974), making estimates of $\partial n/\partial P$ very sensitive to errors in the function of $n(P)$. The advancement in theory and experimental observation of the refractive index of water has led to the improved modeling of density derivative. Zhang and Hu (2009) evaluated three of the latest theoretical estimates of $[\rho(\partial n^2/\partial\rho)_T]$ by Proutiere et al. (1992), Niedrich (1985), and Eisenberg (1965), respectively and found that the use of any of them improved the prediction of pure water scattering, all within an experimental error of 2%. With Proutiere et al.'s (1992) derivation of $[\rho(\partial n^2/\partial\rho)_T]$,

$$\beta_d(\theta) = k_B T \beta_T \left\{(n^2-1)\left[1+\frac{2}{3}(n^2+2)\left(\frac{n^2-1}{3n}\right)^2\right]\right\}^2 f(\lambda_0, \theta, \delta_n) . \tag{7.54}$$

along with the better characterization of other parameters, the prediction using Eq. (7.54), shown in Fig. 7.2, agrees with Morel's measurement with a relative difference of −1%.

Equation (7.54) offers an apparent numerical advantage over Eq. (7.53): the density derivative is represented theoretically as a function of the refractive index of water, which can be measured with relatively high precision (typically 10^{-5} but up to 10^{-7} in Tilton and Taylor, (1938)); on the other hand, with no analytical forms existing, $(\partial n/\partial P)_T$ can only be approximated as $\Delta n/\Delta P$, which is difficult to measure because of very low sensitivity and nonlinearity of function $n(P)$.

7.3.2 Depolarization ratio

Kratohvil et al. (1965) compiled historical values determined for the depolarization ratio of water, which ranged from 0.06 to 0.21. This wide range of values was an indication that the measurements of depolarization are very sensitive to stray light. For example, stray light that is 10% of $\beta(90)$ would create a 50% change in the measured depolarization ratio (Pethica and Smart, 1966). Kratohvil et al. (1965) suggested a value of 0.108, which was the mean of their measurements. Morel (1974) and Shifrin (1988) used a value of 0.091, as measured by Pethica and Smart (1966) at 436 nm. Buiteveld et al. (1994) used a value of 0.051, measured by Farinato and Rowell (1976) at 514.4 nm. Jonasz and Fournier (2007) recommended a value of 0.039, which was determined by Farinato and Rowell (1976) after filtering out stray light using a 0.46 nm bandpass filter.

We recommend using the value of 0.039 for the depolarization ratio because the potential contamination of stray light has been removed. The theoretical results shown in Fig. 7.2 were all estimated using 0.039 for δ_n. No published values for the depolarization ratio for water have been reported since Farinato and Rowell (1976). However, uncertainties in the depolarization ratio value remain. As far as scattering by seawater is concerned, the two main issues are: the spectral dependence of the value, and whether and how the value varies with salinity.

Table 7.2 summarizes the reported spectral values of δ_n and not only are the values different, but their spectral dependence is not consistent either. For example, δ_n reported in Raman and Rao (1923) and in Cohen and Eisenberg (1965) decreases with wavelength, but this spectral behavior is reversed in Kratohvil et al. (1965) and Pethica and Smart (1966). We still do not know how the depolarization varies spectrally. There also no published studies evaluating the effect of salinity on the depolarization ratio. Therefore, until further research is conducted, we must assume a constant value for the range of wavelengths of interest to ocean optics and for the natural range of salinity. This value is 0.039.

Table 7.2. The literature values of δ_n for water determined at different wavelengths

Authors	λ_0 (nm)/δ_n
Raman and Rao (1923)	Violet, blue, green, yellow, red 0.21, 0.155, 0.107, 0.106, 0.094
Cohen and Eisenberg (1965)	436, 546 0.087, 0.076
Kratohvil et al. (1965)	436, 546 0.107, 0.115
Pethica and Smart (1966)	436, 546 0.091, 0.109

7.3.3 Effects of sea salts

From Morel's (1968) measurements, which remain the only experimental determination of scattering by pure seawater, the scattering by pure seawater at a salinity of 38.4 ‰ increases about 30% relative to that by pure water. Based on this, Boss and Pegau (2001) suggested an empirical model adjusting the seawater scattering linearly with salinity,

$$\beta(\theta) = \beta_{pw}(\theta)\left(1 + \frac{0.3}{38.4}S\right), \tag{7.55}$$

where β_{pw} represent the scattering by pure water. Results using the empirical model of Eq. (7.55) and the theoretical models of Eqs. (7.54) and (7.46) are compared in Fig. 7.3(a). These results suggest that scattering does not change linearly with

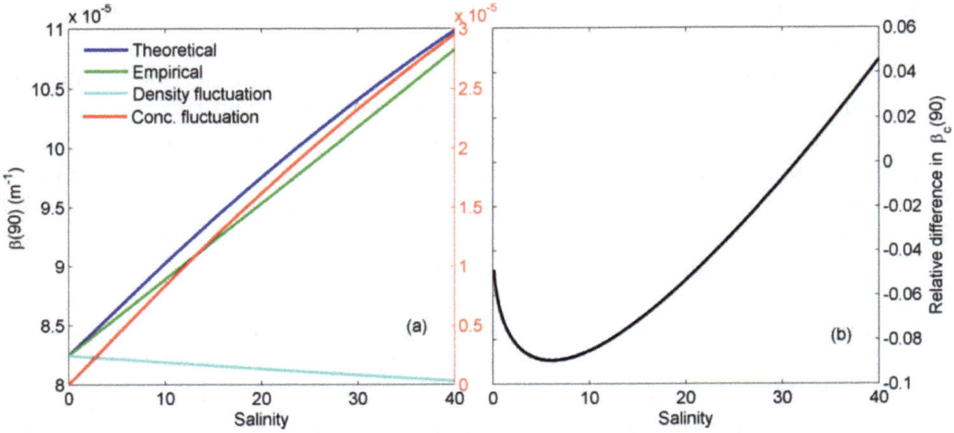

Fig. 7.3. (a) Theoretically and empirically modeled total scattering and its components as a function of salinity; (b) Relative error (Eq. (7.57)/Eq. (7.46) −1) in modeling concentration fluctuations by assuming seawater is ideal. In both (a) and (b), $\lambda = 546$ nm and $T = 20°$C.

salinity, and that Eq. (7.55) would underestimate the theoretical values for a majority of oceanic waters. As validation, the theoretically modeled pure seawater scattering for $S = 38.4$ ‰ is shown in Fig. 7.2 and agrees with Morel's measurements with a relative difference of 1%.

The effects of sea salts on scattering arise from two factors: changes in the density fluctuations (Eq. (7.55)) and additional fluctuations in concentration (Eq. (7.46)). The magnitude of the former is relatively small (about 2.6% decrease as S increases from 0 to 40 ‰), because of contradicting influences of sea salts on β_T (decreasing with S) and on n and ρ (both increasing with S). The effect of the latter is significant. As can be seen from Eq. (7.46) and Fig. 7.3 (a), this increase in scattering is mainly due to the linear term S, modified by nonlinear changes in ρ and a_0. The nonlinearity arises because seawater is an electrolyte solution, which is not ideal, regardless how dilute its concentration.

Assuming that seawater is ideal, $a_0 = 1 - X_{ss}$, where X_{ss} is the molar fraction of sea salts and $X_{ss} = m_{ss}/(m_0 + m_{ss})$, where m represents molar amount and subscript ss represents sea salts, we have

$$-\frac{\partial \ln a_0}{\partial S} = \frac{M_0}{M_{ss}} \frac{1 - X_{ss}}{(1-S)^2} \approx \frac{M_0}{M_{ss}} . \tag{7.56}$$

Inserting Eq. (7.56) into Eq. (7.46), we have

$$\beta_c(\theta) = \frac{M_{ss} S}{N_A \rho} \left(\frac{\partial n^2}{\partial S} \right)^2 f(\lambda_0, \theta, \delta_n) , \tag{7.57}$$

which was proposed by Debye (1944) and used in Morel (1974) and in Jonasz and Fournier (2007) to model the scattering due to concentration fluctuations in pure seawater. Equation (7.57) would underestimate scattering due to concentration fluctuations at salinities <31 ‰ with difference up to −9% and overestimate scattering at higher salinities (Fig. 7.3(b)).

7.3.4 Other relevant issues

7.3.4.1 Spectral dependence

The spectral dependence of scattering by seawater has been characterized by s as in $(\lambda/\lambda_a)^s$. A value of -4.32 has often been used (e.g., Gordon and Morel, 1983) and was first reported by Morel (1974) with an anchor wavelength $\lambda_a = 436$ nm. Twardowski et al (2007) and Shifrin (1988) reported a value of -4.17 based on fitting to Eq. (7.53). The value of s would also change with salinity, ranging from -4.286 to -4.306 for a S from 0 to 40 ‰ with $\lambda_a = 450$ nm based on Eqs. (7.54) and (7.46). Historically, s was used to scale the measured scattering by Morel (1974) (which was only available at 5 wavelengths, e.g., Fig. 7.2) to other wavelengths. Given that the scattering by pure water and pure seawater can now be modeled spectrally, the use of s should be limited.

7.3.4.2 Temperature dependence

Water, the most common molecule on Earth, has several 'anomalous' properties that are relevant to scattering: maximum ρ near 4°C (Vedamuthu et al., 1994), minimum β_T near 46°C (Vedamuthu et al., 1995), and maximum n near 0°C (Cho et al., 2001). It should be expected that through the combined effects of these parameters, the scattering by water may also behave anomalously. For pure water, Cohen and Eisenberg (1965) measured the scattering for T = 5°C to 65°C at two wavelengths of 436 and 546 nm and found a scattering minimum at ~22°C. Modeled scattering by pure water using Eq. (7.54) and Buiteveld et al (1994) equation are compared with the measurements of Cohen and Eisenberg (1965) in Fig. 7.4. Predictions based on Eq. (7.54) agree with the measurements within 1.4% at both wavelengths. The scattering estimated using Eq. (7.54) varies with the temperature nonlinearly with a minimum at 24.64°C (denoted as T_{min} hereafter). The scattering decreases by 4.25% between 0°C and T_{min} and increases by 7% between T_{min} and 70°C. The T_{min} is close to the value of 22°C estimated by Cohen and Eisenberg (1965), however, differs significantly with the estimates by Buiteveld et al. (1994), who found a maximum near 15°C. They used Eq. (7.53), which is valid theoretically, but suffers from the relatively large uncertainty in modeling $(\partial n/\partial P)_T$. Fig. 7.4 reaffirms the advantage in using Eq. (7.54) against Eq. (7.53). The value of T_{min} also increases with salinity, reaching 27.49°C at 40 ‰ (Zhang and Hu, 2010).

7.3.4.3 Polarization

So far, the focus of this review has been on natural, unpolarized light. If the incident light is polarized, the depolarization ratio and the Cabannes factor will be different, as well as the angular dependence. For incident illumination that is vertically polarized, from Eqs. (7.21) to (7.23)

$$E_{total} = E_{VV} + E_{VH} \propto \frac{45\alpha^2+7\beta^2}{45} \tag{7.58}$$

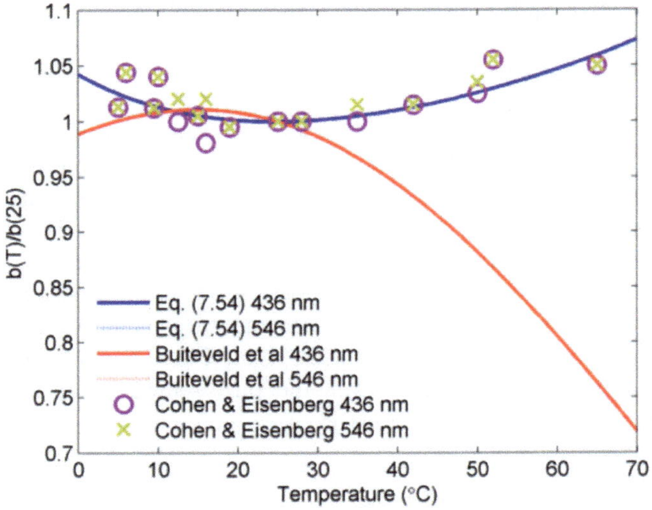

Fig. 7.4. Scattering by pure water normalized at $T = 25°\mathrm{C}$ as a function of temperature.

$$\delta_V = \frac{E_{VH}}{E_{VV}} = \frac{\overline{\alpha_{zy}^2}}{\overline{\alpha_{zz}^2}} = \frac{4\beta^2}{45\alpha^2 + 4\beta^2} \tag{7.59}$$

$$\text{Cabannes factor} = \frac{E_{total}(\beta \neq 0)}{E_{total}(\beta = 0)} = \frac{45\alpha^2 + 7\beta^2}{45\alpha^2} = \frac{3 + 3\delta_V}{3 - 4\delta_V}\,. \tag{7.60}$$

For incident illumination that is horizontally polarized,

$$E_{total} = E_{HV} + E_{HH} \propto \frac{2\beta^2}{15}\left(1 + \frac{45\alpha^2 + \beta^2}{6\beta^2}\cos^2\theta\right) \tag{7.61}$$

$$\delta_H = 1 \tag{7.62}$$

and the Cabannes factor is not defined. Depolarization ratios for different polarization states satisfy the Krishnan relationship (Fabelinskii, 1968),

$$\delta_n = \frac{1 + \frac{1}{\delta_H}}{1 + \frac{1}{\delta_V}} = \frac{2\delta_V}{1 + \delta_V} \tag{7.63}$$

Farinato and Rowell (1976) measured δ_V for pure water, with values of 0.020 or 0.026, with or without a stray light filter, respectively. These values are lower than earlier experimental results, e.g., 0.032 and 0.058 as reported in Kratohvil et al. (1965), who also measured δ_H, with values ranging from 1.00 to 1.05, very close to the theoretical value.

7.4 Conclusions

Light scattering by pure seawater is a physical quantity of critical importance to *in situ* ocean optics and remote sensing. Since the density fluctuation theory was established one century ago by Smoluchowski (1908) and Einstein (1910), a great deal of progress has been made leading to a significant improvement in our understanding of these processes. Recent advancements in both theoretical and experimental characterizations of key thermodynamic parameters have allowed us to model scattering within the experimental error of measurements (Zhang and Hu, 2009; Zhang et al., 2009). The measurements of Morel (1966, 1968) for pure water and pure seawater scattering at one salinity still are the best values for validating theory. Despite our progress, there are still uncertainties that need to be further constrained. We still do not know whether and how the depolarization ratio varies spectrally and/or with salinity. It seems that additional experimental efforts, particularly those addressing the uncertainty with respect to the depolarization ratio, are needed. The Matlab codes used to generate Figs. 7.2 to 7.4 can be downloaded from the author's website, http://www.und.edu/instruct/zhang.

References

Austin, R. W., and G. Halikas (1974), The index of refraction of seawater *Rep. SIO Ref. No. 76-1*, 121 pp, Scripps Institute of Oceanography, La Jolla.

Beysens, D., and P. Calmettes (1977), Temperature dependence of the refractive indices of liquids: Deviation from the Lorentz–Lorenz formula, *J. Chem. Phys.*, **66**, 766–771.

Boss, E., and W. S. Pegau (2001), Relationship of light scattering at an angle in the backward direction to the backscattering coefficient, *Appl. Opt.*, **40**(30), 5503–5507.

Brinkman, H. C., and J. J. Hermans (2001), The effect of non-homogeneity of molecular weight on the scattering of lght by high polymer solutions, *J. Chem. Phys.*, **17**(6), 574–576.

Buiteveld, H., J. H. M. Hakvoort, and M. Donze (1994), The optical properties of pure water, *SPIE, 2258* (Ocean Optics XII), 174–183.

Cabannes, J. (1920), Relation entre le degré de polarisation et l'intensité de la lumière diffusée par des molécules anisotropes. Nouvelle détermination de la constante d'Avogadro, *J. Phys. Radium*, **1**(5), 129–142.

Cabannes, J. (1922), Considérations théoriques sur la diffusion de la lumière par les liquides transparents. Polarisation de la lumière diffusée latéralement, *J. Phys. Radium*, **3**(12), 429–442.

Cho, C. H., J. Urquidi, G. I. Gellene, and G. W. Robinson (2001), Mixture model description of the T-, P dependence of the refractive index of water, *J. Chem. Phys.*, **114**(7), 3157–3162.

Cohen, G., and H. Eisenberg (1965), Light scattering of water, deuterium oxide, and other pure liquids, *J. Chem. Phys.*, **43**(11), 3881–3887.

Debye, P. (1944), Light scattering in solutions, *J. Appl. Phys.*, **15**(4), 338–342.

Einstein, A. (1910), Theorie der Opaleszenz von homogenen Flüssigkeiten und Flüssigkeitsgemischen in der Nähe des kritischen Zustandes, *Annalen der Physik*, **338**(16), 1275–1298.

Eisenberg, H. (1965), Equation for the refractive index of water, *J. Chem. Phys.*, **43**(11), 3887–3892.

Fabelinskii, I. L. (1968), *Molecular Scattering of Light*, Plenum Press, New York.

Farinato, R. S., and R. L. Rowell (1976), New values of the light scattering depolarization and anisotropy of water, *J. Chem. Phys.*, **65**(2), 593–595.

Gordon, H. R., and A. Morel (1983), *Remote Assessment of Ocean Color for Interpretation of Satellite Visible Imagery, a Review*, Springer-Verlag, New York.

Jonasz, M., and G. R. Fournier (2007), *Light Scattering by Particles in Water: Theoretical and Experimental Foundations*, Academic Press, New York.

Kerker, M. (1969), *The Scattering of Light and Other Electromagnetic Radiation*, Academic Press, New York.

King, L. V. (1923), The complex anisotropic molecule in relation to the theory of dispersion and scattering of light in gases and liquids, *Nature*, **111**(2794), 667.

Kirkwood, J. G., and R. J. Goldberg (1950), Light scattering arising from composition fluctuations in multi-component systems, *J. Chem. Phys.*, **18**(1), 54–57.

Kratohvil, J. P., M. Kerker, and L. E. Oppenheimer (1965), Light scattering by pure water, *J. Chem. Phys.*, **43**(3), 914–921.

Kraut, J., and W. B. Dandliker (1955), Light scattering by water, *J. Chem. Phys.*, **23**, 1544–1545.

Martin, W. H. (1923), The scattering of light by liquids: Effect of direction on polarization and intensity, *Proceedings and Transactions of The Royal Society of Canada*, XVII, 151–155.

Millero, F. J., R. Feistel, D. G. Wright, and T. J. McDougall (2008), The composition of Standard Seawater and the definition of the Reference-Composition Salinity Scale, *Deep-Sea Res. I*, **55**(1), 50–72.

Morel, A. (1966), Etude Experimentale de la diffusion de la lumière par l'eau, les solutions de chlorure de sodium et l'eau de mer optiquement pures, *Journal de Chimie Physique*, **10**, 1359–1366.

Morel, A. (1968), Note au sujet des constantes de diffusion de la lumière pour l'eau et l'eau de mer optiquement pures, *Cahiers Oceanographiques*, **20**, 157–162.

Morel, A. (1974), Optical properties of pure water and pure sea water, in *Optical Aspects of Oceanography*, edited by N. G. Jerlov and E. S. Nielsen, pp. 1–24, Academic Press, New York.

Mysels, K. J., and L. H. Princen (1959), Light scattering by some laurylsulfate solutions, *J. Phys. Chem.*, **63**(10), 1696–1700.

Niedrich, Z. (1985), Dispersion interactions and the refractive index of liquids, *Physica*, **128B**, 69–75.

Parfitt, G. D., and J. A. Wood (1968), Rayleigh ratios of pure liquids; comparison of experiment with theory, *Transactions of the Faraday Society, 64*, 805–814.

Pethica, B. A., and C. Smart (1966), Light scattering of electrolyte solutions, *Trans. Faraday Soc.*, **62**, 1890–1899.

Prins, J. A., and W. Prins (1956), Dépolarisation et intensité de la lumière diffusée par les liquides à molécules anisotropes, *Physica*, **22**(6–12), 576–578.

Proutiere, A., E. Megnassan, and H. Hucteau (1992), Refractive index and density variations in pure liquids: A new theoretical relation, *J. Phys. Chem.*, **96**, 3485–3489.

Raman, C. V., and K. S. Rao (1923), LXIII. On the molecular scattering and extinction of light in liquids and the determination of the Avogadro constant, *Philosophical Magazine Series 6*, **45**(267), 625–640.

Shifrin, K. S. (1988), *Physical Optics of Ocean Water*, American Institute of Physics, New York.

Smoluchowski, M. v. (1908), Molekular-kinetische Theorie der Opaleszenz von Gasen im kritischen Zustande, sowie einiger verwandter Erscheinungen, *Annalen der Physik*, **330**(2), 205–226.

Stockmayer, W. H. (1950), Light scattering in multi-component systems, *J. Chem. Phys.*, **18**(1), 58–61.

Strutt, R. J. L. R. (1918), The light scattered by gases: its polarisation and intensity, *Proceedings of the Royal Society of London. Series A*, **95**(667), 155–176.

Tilton, L. W., and J. K. Taylor (1938), Refractive index and dispersion of distilled water for visible radiation at temperature 0 to 60°C, *J. Res. Natl. Bur. Stand.*, **20**, 419.

Twardowski, M. S., H. Claustre, S. A. Freeman, D. Stramski, and Y. Huot (2007), Optical backscattering properties of the 'clearest' natural waters, *Biogeosciences*, **4**(6), 1041–1058.

Vedamuthu, M., S. Singh, and G. W. Robinson (1994), Properties of liquid water: Origin of the density anomalies, *J. Phys. Chem.*, **98**(9), 2222–2230.

Vedamuthu, M., S. Singh, and G. W. Robinson (1995), Properties of liquid water. 4. The isothermal compressibility minimum near 50°C, *J. Phys. Chem.*, **99**(22), 9263–9267.

Zhang, X., and L. Hu (2009), Estimating scattering of pure water from density fluctuation of the refractive index, *Opt. Express*, **17**(3), 1671–1678.

Zhang, X., and L. Hu (2010), Effects of temperature and salinity on light scattering by water, paper presented at SPIE – Ocean Sensing and Monitoring, Orlando, FL, USA.

Zhang, X., L. Hu, and M.-X. He (2009), Scattering by pure seawater: Effect of salinity, *Opt. Express*, **17**(7), 5698–5710.

Zimm, B. H. (1945), Molecular theory of the scattering of light in fluids, *J. Chem. Phys.*, **13**(4), 141–145.

Index

MIX
Papier aus verantwortungsvollen Quellen
Paper from responsible sources
FSC® C105338

If you have any concerns about our products,
you can contact us on
ProductSafety@springernature.com

In case Publisher is established outside the EU,
the EU authorized representative is:
Springer Nature Customer Service Center GmbH
Europaplatz 3, 69115 Heidelberg, Germany

Printed by Libri Plureos GmbH
in Hamburg, Germany